INTO THE BLACK

The Extraordinary Untold Story of the
First Flight of the Space Shuttle
and the Men Who Flew Her

Rowland White

CORGI BOOKS

TRANSWORLD PUBLISHERS
61–63 Uxbridge Road, London W5 5SA
www.penguin.co.uk

Transworld is part of the Penguin Random House group of companies
whose addresses can be found at global.penguinrandomhouse.com

Penguin
Random House
UK

First published in Great Britain in 2016 by Bantam Press
an imprint of Transworld Publishers
Corgi edition published 2017

A CIP catalogue record for this book
is available from the British Library.

ISBN
9780552160223

Typeset in 10/14pt Stone Serif by Falcon Oast Graphic Art Ltd.
Printed and bound by Clays Ltd, Bungay, Suffolk.

Penguin Random House is committed to a sustainable
future for our business, our readers and our planet. This book is made
from Forest Stewardship Council® certified paper.

MIX
Paper from
responsible sources
FSC® C018179

1 3 5 7 9 10 8 6 4 2

'Rowland White has written an unforgettable book, one **destined to become a classic** in the emerging field of space history. Moving effortlessly from planning meetings in obscure offices within NASA, the Pentagon, and industry to the unforgiving environment of high-risk flight test and flight into space, this book traces a remarkable journey and, along the way, introduces his readers to a courageous group of far-seeing engineers and astronauts whose bold vision and tenacious work gave humanity its first reusable space transportation system. A work of transcending importance, it will inform specialists and laypersons alike.'
Dr Richard P. Hallion, Former US Air Force Historian and Chief Historian, Air Force Flight Test Centre, Edwards Air Force Base

'A truly edge-of-your seat ride into space. In this **superbly researched**, painstakingly pieced together book, Rowland White has written a sensational account of the Space Shuttle story. With a cast of extraordinary characters, incredible science and the kind of cutting-edge technology that almost defies comprehension, he recounts all the drama, jeopardy, intense beauty and astonishing adventure of what must be one of man's most incredible achievements.'
James Holland, bestselling author of *Dam Busters* and *The War in the West*

'Impeccably, painstakingly researched from start to finish. **The level of access Rowland White has secured here is nothing short of breath-taking**. The story is therefore woven through with revealing testimony straight from the mouths of the astonishing cast of characters behind the Space Shuttle – heroes all, but made real flesh and blood within these pages. To hear so much of it in their own words means the whole thing feels incredibly personal and revealing. Even more affecting is the way the storytelling reveals such boyish fascination with flight, so that *Into the Black* reads like the very best fiction – a real page-turner. I loved it.' Neil Oliver

'Rowland White's account of *Columbia*'s inaugural flight in 1981, and all the preparations that led up to it, could not be a more timely reminder of what it takes to design, launch and fly a complex manned space vehicle. White's research is thorough, his writing style is superb, and he has a gripping and fresh story to tell. This is a **genuine "must-have" book** for anyone fascinated by the sharp end of space flight.'
Piers Bizony, author of *Starman*, *The Space Shuttle*, and *The Making of Stanley Kubrick's 2001: A Space Odyssey*

Rowland White is the author of three critically acclaimed works of aviation history: *Vulcan 607*, *Phoenix Squadron* and *Storm Front*. All three have been *Sunday Times* top ten best-sellers. His writing has appeared in a variety of national magazines and newspapers including the *Guardian*, *Daily Mail* and *Esquire* magazine.

Born and brought up in Cambridge, Rowland studied Modern History at Liverpool University. He has held a private pilot's licence since 1998 and lives near Cambridge with his wife and three young children.

For more information on Rowland White and his books, see his website at www.rowlandwhite.com or @RowlandWhite

Also by Rowland White

VULCAN 607
PHOENIX SQUADRON
STORM FRONT
THE BIG BOOK OF FLIGHT

'*The Right Stuff* of our times.' *Daily Mail*

'Beautifully researched and written, **Into The Black** tells the true, **complete story of the Space Shuttle better than it's ever been told before**.' Col. Chris Hadfield, astronaut, Space Station commander, author of *An Astronaut's Guide to Life on Earth*

'**A work of great importance** – knitting together separate strands of what has so far been a hidden story worth the telling but largely ignored. I am impressed by the thoroughness of scholarship and the crafting of an important piece of research into a highly readable narrative. *Into the Black* illuminates a complex piece of space history left in the shadows until now. Impossible to put down.' Dr David Baker, Editor of *Spaceflight* magazine

'A remarkable book.' David Scott, Commander of Apollo XV, Moonwalker

'**White is the Robert Caro of the Space Shuttle** . . . *Columbia* showed that if they put their heart and soul into it, humans could glide back from space and land on a runway. When it worked it looked easy. *Into the Black* shows it was really a white-knuckle ride, every time.' *The Times*

'Meticulously researched . . . the astronauts are, not surprisingly, the principal heroes of White's book, although there's a lot of credit given to the visionaries on the ground who made the whole thing happen. But **it's the stuff about flying that will grip most readers' imaginations**, and it's here, particularly in the tense final section of the book, that White really comes into his own.' *Mail on Sunday*

'I grew up in the world of experimental test flying which spawned the first astronauts, some of whom I actually served with, and they popped out at me from almost every page of this book. **An astonishing amount of research has gone into this splendid work**. I found it totally absorbing.' Captain Eric Brown, former UK Chief Naval Test Pilot

'White captures the astronauts' yearning for space as well as its stark dangers . . . As White shifts from budgets and politics on the ground to the mission itself, the story becomes **thrilling**.'
Financial Times

'The release of a new title by Rowland White is something to be eagerly anticipated. Not only is his research meticulous, but he also knows how to write. *Into the Black*'s *pièce de résistance* is its description of the first orbital spaceflight by *Columbia* . . . **never before explored in such depth** and bound to engage even those for whom spaceflight is not a primary interest.'
Aeroplane magazine

'Brilliantly revealed, *Into the Black* is the finely tuned true story of the first flight of the Space Shuttle *Columbia*. Rowland White has magnificently laid bare the unknown dangers and unseen hazards of that first mission. He has also given us amazing insight into a world of science and engineering, the victories and defeats, for the first time. **It's the perfect tale that educates as it entertains.** Once read, not forgotten.'   Clive Cussler

'*Into the Black* isn't just spectacularly researched, it's **told like a thriller**, unfolding the edge-of-death tale of the Space Shuttle *Columbia*'s maiden voyage in riveting fashion. Rowland White performs a rare feat here, stitching together comprehensive research – countless interviews, declassified files, flight documents – into a tale of courage and daring as streamlined and elegant as the spacecraft herself. Buckle in and hold on tight – this thing's got rocket propulsion.'   Gregg Hurwitz, best-selling author of *Orphan X*

'White has a skilled eye for flight engineering and the culture of pilots . . . He packs his latest effort with rich detail . . . *Into the Black* captures the aeronautical challenges of shuttle flight and the heroism of astronauts John Young and Robert Crippen, who boarded *Columbia* for the initial mission even though there had been no unmanned test launch first . . . **a volume no aviation buff will want to miss.'**   *Wall Street Journal*

'Told with vivacity and aplomb . . . **this is the definitive word on Columbia.**'   *How It Works* magazine

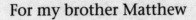

For my brother Matthew

CONTENTS

FOREWORD

Almost two decades into the twenty-first century, with the Space Shuttle programme now shut down and firmly in the history books, we all know how it turned out. In over 130 flights more than 350 men and women from many nations flew into orbit, the International Space Station was built, the enormously successful Hubble Space Telescope was placed on orbit, planetary explorers such as *Cassini*, *Galileo* and *Magellan* were sent to the planets, entire communication satellite constellations were created, important defence payloads were deployed. We also remember the high costs and of course the dramatic failures as well, the tragic loss of two crews and the *Challenger* and *Columbia* spaceships. Today *Atlantis*, *Discovery*, *Endeavour* and *Enterprise* all reside in museums across the USA.

But Rowland White's book *Into the Black* goes way back before all that.

This book is actually several fascinating narratives in one, some of which have never been told before, set in a time when Apollo's triumphal flights to the moon were still very fresh in the world's mind. America was setting out, in the very depths of the Cold War, to develop an unheard-of capability: an enormous, mighty flying machine whose size was driven by classified national security payloads, and which would launch from Earth over and over again. She would operate in orbit doing many different jobs, first as a delicate spacecraft. Then, during a re-entry that would begin at Mach 25, she would slowly morph like a butterfly from a spaceship into an airplane, a gigantic glider; there would be one opportunity, and only one, for a nice landing. Nevertheless, there would be no crew escape capability: survival of the crew would rely on her ability to land safely on a runway. This new 'Space Shuttle' was to operate for several decades, and she would be built using many utterly new technologies, from complex computer systems to powerful but compact engines and lightweight thermal protection systems; the list went on and on.

And this enormous system, for the first time in the history of spaceflight, would launch on its very first mission with astronauts aboard at her controls, without the benefit of all the unmanned test flights that had preceded previous new manned space vehicles.

Of course this book is a personal story of the test pilots, the engineers, the managers and, yes, the politicians who came together to pull off the first flight of this iconic programme. Before the maiden flight, thousands and thousands of Americans all across the country, in manufacturing facilities and simulators and control centres, worked on the programme.

But *Into the Black* also tells the intriguing story of how two entire national space cultures meshed with each other, and sometimes collided. One, NASA, prided itself on operating in, and even inviting, the searing light of public visibility; the other was completely invisible to both the American and the Soviet public, operated by the National

Reconnaissance Office. Although the NRO was cloaked in secrecy and its very existence was totally hidden from public view, it had played a vital daily role since the beginnings of the space programme in the strategic struggle with the Soviet Union and was developing and flying billion-dollar spacecraft with secret codenames such as GAMBIT, HEXA-GON, DORIAN. It was out of this classified netherworld that several astronauts came to NASA from the Air Force's Manned Orbiting Laboratory, after MOL was cancelled by President Nixon just weeks before the first Apollo landing on the moon. Eventually, these 'MOL guys' would become an important part of the backbone of the early Shuttle flights, most especially the first.

But most of all, this is a bang-up flying story. On a brilliantly bright, sunny Florida morning in April 1981, after years of training and waiting, two astronauts rode up the launch pad elevator, climbed aboard the mighty *Columbia* on launch pad 39A and proceeded to attempt what no one had done before. One, a grizzled space veteran of both Gemini and Apollo who had previously landed on the moon, the other a space rookie who had emerged from MOL. Both were veterans of at-sea carrier operations and were experimental test pilots. They had ridden this very elevator before, only to be frustrated by a launch scrub. This day had eventually come following years of delays due to failing test engines, thermal tiles and software redesigns. But finally, to them, this looked like the day.

With thousands of support personnel monitoring their every move and examining every bit of the telemetry streaming from the Shuttle on the pad, John Young and Bob Crippen were about to fly the first flight. For all their experience, neither had ever done anything like the first flight of the Shuttle was going to demand. For those of us who were somehow involved in that endeavour, we would eventually regard the Shuttle as the world's greatest flying machine. At the time, we, and all of America's friends and foes, were watching, but not a soul knew how it was going to turn out!

Richard H. Truly
Vice Admiral, US Navy (Ret.)

AUTHOR'S NOTE

Like most large organizations founded for a specific purpose, NASA loves an acronym and is fond of a contraction. Mostly this makes sense. It's quicker and easier to talk about SRBs instead of solid rocket boosters, FRSI rather than felt reusable surface insulation, TPS as opposed to thermal protection system, and CapCom not Capsule Communicator. Sometimes there's humour in an acronym: TFNGs was more polite than describing the 1978 astronaut intake as The F**king New Guys. Sometimes it can get a little excessive. PLT instead of Pilot is the one that springs to mind. To a general reader it can quickly become overwhelming. And so, while I've included a glossary, I've also made an effort not to use acronyms as freely as they appeared both in NASA literature and in the conversations I've had with NASA personnel.

Halfway through the story that follows, the home of NASA manned spaceflight in Houston changed name from Manned Spacecraft Center to Johnson Space Center. Needless to say, that meant it also changed from MSC to JSC. As often as possible I've done my best simply to refer to it as Houston.

Similarly, there is often confusion around Kennedy

Space Center. Geographically, KSC is located on Merritt Island and not on Cape Canaveral which lies adjacent. An attempt to change the name of Cape Canaveral to Cape Kennedy never took root, but further confuses the issue. And yet the popular term 'the Cape' has become so synonymous with NASA and rocket launches that, although technically incorrect, I've chosen to use it and terms Kennedy, Kennedy Space Center and KSC fairly interchangeably. Apologies to those readers for whom this feels a little fast and loose.

Students of the Space Shuttle programme will also be aware that technically speaking the term 'Space Shuttle' refers not just to the familiar white and black spaceplane but to the whole launch assembly including the big fuel tank and twin rocket boosters. The spaceplane alone is correctly labelled the Orbiter. However, given that pretty much everyone understands the Space Shuttle to be the spaceplane with wings that lands on a runway at the end of the mission, I felt that maintaining this distinction was both pedantic and confusing, and have used the terms 'Space Shuttle' and 'Orbiter' interchangeably.

I've taken the same approach to dialogue as in my previous books. Where it appears in quotation marks it's either what I've been told in interviews was said by those who were there, from contemporary recordings, or what's been reported in previous accounts, official, unofficial, published and unpublished. Where speech is in italics – often drawn from places like standard operating procedures – it represents genuine dialogue taken from a general source to lend authenticity to a scene. Given the ceaseless training that ensured procedures were followed, I hope those involved will agree that it's an accurate reflection of what was said at the time. Also in italics are characters' internal thoughts. As with speech, these represent direct transcriptions of what I've been told by participants – or have read – that they were thinking at the time.

Much of the material in the latter part of the book remains deeply classified. Those with the required security clearances to tell the story are not permitted to talk about it. I've pieced together events from a wide variety of sources to produce an account that is the fullest and most accurate yet published. I've been reassured that this is the case; however, any mistakes are my own.

R.W.

Nant-y-Feinen, 2015

ACKNOWLEDGEMENTS

There's a long list of people to thank. As ever, the first challenge was just trying to make contact with the people I needed to talk to. When I began to contemplate writing about the inaugural flight of the Space Shuttle – another subject that first captured my imagination as a boy who'd been born too late to share in the excitement of the Apollo moon missions – it felt as if this was going to be a good deal harder than it had been with my first three books about some of Britain's post-war military endeavours. Literary agent Ed Victor helped me get my foot in the door by talking to one of his clients, moonwalker Dave Scott, on my behalf. Dave was kind enough to do the same with Bob Crippen, one of the two-man crew on *Columbia*'s maiden voyage. When Bob said he was willing to meet me and talk about his memories of the time, I knew I was in business. Bob also put me in touch with his friend and contemporary in the Astronaut Office, Richard Truly.

Richard – and his wife, Cody – were extremely welcoming and generous with their time. Looking back on it now, I'm astonished by how ignorant I was of both how much I had to learn and the shape of the overall story – beyond trying to reclaim the sense of excitement that surrounded that first flight after six years in which American astronauts had been grounded without a spacecraft. Talking to Richard helped fire my imagination and crystallize the story that follows.

Bob and Richard enjoyed exceptionally illustrious careers at NASA as astronauts then administrators and both provided help that was way beyond the call of duty. I'm not sure that writing *Into the Black* would have been possible without them. I'm hugely indebted to both of them.

Thanks to Richard I was lucky enough to talk to three of the 1978 astronaut intake whose recollections helped provide a different perspective on the first Shuttle mission. I'm grateful to Jon McBride, Rick Hauck and Dan Brandenstein for their time and willingness to help. All had fascinating stories to tell. Richard also introduced me to Simulator Supervisor Jerry Mill. I was really pleased that the two of us got to talk – not least because I learned from him that if you're handed a jacket by the Prince of Wales you don't put it down, however interesting a Space Shuttle simulator might be.

Bob Crippen put me in touch with Fred Haise. It's not every day you have the opportunity to interview a member of the crew of *Apollo XIII*. That was a real high point. So too was talking at length to Joe Engle, whose career in aviation, as a visitor to space in both the X-15 and the Shuttle, must rate as one of the most remarkable there's been. Fred and Joe were patient and forthcoming in answering my questions. It was a pleasure and privilege to speak to them both.

For reasons that will be obvious, Dr Hans Mark's story was central to the one I wanted to tell. I hadn't expected to have the opportunity to talk to him but he was gracious enough to respond to a completely unsolicited email and to answer my questions on two occasions. I'm grateful to him.

I'd mentioned on my website that I was researching a book on the Shuttle and Steven Quayle kindly dropped me a line to suggest that I talk to his friend Christopher Stott, who in turn introduced me to Jay Honeycutt, the ex-Director of Kennedy Space Center. I'm glad he did. Jay and his wife, Peggy, were warm and generous hosts during my time in Cocoa Beach. Jay also helped put me in touch with others I was to have valuable conversations with, most notably ex-Flight Operations boss George Abbey and ex-Kennedy Space Center Public Affairs Officer Hugh Harris.

Closer to home, a big thank you to Dr David Baker, ex-NASA engineer, space writer and editor of the British Interplanetary Society's magazine *Spaceflight*. David has been enormously encouraging as well as being an extra-ordinarily well-informed sounding board. I also got the chance to talk to British test pilot Captain Eric 'Winkle' Brown, which was as great a privilege as talking to those who flew the Shuttle. I'm certain that had he not been born in Edinburgh, Eric would have added 'astronaut' to his incredible list of accomplishments.

Thanks to Charlotte Surtees of Darlow Smithson, and Jeremy Hall, who directed that company's excellent docu-mentary *The Last Flight of the Space Shuttle*. Jeremy helped me make contact with NASA engineer Tom Moser. Tom's experience proved key, and I'm grateful to him for the time he spent talking to me.

Other writers who had absolutely nothing to gain from sharing their insights and reflections but were happy to do so include Andrew Chaikin, author of the definitive account of the Apollo programme, *A Man on the Moon*; Andrew Smith, author of *Moondust* (which might have been the thing responsible for getting me thinking about space again); and Bill Sweetman, *Aviation Week*'s senior international defence editor. Thank you all.

Thanks to friends Nick Cook, Rob Lenaghan (who, fortuitously, is a real rocket scientist) and, as ever, James Holland, who never ceases to be a source of encouragement, inspiration, and very good company.

At Transworld Publishers, I'd like to thank Viv Thompson, Phil Lord and Richard Shailer. Viv and Phil ensured my scribblings were as handsomely presented as they are inside the book, while Richard was responsible for *Into the Black*'s beautiful dust jacket. Alongside them, copy-editor Dan Balado has done an incredible job of sorting out the text, sparing my blushes on many occasions through his care and thoroughness. Behind it all, my publisher, Bill Scott-Kerr, has kept the faith despite – and I see this now – my delivering a first draft which he says he didn't really understand. My agent Mark Lucas has been similarly enthusiastic about the story's potential and careful in his editorial advice. I can't really imagine doing this with anyone but you two. Thank you.

At Touchstone in New York, I'd like to thank my editor, Matthew Benjamin, for his enthusiasm for the book and sound advice on the manuscript.

Once again, Lalla Hitchings has done a wonderful job of transcribing the interviews and in doing so has helped make the book possible, providing me with a little more time to write when there too often appeared to be none.

And on that subject I must thank Michael Joseph managing director, Louise Moore. Lou has been unbelievably kind, patient and understanding when she must have been feeling anything but. At a time when the writing was at its most intense, I felt nearly overwhelmed by the twin demands of the book and the day job. Somehow, Lou managed to be supportive when she had every right to be pretty fed up.

Again, a mention of Rory, Jemima and Lexi, with whom I've not spent nearly enough time for too long now. I'd like that to change. The ball's in my court, of course.

Lastly, there's my amazing wife, Lucy. You've endured with good grace, selflessness and strength a year that I suspect might have seen other people curling up in a little ball and twitching in a corner. Somehow you got through 2014 and still managed to be you. And I love you for it. I'm very lucky. Thank you. We should have a party. Celebrate. Something like that.

STS-1 LOCATION MAP

CANADA

USA

CALIFORNIA

•1, 2

3• •4

TEXAS

ALABAMA
•6

•5

Gulf
of
Mexico

FLORIDA
•7

ATLANTIC

OCEAN

1 NASA Ames Research Center
2 Sunnyvale Air Force Base – the Blue Cube
3 Vandenberg Air Force Base
4 Edwards Air Force Base
5 NASA Johnson Space Center, Houston
6 NASA Marshall Space Flight Center, Huntsville, Alabama
7 NASA Kennedy Space Center

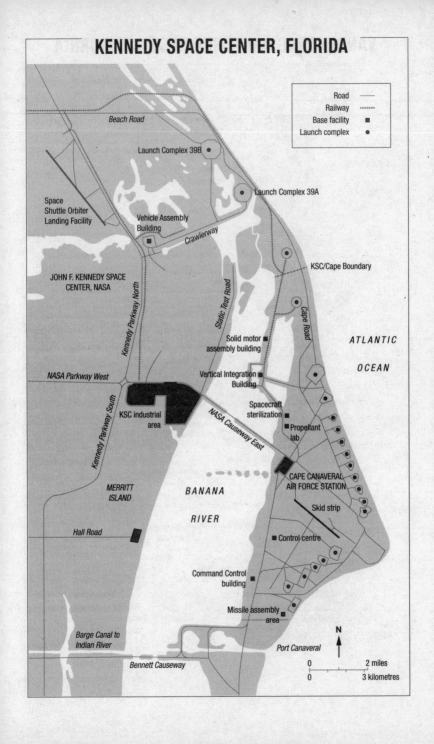

KENNEDY SPACE CENTER, FLORIDA

Road
Railway
Base facility ■
Launch complex ●

Beach Road

Launch Complex 39B ●

Launch Complex 39A ●

Space
Shuttle Orbiter
Landing Facility

Vehicle Assembly
Building

Crawlerway

KSC/Cape Boundary

JOHN F. KENNEDY SPACE
CENTER, NASA

Kennedy Parkway North

Static Test Road

Cape Road

ATLANTIC

OCEAN

NASA Parkway West

Solid motor
assembly building

Vertical Integration
Building

Spacecraft
sterilization

Propellant
lab

Kennedy Parkway South

KSC industrial
area

NASA Causeway East

CAPE CANAVERAL
AIR FORCE STATION

MERRITT
ISLAND

BANANA

RIVER

Skid strip

Hall Road

Control centre

Command Control
building

Missile assembly
area

Barge Canal to
Indian River

Port Canaveral

N

Bennett Causeway

0   2 miles

0   3 kilometres

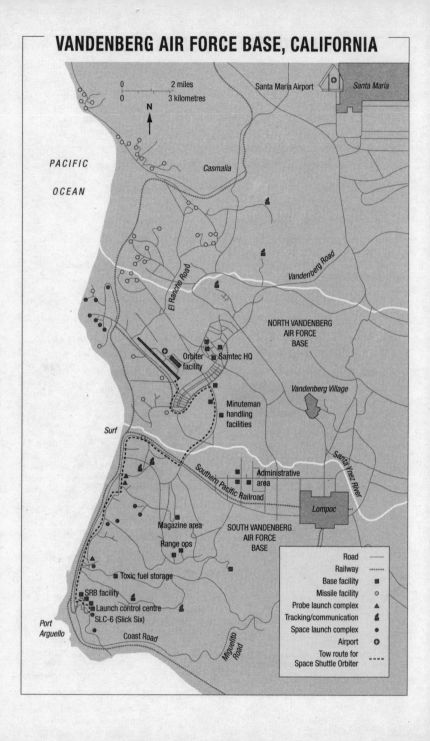

VANDENBERG AIR FORCE BASE, CALIFORNIA

PACIFIC OCEAN

Santa Maria Airport

Santa Maria

0 — 2 miles
0 — 3 kilometres

N

Casmalia

Vandenberg Road

El Rancho Road

NORTH VANDENBERG AIR FORCE BASE

Vandenberg Village

Orbiter facility ■ Samtec HQ

Minuteman handling facilities

Surf

Southern Pacific Railroad

Santa Ynez River

Administrative area

Lompoc

SOUTH VANDENBERG AIR FORCE BASE

Magazine area

Range ops

Toxic fuel storage

SRB facility

Launch control centre

SLC-6 (Slick Six)

Port Arguello

Coast Road

Miguelito Road

Road	—
Railway	+++++
Base facility	■
Missile facility	○
Probe launch complex	▲
Tracking/communication	≦
Space launch complex	●
Airport	⊕
Tow route for Space Shuttle Orbiter	---

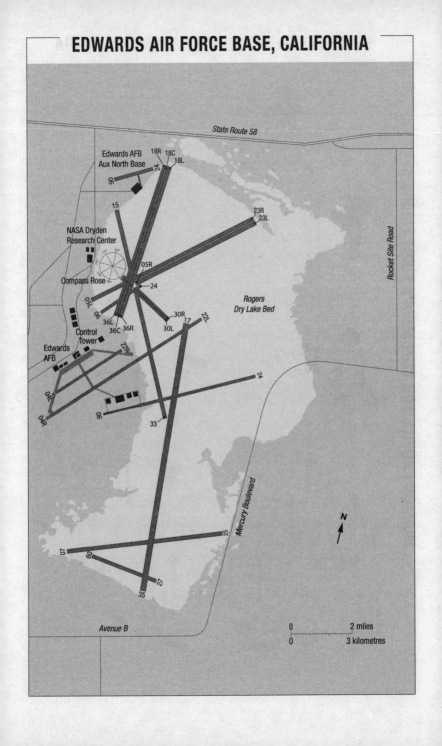

EDWARDS AIR FORCE BASE, CALIFORNIA

State Route 58

Edwards AFB
Aux North Base

18R 18C
18L
24
90

15

NASA Dryden
Research Center

05R

Compass Rose

24

05L
90

36L
36C 36R

30R

30L

17 22L

Control
Tower

Edwards
AFB

22R

24

04L

90

04R

33

23R
23L

Rogers
Dry Lake Bed

Rocket Site Road

Mercury Boulevard

25

07
09

27

35

Avenue B

N

0   2 miles
0   3 kilometres

MISSION PROFILE FOR *COLUMBIA'S* FIRST FLIGHT

Orbital Altitudes:
132.7 x 133.7 n.mi.
147.9 x 148.0 n.mi.
Inclination:
40.3 degrees

OMS-1
(0:10:34.1)

OMS-2
(0:44:02.1)
132.7 x 133.7 n.mi.

OMS-3
(6:20:46.5)

OMS-4
(7:05:32.5)
147.9 x 148.0 n.mi.

On-Orbit Operations

Manoeuvre for Re-Entry
(53:21:31.1)

Re-Entry
(53:49:01.1)
400,000 ft.
65.8 n.mi.

External Tank Separation
(0:08:58.1)

Main Engine Cut-Off
(0:08:34.4)
383,000 ft

Solid Rocket Booster Separation
(0:02:10.4)
163,000 ft

External Tank Impact
Indian Ocean

SRB Recovery
Atlantic Ocean
(0:07:10)

Launch
Kennedy Space Center
(0:00:00.3)

Glide Path
4,300 n.mi. from
Edwards Air Force Base

Landing Speed
180 knots
Rollout: 8,993 ft

Landing
Edwards Air Force Base
(54:20:54.1)

Event times are in actual Mission Elapsed Time (MET) in hours, minutes, seconds from launch.

ORBITAL INCLINATIONS DURING STS-1

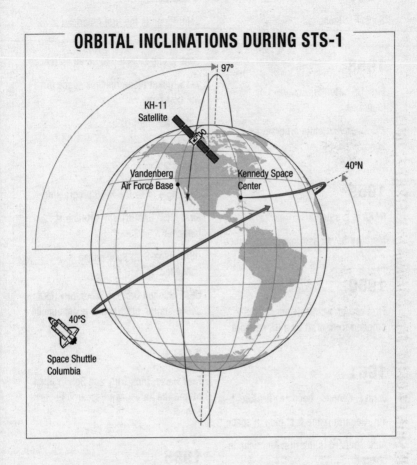

TIMELINE

1957

Sputnik is launched

1958

First US satellite, *Explorer 1*, is launched

Discoverer satellite programme announced

1959

NASA is created

Mercury Seven astronauts announced

1960

First re-entry capsule successfully dropped from orbit by a US satellite

1961

John F. Kennedy becomes President

Yuri Gagarin is the first man in space

Alan Shepard is the first American in space

Aerospace Research Pilots School (ARPS) opens at Edwards Air Force Base

National Reconnaissance Office (NRO) is created

1962

John Glenn is the first American in orbit

John Young selected to be an astronaut

X-15 rocket plane reaches space for the first time

Cuban Missile Crisis

1963

Joe Engle joins the X-15 programme

First KH-7 GAMBIT spy satellite is launched

USAF Dyna-Soar spaceplane is cancelled

USAF Manned Orbiting Laboratory (MOL) space station programme is announced

1964

Fred Haise, Dick Truly and Bob Crippen attend the Aerospace Research Pilots School

1965

John Young flies his first space mission aboard *Gemini III*

Joe Engle earns his astronaut wings in the X-15

First group of MOL astronauts – including Dick Truly – announced

1966

Joe Engle and Fred Haise join NASA

Bob Crippen joins the MOL programme

John Young flies his second space mission aboard *Gemini X*

1967

Apollo I fire

Six-Day War in the Middle East

Ex-MOL pilot Mike Adams is killed flying the X-15

MOL pilot Bob Lawrence is killed in a plane crash at Edwards

1968

2001: A Space Odyssey is released

Soviet tanks roll into Czechoslovakia

Apollo VIII flies around the moon

1969

Space Task Group make recommendations for NASA post-Apollo

Hans Mark becomes Director of NASA's Ames Research Center

First KH-8 GAMBIT spy satellite is launched

John Young orbits the moon on board *Apollo X*

USAF Manned Orbiting Laboratory programme is cancelled

Apollo XI – Neil Armstrong becomes the first man to walk on the moon

Bob Crippen, Dick Truly and five other MOL astronauts join NASA

1970

Apollo XIII survives an in-flight explosion on its way to the moon

NASA cancels all lunar missions beyond *Apollo XVII*

NASA and USAF sign their first agreement on the Space Shuttle

1971

Apollo XIV flies

First KH-9 HEXAGON spy satellite is launched

Soviet Union's *Soyuz 11* crew are killed on re-entry

Rocketdyne wins contract to build the Space Shuttle's main engines

Apollo XV flies

Joe Engle is replaced on the crew of *Apollo XVII* by geologist Harrison 'Jack' Schmitt

The Space Shuttle's basic layout is settled

Development of the KH-11 KENNEN digital spy satellite is approved

1972

President Nixon approves the Space Shuttle programme

Ames Director Hans Mark warns of the diffculties in developing the Shuttle's heatshield

Deep space probe *Pioneer 10* is launched

John Young commands *Apollo XVI* moon mission

Wernher von Braun retires as Director of Marshall Space Flight Center

North American Rockwell wins the contract to build the Space Shuttle

Apollo XVII – the last lunar mission – flies

1973

Apollo-Soyuz crew announced

Manned Spacecraft Center at Houston is renamed Johnson Space Center

Skylab is launched

National Reconnaissance Office uses a spy satellite to take pictures of *Skylab* in space

Pioneer 11 is launched

1974

Alan Shepard, the first American in space, retires

John Young becomes head of the Astronaut Office

Construction of the first Space Shuttle prototype begins

1975

First ignition of the Space Shuttle main engines

Joint US/USSR *Apollo-Soyuz* mission flies

1976

George Abbey becomes NASA's Director of Flight Operations

Kremlin authorizes development of a Soviet Space Shuttle

Crews for Space Shuttle Approach and Landing tests at Edwards AFB are named

NASA's first Gulfstream Shuttle Training Aircraft is delivered

Roll-out of *Enterprise*, the first Space Shuttle prototype

First flight of NASA's Boeing 747 Shuttle Carrier Aircraft

First NRO KH-11 KENNEN digital spy satellite is launched

1977

Jimmy Carter becomes President

First run of Shuttle main engine at 100%

Hans Mark becomes Under-Secretary for the Air Force and Director of the NRO

Star Wars movie is released

Deep space probes *Voyager 1* and *Voyager 2* are launched

First flight of Space Shuttle *Enterprise* at Edwards AFB

1978

John Young and Bob Crippen announced as crew for STS-1, the first Space Shuttle flight into space

Thirty-five new astronauts join NASA

Skylab rescue mission cancelled

Massive fire at Space Shuttle main test stand

1979

Roll-out of Space Shuttle *Columbia*

SALT II treaty signed

James Bond film *Moonraker* is released

Fred Haise retires from the Astronaut Office

Skylab burns up on re-entry

Columbia is delivered to Kennedy Space Center

1980

NASA awards contract for on-orbit tile repair kit to Martin-Marietta

Iranian hostage rescue attempt fails

Columbia is mated to the external tank and solid rocket boosters

Bob Crippen undergoes training for on-orbit tile repair

Columbia is moved to the launch pad

1981

Ronald Reagan becomes President

Iranian hostages are released

Hans Mark nominated as NASA's Deputy Administrator

Two pad-workers die from nitrogen poisoning at Kennedy Space Center

Assassination attempt on Ronald Reagan

First flight of Space Shuttle *Columbia*

PROLOGUE

The Next Generation

'The Shuttle is to space flight what Lindbergh was to commercial aviation.'

Arthur C. Clarke

Houston, 1969

DOTTIE LEE'S MOTHER thought it was unusual for her ten-year-old daughter to display such a love of classical music. That, though, seemed positively conventional next to Dottie's keen interest in astrophysics. Growing up in New Orleans in the 1930s, Dottie had known that there would one day be men on the moon. Now, thirty years later, she was working as an engineer at Houston's Manned Spacecraft Center and the prospect of Neil Armstrong and Buzz Aldrin realizing that goal was just months away. After graduating with a degree in Mathematics in 1948, Lee had trained on the job under Houston's Chief Engineer, Max Faget. Part of the team working on the heatshield that would protect the *Apollo XI* crew as they returned to Earth, she was already anticipating the celebrations that would mark their safe return.

On 1 April 1969, Lee took a phone call from Faget's office asking her to report to a room on the third floor of Houston's Building 36. She was told not to tell anybody where she was going. Lee arrived to discover around twenty other people gathered in a large room untidily stacked with chairs, boxes and other bits and pieces it seemed no one knew what else to do with. Alongside Lee, with her discipline of aerothermodynamics, there were representatives of all the centre's different engineering departments. No one from this team, which had, with *Apollo VIII*, already successfully flown a manned mission around the moon, had any idea what they'd been invited for. The best guess was that it was some kind of ambitious April Fool's Day practical joke.

The group weren't waiting for long before Max Faget walked into the room. Carrying a garment bag, the impish Chief Engineer, 5ft 6in in his bow tie, jumped up on to a desk in front of them. Lee had seen this kind of thing before from her boss. Faget was prone to performing headstands whenever and wherever he thought he needed to get the blood flowing to his brain, seemingly unconcerned about his pockets emptying noisily on to the floor. Standing on the desk, Faget unzipped the bag and pulled out a hand-made balsawood and paper model aeroplane. Hand-finished spars ribbing the cigar-shaped fuselage and straight wings were visible beneath the model's translucent pale brown skin. At the front was an upturned snub of a nose; at the back a pair of vertical fins mounted at each end of the horizontal tail. It was a shape unfamiliar to those watching the performance, and Faget was keen to explain it to them.

'It's stable in two attitudes,' he declared in his syrupy Cajun accent. 'Zero degrees angle of attack . . .'

The designer raised the model in his hand then launched it from on top of the desk, watching it fly gently across the room before skittering over the hard floor. One of the audience retrieved the model and returned it to Faget. So far, so unremarkable.

'And,' Faget continued, lining up his creation by eye before launching it again, nose high and tail low, 'sixty degrees angle of attack.'

Instead of flying horizontally, nose first, the glider presented its underside to the direction of travel, not so much flying as falling forward through the air. But it was doing so in as settled and undramatic a fashion as it had when it was flying more conventionally. To a gathering of NASA engineers, Faget's point was clear: this was a winged aeroplane capable of re-entering the atmosphere from space as safely as the blunt-bodied capsules that had so far returned America's astronauts to Earth.

'We are going to build the next-generation spacecraft,' Faget told them.

Space Shuttle *Columbia*, 1981

After the violence, fire and thunder of launch, there was no wind noise, nor engine noise, no force nor acceleration. But *Columbia*, the world's first reusable spacecraft, was travelling three times faster than any other winged flying machine in history, orbiting over 100 miles above Earth at a speed of 17,500 mph.

Inside it was quiet, except for the reassuring hum of cooling fans. Bob Crippen, the Space Shuttle's Pilot, made his way back to the aft crew station, the control panel built into the rear bulkhead of the flight deck, to test the mechanism of the big payload bay doors that ran the length of the spacecraft's back from flight deck to tail. *Here we go again*, he thought as he anticipated a job he'd repeated to the point of tedium in training; but then, slowly propelling himself backwards, careful to keep his head level and feet on the deck, he realized he wasn't really walking at all. He reached his position on the starboard side of the crew station and turned to John Young, *Columbia*'s Commander, standing next to him. 'You know,' Crippen said, 'this feels like every time I've done it in the simulator, except my feet aren't on the floor . . .'

Young smiled. A veteran of four previous spaceflights including command of the *Apollo XVI* moon mission, Young may have been NASA's most senior astronaut but he'd lost none of his enthusiasm for space travel. He loved the feeling of weightlessness. *Who wouldn't?* he thought, convinced that if there was no danger of breaking bones, everyone would be whirling around their living rooms.

Young floated forward again as Crippen monitored the slow progress of the payload bay doors on a pair of fore-and-aft-pointing closed-circuit TVs. Nearly a minute after the doors cracked uncertainly into action, Crippen was afforded a clear view aft through the big windows in the rear bulkhead of the flight deck. Looking across the payload bay towards the back of the Orbiter, he noticed an ugly peppering of black on the otherwise smooth white curve of the rocket pod that flanked *Columbia*'s vertical tail. *Uh-oh,* he thought.

'Hey, John,' he called to the mission Commander, 'come and take a look at this.'

To protect her from the blast furnace heat of re-entry, *Columbia* was covered in a dense mosaic of lightweight silica tiles, each unique, each fitted tightly against its neighbours. The integrity of the tiled heatshield was critical to the spacecraft's survival. If it was lost or compromised, there was no way home. When they plunged back into Earth's atmosphere, neither *Columbia* nor her two-man crew could hope to survive temperatures of nearly 3,000°F – hot enough to melt steel.

And Crip had just discovered there were tiles missing from the heatshield.

In Houston's Building 13, home to the Structures and Mechanics Division responsible for *Columbia*'s heatshield, Dottie Lee was watching the first television images beamed back from the new spacecraft with a small group of colleagues. They stared at the black scars on the Shuttle's rocket pod. At first people were too shocked to say anything. Then there was just one question on people's lips: 'Can they come back?'

And Dottie realized she was crying.

PART ONE

In and Out of the Shadows

'There wasn't going to be a British space programme. The Russians weren't going to put people on the Moon, and it was putting people out there that was important. We understand environments through experience, not robot telemetry. All my hopes for a science-fiction future, for performing the exploration the human being is hardwired for, were pinned on the American space programme.'

Warren Ellis

ONE

21 February 1962

LIEUTENANT COMMANDER JOHN Young was strapped into the cockpit of the McDonnell F-4H-1 Phantom II. His breath condensed into thick clouds as he exhaled into the cold, dry air. It was just 8°F. The bitter temperature was no accident, however. The average in Brunswick, Maine, in February sat well below freezing. Cold air, though, meant denser air. And that gave the thin wings and spinning turbines of the big Phantom's two turbojet engines more to bite on. Cold air gave Young the leg-up he needed if he was going to be able to add to the clutch of world records already held by the Navy's impressive new interceptor. Not that he was relying solely on the weather. To save weight he was flying the two-seat jet alone. The weapons systems had been stripped out along with many of the cockpit instruments. Parts of the aluminium airframe – including the speed brakes – had been replaced with balsawood, and the F-4's two General Electric J79 engines had been uprated to provide extra thrust. That, though, came at a cost in fuel consumption. And to further save weight, Young would begin his take-off roll with just 1,800lb of fuel in the tanks. In full afterburner, those two J79s were going to guzzle that in barely thirty seconds. But that was all the time Young needed to streak to a height of 3,000 metres (a little under 10,000 feet) faster than any aircraft had ever done before.

After closing the canopy, Young started the engines, scanning what remained of the instruments for any signs of trouble. It all looked good. Ahead of him, through the cockpit glass, was a mile and a half of asphalt runway more used to the comings and goings of the Navy's lumbering propeller-driven patrol aircraft. He was barely going to need a fraction of it.

With his gloved left hand, he advanced the two throttle levers to full military power, then on through the gate to light the afterburners. Cold fuel sprayed into the exhausts and the Phantom, spears of flame shooting backwards from beneath the tail, strained against the holdback bolt. Instead of wasting fuel as she accelerated down the runway, the Phantom was being held until, with the engines throwing out a maximum 32,500lb of thrust, the fuel levels fell to 1,800lb. Young watched the fuel gauge until, when the needle hit 1,800, the jet, released, leapt forward like a bullet from a gun.

Thirty-three and a half seconds later, as the Phantom soared through the 3,000-metre mark, still climbing steeply at an angle of 60°, the thirty-one-year-old US Navy test pilot in the cockpit wrote his name into history for the first time.

Young was brought up poor. His father, Hugh, a civil engineer, was laid off during the Great Depression and moved his family back to Georgia where he pumped gas at a filling station for a few dollars a day. When John was five the family moved on to Orlando after his father found new work. But then, a family that had already endured its fair share of hardship was struck again.

Halfway through his first school year in Florida, John's mother, suffering from schizophrenia, was taken away in a straitjacket to Florida State Hospital in Chattahoochee. He hadn't even known she was unwell, but she never returned from the asylum. Yet despite permanently losing his mother to illness, Young prospered, inheriting an intuitive talent for engineering from his father. And he lived and breathed aeroplanes, reading obsessively about heroes like Charles Lindbergh, Eddie Rickenbacker and Jimmy Doolittle, augmented by science fiction such as Flash Gordon, Buck Rogers and Edgar Rice Burroughs' influential *John Carter of Mars*. A straight-A student at Orlando High School, Young seemed to excel in whatever he put his mind to, from athletics to physics. It was to take him to Georgia Tech on an ROTC scholarship and a BSc in Aeronautical Engineering, before he joined the Navy which, in its wisdom, instead of sending him to flight training, informed him he'd be joining the crew of the destroyer USS *Laws* as a gunnery officer.

Bitterly disappointed, he endured a frustrating tour aboard the *Laws* firing 5-inch shells into North Korean positions before finally being assigned to flight school in 1953. After earning his wings of gold as a naval aviator he joined a fleet fighter squadron before going to test pilot school at Naval Air Station Patuxent River, Maryland – a site, as a naval aviator, he regarded as hallowed ground. When he graduated second in his class in 1959 he was singled out for the Navy's new F-4 Phantom. A hot ship, the F-4 had already been setting new speed and altitude records for a couple of years before Young was chosen as one of the five pilots for Project HIGH JUMP, the US Navy's attempt to seize a raft of time-to-climb world records from the Air Force. He thought such an impressive demonstration of the fighter's performance might give the Russians pause for thought about invading Western Europe. Soon HIGH JUMP would give him an opportunity to assert his country's superiority in an entirely different arena.

With the 3,000-metre record in his pocket, Young landed back at Naval Air Station Brunswick and climbed out of the cockpit. Afterwards, the range trackers told him that the whole thing had happened so quickly they thought it had probably been easier to fly the Phantom than try to track it. In a southern-country-boy drawl, Young told them: 'It's not like flying as much as it's like riding a rocket!'

It was a notion fresh in the minds of everyone involved in Project HIGH JUMP as they worked through the bitter cold to prepare the Phantom for a second flight, an attempt on the 6,000-metre record. Just a day earlier, on 20 February, a Marine Corps fighter pilot called John Glenn had blasted off from Cape Canaveral in Florida on top of a man-rated Atlas missile to become the first American to orbit the Earth. The space race was in full cry. And, despite the success of Glenn's mission, the United States was playing catch-up.

The Soviets had been first out of the blocks when, in October 1957, a Soviet R-7 intercontinental ballistic missile launched a small satellite called *Sputnik* into an orbit that saw it fly over the continental United States several times a day. To the American public it was both a humiliation and a threat. It felt as if the race had already been lost. And what had been a relatively leisurely pursuit of spaceflight by all three branches of the country's armed forces became a national priority. On 1 October 1958, the National Advisory Committee for Aeronautics (NACA) became NASA, the National Aeronautics and Space Administration. Before the end of the year the United States launched an orbiting satellite of its own and a Space Task Group was set up within NASA at Langley; their job was to direct a manned space-flight programme. With thirty million dollars' worth of funding they began work on developing a small, one-man space capsule they christened *Mercury*. In April 1959, John Glenn and six other military test pilots were introduced to the world as the men who would be America's first astronauts. The Mercury Seven. A half-million-dollar deal with *Life* magazine was to see them become some of the most famous people in the country. It seemed as if, after the disappointment of *Sputnik*, the American programme was back on track. Until, in April 1961, Soviet cosmonaut Yuri Gagarin became the first man in space, completing a single orbit of the Earth inside his tiny *Vostok 1* capsule.

In reply, NASA just managed to poke its nose into space by firing Mercury astronaut Alan Shepard above the atmosphere in a brief, fifteen-minute sub-orbital flight before, on 25 May 1961, the new President, John F. Kennedy, stood up in a special joint session of Congress to announce the goal of putting a man on the moon and returning him safely to Earth before the end of the decade. The excitement that surrounded John Glenn's orbital flight in February 1962, followed by the President's visionary speech at Rice University later that year, in which he again sold the idea of a moonshot, seemed to reassure the public that NASA was on its way.

* * *

John Young wanted to be part of it. Not only did he believe

that joining the space programme would be the best way to make use of his considerable skills as a test pilot, he was also curious about what they'd find when they got there. He applied to NASA and was called to Houston for interview in June 1962. Asked by the selection panel what he thought astronauts needed to study before travelling to the moon, he answered, 'Geology.' It was the kind of response that illustrated, to those who knew him, that there was a good deal more to Young than that laconic drawl might suggest.

Three months later he got a call from NASA asking if he was still interested in becoming an astronaut.

'Yes, sir, I am,' he said.

TWO

WHEN *SPUTNIK* CAPTURED the world's imagination, Robert Laurel Crippen couldn't help but feel a twinge of envy. He was in his sophomore year at Sam Houston State University in Huntsville, Texas. Crippen was fascinated by the Soviets' achievement. As a freshman, he'd written an English paper on rockets and space, but now, still at college, he was frustrated at the timing of the little satellite's launch. Manned spaceflight would surely follow, and there were others just a few years older than he was, he realized, who would get to fly in space before he was ready.

Crippen's father was an oil worker who, after losing a couple of fingers in an accident, used his insurance money to buy a chicken ranch in the small town of Porter, Texas. In time, he and his wife added a service station and beer joint, called it Crippen's, and hung a neon sign outside. He'd been up in an aeroplane once, but decided it wasn't for him. His reluctance didn't stop him driving his five-year-old son out to the perimeter fence at Houston's Hobby Airport to watch the airliners take off and land, however. On one occasion a stewardess spotted them and asked, 'Do you want to take a look?' She invited father and son through a gate to a parked-up DC-3. They climbed through the rear door and walked up the taildragger's sloping floor to the cockpit. The boy sat in the captain's chair and took it all in. Bob Crippen wondered if it had been that first encounter that spurred his interest in aviation.

After high school, the young Crippen worked as a railroad switchman and served meals in a boarding house to help pay his way through college. But by now he had very clear goals in mind. He was studying for a degree in Aerospace Engineering at the University of Texas when the names of NASA's Mercury Seven astronauts were announced. *I'd like to do that*, he thought and, noting that all were test pilots, decided to join the Navy because he believed it would make him the best pilot he could possibly be. *At least*, he figured, *if I can land on one of those aircraft carriers, I certainly wouldn't be* too *bad*.

He climbed aboard an aeroplane for only the second time in his life when he took a seat on a Trans Texas DC-3 going from Austin to Dallas to join the Navy. This time, though, it was going to take off. He'd imagined it often enough, now Crippen was going to fly.

While *Sputnik* had inspired Bob Crippen's thoughts of space-flight, the reaction of the Air Force to the Soviets' pioneering efforts in space was rather more belligerent. As the Navy taught Crippen to fly in a little piston-engined Beech T-34B Mentor with a maximum speed of just 188 mph, the Air Force, stung by *Sputnik*, was issuing the country's aircraft manufacturers with a flurry of increasingly ambitious requests for proposals – RFPs. Among the first was one for a Strategic Orbital System. The study was completed and submitted before quickly being followed by an invitation to tender for a Strategic Lunar System. Again, competing bids were submitted. But when that was followed with an RFP for a Strategic Interplanetary System, engineers at the Hughes Aerospace Company were incredulous. The whole idea of projecting power from deep space seemed so absurd that Hughes chose not to produce a proposal, only to prompt a phone call from the Pentagon demanding that they take part. Across the Atlantic, even the Royal Air Force were discussing their own plans for Near Space Control and Far Space Control. As far as the free world's military were concerned, *Sputnik* had created a new battleground upon which they must now be ready to fight.

They were swimming against the tide. In January 1958, three months after the launch of *Sputnik*, President Eisenhower suggested a bilateral agreement between the United States and the Soviet Union 'that outer space should be used only for peaceful purposes'. Then, in February the same year, he appointed a Science Advisory Committee to define the nation's space programme. In the committee's report, 'An Introduction to Outer Space', while there was an acceptance that defence was part of the jigsaw, the emphasis was on exploration, science and prestige. NASA, a civilian agency 'devoted to peaceful purposes for the benefit of mankind', was the result.

But the creation of the new agency did nothing at all to dampen the ambition of the Air Force to carve out its own place in space. And if *Sputnik* had sparked the flying branch's nascent space efforts, Yuri Gagarin's first manned space-flight in April 1961 fanned the flames. In the same month that Kennedy first talked of putting a man on the moon, the Air Force published a detailed 200-page proposal for their very own military moon mission, dubbed LUNEX. Inside were designs for lunar landers and re-entry vehicles, flowcharts mapping out the route from 'Preliminary Design' to 'First Operational Launch', flight-testing schedules, comprehensive equipment manifests, and personnel requirements from aerospace pilots to diet supervisors. The training of the crew required a programme 'unique to the Air Force'. The foreword to the Lunar Expedition Plan pointed out the 'dire need for a goal for our national space program'. The country now had that. And it didn't need the threat drawn by the Air Force that by the early seventies Soviet 'military facilities may have been established on or in orbit around the moon' for it to embrace that goal.

Yet the Air Force's interest in space was not entirely unreasonable. In 1961, it remained essentially uncharted territory. And, in a speech to the American Ordnance Association in October that year, the cigar-bothering Air Force Chief of Staff, General Curtis LeMay, explained why:

> Looking back at the history of airpower, you will recall the first use of the airplane in World War I was for reconnaissance. For a time air operations were conducted politely and with chivalry. Opposing pilots waved and nodded to each other as they passed. Both sides had equal access to the sky. But once reconnaissance began changing the course of battles, the rules changed. It didn't take long before commanders realized that it was necessary to deny the opposition this aid from the sky. Soon opposing airmen were engaged in battle. First it was air-to-air bombs and small arms. Then they graduated to the machine gun. After this came bombers and aerospace had become another area of conflict. I think we will be very naive if we don't expect and prepare for the same trends in space.

* * *

Thoughts of war in space were far from Bob Crippen's mind when he joined attack squadron VA-72 'Blue Hawks' in the summer of 1962. It was enough of a challenge closer to Earth. Not long after joining the unit, the young naval aviator took off wearing a thick rubberized immersion suit for the first time. The 'poopy suit' was designed to protect pilots who were forced to eject over cold water. The four Douglas A-4 Skyhawks climbed out from Naval Air Station Oceana with ordnance slung under their wings and headed out to sea – the division was practising its bombing. Crippen rolled into a dive towards the floating target. Tracking the pipper in the little jet's bombsight, he pressed the bomb release and pulled up, but, made ham-fisted by the heavy, unfamiliar poopy suit, he snatched back on the stick too forcefully. He was immediately mashed back into his ejection seat, the Skyhawk pulling 8.5gs as her nose came up over the horizon. It was too much, ripping a slat off the starboard wing. He nursed the shaken jet back to base, but got saddled with the nickname 'Slats'. (He was luckier than 'Leaky', the squadron mate who, caught short in the cockpit, managed to pee into his leather flying glove without spilling a drop.) In the end, though, 'Slats' didn't stick. Everyone just knew the tanned, brown-eyed Texan with the crewcut as Crip.

After his first night landing aboard an aircraft carrier he thought, *I can do just about anything*. For a while, as he flew from the deck of the Navy's Forrestal Class aircraft carrier, USS *Independence*, during the Cuban Missile Crisis, it looked like that might include war. Test-flying, though, remained his goal, and he filled in the application form in 1964. It offered him a choice: do you want to go to the Navy's test pilot school at Patuxent River? Or would you be prepared to go to the Empire Test Pilots School in the UK, or train with the Air Force instead? He figured the odds of getting in were better if he said yes to all three of them. When his orders came through telling him he'd been posted to Edwards Air Force Base he was surprised to discover he'd be a student at an outfit called the Aerospace Research Pilots School – *I thought I was going to test pilot school*. The syllabus, he learned, included spaceflight. And one of the instructors on the space dynamics and control course was a young Navy pilot who, like Crip, had only just over a year before been flying off carriers during the Cuban Missile Crisis. His name was Lieutenant Dick Truly.

THREE

Edwards Air Force Base, 1963

WHEN DICK TRULY arrived at Edwards with his wife, Cody, two young sons and the family dog after a long cross-country drive, he felt *like a cowboy just arrived at IBM*. There were few places in the world that were more glamorous than Edwards. *Like Peyton Place*, thought Truly. The vast air base, scratched on to the Mojave Desert alongside the endless runways provided by Rogers Dry Lake Bed, was teeming with young couples glowing in the face of their illustrious profession: flying and testing the world's most advanced, high-performance aircraft. Instead of training at the Navy's own test pilot school at Pax River, an exchange programme saw Truly chosen as one of two Navy students at the Air Force's new Aerospace Research Pilots School. After feeling top of the heap flying supersonic F-8 Crusaders off pitching carrier decks, Truly knew he was back at square one; a nugget, as the Navy put it. Even putting aside the exotic flying machines – nearly every one of which had at some point claimed a world record – everything at Edwards was newer and more state-of-the-art, from the blue trucks that ferried people around to the 'bricks', walkie-talkies possessed by everyone on the flightline. The F-8 had been the best the Navy had to offer, but this was something else. He had come a long way from Fayette, Mississippi, the segregated town of just 1,000 people where he'd grown up.

Dick, Cody and the rest of the students of Class 64A and their wives were greeted in the school's lecture theatre by the Commandant. It wasn't a sight to inspire confidence in choosing 'Test Pilot' as a vocation. Colonel Chuck Yeager strode on stage wearing full dress uniform, his chest adorned with coloured ribbons in recognition of an extraordinary flying career. Beyond that, though, he was looking pretty beat up. America's most famous test pilot welcomed the group to Edwards while sporting a fresh bandage around his neck. The left side of his face was still raw from a month spent in hospital having scabs scraped off it every four days to prevent scarring. His left arm was suspended in a sling. His audience soon got to see what had done the damage.

From the auditorium, they were taken to the school's hangar. On the floor lay the crumpled wreckage of the Lockheed NF-104 AeroSpace Trainer in which, just weeks earlier, Yeager had lost control while trying to add the world altitude record to his résumé, alongside 'WWII ace' and 'first man to break the sound barrier'. After he'd punched out, Yeager's rocket ejection seat had smashed through his helmet visor, igniting the oxygen he was breathing. Suffocated by smoke and well-nourished fire, he hung beneath his parachute trying to shovel air in through the broken faceplate using his gloved hand. He succeeded only in setting that on fire too. Not until he managed to push open the visor was he able to shut off the oxygen and extinguish the flames. But by the time he hit the ground he was a mess. When the rescue helicopter reached him, the black rubber lining of Yeager's helmet had bonded to his face, and in removing the remains of the glove from his burned hand with a penknife, he had also managed to pull off the broiled tips of two fingers.

America's most celebrated test pilot had been attempting to soar beyond 120,000 feet, a height at which a normal aeroplane has long since ceased to be controllable because of the thinness of the air. Built specially for the school, the AeroSpace Trainer was half aircraft, half spacecraft, though. A standard F-104 Starfighter, fitted with a 6,000lb-thrust LR-121 rocket in the tail to provide power as the jet engine struggled for breath, and small puffer jets to provide attitude control in the stratosphere, the NF-104 could be flown where aerodynamic flight controls no longer functioned. But then it didn't *feel* like flying any more, and Yeager's accident had highlighted that. The great aviator's legendary feel for an aeroplane was no longer of any use to him. To equip them to succeed in test-flying's emerging new world, Yeager's students needed to be taught new skills. The syllabus of the year-long course was designed first to teach the students to be test pilots. Then to be astronauts.

The Aerospace Research Pilots School had come into being in 1961, just months after Alan Shepard and Gus Grissom had between them accumulated a grand total of just fifteen minutes in sub-orbital space. There were reminders everywhere that this was still largely an unknown arena. The students had no printed textbooks. Instead, typed, sometimes handwritten, study notes were collected in ring-bound files. Much focus was placed on biomedical research and the potential effects of spaceflight on the human body. Before being accepted by the school, each of the students had spent a week at Brooks Air Force Base near San Antonio, Texas, being prodded, probed and measured. As students they returned to Brooks to endure capillary-bursting rides in a centrifuge designed to replicate the g-loads astronauts were expected to be subjected to on re-entry. They began with two minutes at 8g, then thirty seconds at 12g, during which each part of their body weighed twelve times what it should – forces that would rip the wings off an aeroplane. As they spun, the students used a hand controller to move lights on the instrument panel to the centre of a display to prove they were still conscious – a task complicated by the effort to make sure they didn't swallow their tongues. They finished with a run to 15g and back. It earned each of them membership of the Order of the Elephant and a certificate illustrated with a picture of an elephant standing on a man's chest. That gave a pretty good idea of what it had felt like.

They trained at the other extreme, too, in a zero-g environment, for the weightlessness they would experience in space. Inside a modified Boeing KC-135 tanker-transport, twenty airline-style seats were separated with cargo netting from an empty section of cabin at the back lined with thick crash pads. The big four-engined jet was flown in a series of climbs and dives, and as it bunted over the top of each parabola the passengers in the back would enjoy the sensation of zero gravity for thirty to forty seconds before a series of flashing lights triggered by the pilot signalled that the KC-135 was pulling out of the dive. Before their first flight the instructor warned them that they would fall into two groups: those who'd been sick and those who would be. Not for nothing was the jet known as the 'Vomit Comet'. In the end, though, they were having too much fun floating around, bouncing off the walls and each other, to remember to puke. Since arriving at Edwards, flying as a passenger had become a novel experience for Class 64A.

When he arrived at Edwards, despite being certain that the F-8 Crusader was the best flying machine there ever was, Dick Truly soon discovered there was a good deal more on offer. The rule was that you could only be current on five types of aeroplane. The reality was that if there was another machine you wanted to fly, the paperwork could be done the night before: one in, one out, though the total never exceeded five. Truly and the other students flew and flew. During the first half of the course it was often T-33 and T-38 trainers, but Truly also added powerful Century series fighters like the F-100 Super Sabre, F-101 Voodoo and F-106 Delta Dart to his burgeoning logbook. There was bigger iron on offer too in the form of a twin-engined NB-57 Canberra bomber, a piston-engined B-26 for stability and control training, and even the C-130 Hercules four-engined turboprop cargo plane. By the end of six months' flight-testing the F-8 still had its claim on Truly's affections, but he now understood its strengths and weaknesses. Then, as he and his fellow students moved from test-flying to space training, there was one machine which, while it may have failed to impress the USAF as an interceptor, was practically tailored to the requirements of training Air Force astronauts: the Lockheed F-104 Starfighter, the only aircraft ever to have simultaneously held the world records for both speed and altitude.

On the ground, the class studied orbital mechanics, the mathematics of calculating satellite movements, bio-medicine, spacecraft design and aerothermodynamics. They also 'flew' the school's ground-based GPS T-27 simulator, in which they could practise orbital manoeuvring, rendezvous and docking in a facsimile of a spacecraft.

It seemed to be doing the trick. Ten graduates from the Edwards test pilot school had already been accepted into the ranks of NASA's astronaut corps. Another, Captain Joe Engle USAF, would soon be earning his astronaut wings elsewhere.

FOUR

Joe Henry Engle loved airplanes. *Really* loved airplanes.

His first memory of growing up in Chapman, Kansas, in the 1930s was of his sister building him a model plane out of tin can. He dug sandpits for cockpits, kitting them out with cans for instruments and a branch for a control column. His weapon of choice was the Curtiss P-40 Warhawk. After watching a movie about the exploits of the Flying Tiger Squadron in China he decided that was all he ever wanted to do. Joe was still a boy when the war ended and, as happy as he could see that made people, tearing at him was the thought that he'd never get to be a fighter pilot and shoot down Japanese Zeroes. But he could still fly.

Fair-haired and hazel-eyed, Engle chose to study Aeronautical Engineering at the University of Kansas while during the summers he worked as a draughtsman at the Cessna Aircraft Company in Wichita. That was where he learned to fly, buying an hour of flying time for every eight hours' work he did clearing up at the local airport after he'd finished his day at Cessna. After graduating, Engle had to wait nine months before getting a slot at Air Force flying school. He used the time predictably enough: building an aeroplane with the Cessna supervisor who'd taught him to fly. They constructed the wings in the garage, the fuselage in the living room and the tail in the spare bedroom; the kitchen was used for engine testing. When the whole thing was finished, they rolled the fuselage outside, attached the wings, and took off from the street outside.

By the end of the fifties, Engle, a lean, loose-limbed six-footer, was a fighter pilot. Assigned to a new squadron, the 474th, based at George Air Force Base in the Mojave Desert, he was flying a former world air speed record holder in the shape of the North American F-100 Super Sabre, the first fighter in the world capable of travelling faster than sound in level flight. The squadron's primary mission was air defence and, even as a young Second Lieutenant, Engle's talent was sufficiently evident that he was made a designated flight leader. Whenever he could he'd lead a flight of F-100s up into the skies over Death Valley to practise dogfighting. He might have missed out on the Flying Tigers but Joe Henry Engle was doing the things he'd dreamt about day and night his entire life.

The icing on the cake was that the 474th shared the George flightline with another F-100 squadron commanded by a pilot Engle had previously only admired from a distance: Colonel Chuck Yeager. Engle would sit in the squadron ops block just watching the way Yeager walked out to his jet with his parachute. He tried to emulate Yeager's fearless rolling gait, and if he knew Yeager was briefing his squadron he'd slip into the back of the room just to hear his distinctive drawl. *Nobody*, thought Engle, *can talk like a West Virginian*, but it didn't stop him trying. Growing increasingly confident, Engle once made the mistake of attempting to bounce an unsuspecting Yeager in the air. Spotting two Super Sabres returning to George from the north-east, the young fighter pilot set up his own four-ship flight to attack. Following his lead, the four 474th jets rolled in, only to have their asses handed back to them by Yeager and his wingman. It was a lesson learned, but it got Engle noticed.

Engle's logbook needed to show 1,500 hours of flying before he was eligible for test pilot school. On days off, at weekends, he made it up any way he could, pulling gunnery targets in P-80 Shooting Star jets and prop-driven B-26 Invaders, or ferrying gunnery range officers to and from their observation posts at dawn in George's old high-winged De Havilland Beaver bush plane. In Engle, Yeager recognized something of a kindred spirit. 'For the best pilots,' he maintained, 'flying is an obsession, something they must do continually.' Yeager recommended Engle for test pilot school at Edwards.

After graduating, Jovial Joe stayed on at Edwards, assigned to the Fighter Test Operations division where he took part in an evaluation of a new subsonic attack jet designed for close air support: Cessna's YAT-37 Dragonfly. He was loving every minute, but in the wake of Gagarin's first spaceflight it seemed clear where the future of the Air Force lay and so Engle applied for a place at Chuck Yeager's Aerospace Research Pilots School. And when NASA announced a new astronaut selection, he submitted his application.

Soon afterwards, he was called in to see Major-General 'Twig' Branch, Edwards' Commanding Officer, who told him, 'I'm pulling your application to NASA.' Engle wondered what he'd done wrong. Then Branch asked him, 'Did you really want to go?' The truth was, Engle *did* have mixed feelings about it. The idea of exchanging real stick-and-rudder flying for the controls of a capsule didn't thrill him.

'Well, I thought it would be a career thing,' Engle replied weakly.

'We've got something else in mind for you,' Branch said, 'but I can't tell you right now . . .'

Engle soon discovered that the gods had blessed him with a slot on the X-15 programme: a joint USAF/NASA hypersonic flight research project. He couldn't have been more thrilled. The rocket-powered North American X-15 was the ultimate winged flying machine; the pinnacle of stick-and-rudder flying. For the next three months the training was as intense as he had ever known. By day he'd replicate the high-speed gliding landings made by the X-15. In the evenings, at weekends, whenever he could persuade any of the operators to come in, he'd ride his Lambretta scooter across the Mojave back roads to NASA's Dryden Flight Research Center and sit in the simulator, imagining the flight just as he had as a kid sitting in a sandpit in Chapman.

NASA's Mercury capsule was not the first American manned spacecraft to fly. That honour belonged to the X-15 (even if, ultimately, the capsule was first into space). In September 1959, North American's experimental rocket plane lit its engines for the first time and crested at a height of 52,341 feet. It was still a long way short of spaceflight, but it marked the beginning of perhaps the most successful flight test programme ever undertaken. It was December of the following year before NASA successfully lofted an unmanned Mercury capsule into sub-orbital space atop a Redstone missile launched from Cape Canaveral. On 5 May 1961, NASA's Alan Shepard was strapped into the same combination to become the first American in space. He was followed by Gus Grissom, John Glenn and, in May 1962, Scott Carpenter. Then on 17 July 1962, USAF test pilot Robert White flew the X-15 to a height of 314,000 feet – above the 50-mile limit where the Air Force considered Earth's atmosphere ended and space began. The flight earned White his astronaut wings from the Air Force, but, although he was the fifth American in space, his achievement was entirely overshadowed by the fanfare and circus surrounding NASA's Mercury Seven.

Engle's first X-15 flight was just for familiarization. But even using only 50% of the available engine power, the North American rocket plane was still going to fly him twice as fast as he'd ever travelled before. He nearly didn't launch at all though. Engle had already called for an abort when an instrument failure corrected itself and allowed him to go. The stub-winged black needle dropped away from beneath the wing of NASA's NB-52 mothership, 'Balls 8', and Engle lit the Reaction Motors XLR99 engine. As he accelerated away, an electrical malfunction caused him to lose all his flight instruments bar the g-meter. Having practised and memorized the flight profiles in the simulator, though, Engle was completely unfazed by the inconvenience. In fact, being forced to fly the world's hottest ship by the seat of his pants was Engle's idea of fun.

As he gained altitude, he kept a close eye on the g-meter spooling up – his only way of measuring the accuracy of his flying – but pushing over the top of the climb at 77,000 feet and four times the speed of sound, he looked out of the flat plate glass of the cockpit to see the lake bed he had to land on disappearing behind him. Instinctively, Engle rolled the rocketship on to its back and gently pulled the nose down to head for denser air at lower altitude and avoid over-shooting. As the X-15 descended, Engle completed the 360° roll, levelled the wings, and opened the speed brakes to bleed off his excess velocity. The X-15 felt satisfyingly crisp and responsive. *Boy*, he thought to himself, *this thing really is a nice flying airplane*.

On the ground, the engineers monitoring the telemetry in the control room thought that Engle had lost control. After a safe landing, one of them quizzed him at the debriefing, 'Hey, you didn't roll that airplane, did you?'

'Who, me?' Engle replied, all innocence.

'I didn't think so.' It was inconceivable to the technician that a rookie would perform aerobatics in a machine that had, in just its previous mission, flown a man into space and back. *On his first flight*.

Now he was flying rocket planes at Edwards, Engle became good friends with Chuck Yeager. The ARPS Commandant invited his protégé on long hiking trips he led up in the Sierra Nevada mountains. The two men shared a love of hunting. Fishing too. When, because of a change to the X-15 flying programme, Engle was unable to join Yeager on an expedition up to Kern River to catch golden trout, he thought he'd give the camping party a little variation to their otherwise all-fish diet. He packed thick steaks into a helmet bag and tucked them behind one of the airbrakes of a Starfighter. Beneath the other he filled a bag with frozen fishsticks. As he flew low over the camp, Yeager flashed him with a signal mirror. Engle pulled the F-104 round the corner for another pass, low and slow. He popped the airbrakes, and one of the bags dropped to the ground; the other got caught up for a few seconds before ripping loose and tumbling down a ravine. Yeager opened the bag with the steaks, and, unable to resist the temptation to double up, climbed down the canyon to search for the other bag. After a long hike, he found it. And the fishsticks. It was a gotcha Engle couldn't have pulled off any better if he'd tried.

Joe Engle earned his astronaut wings flying the X-15. Three times in 1965 he flew beyond the 50-mile limit, the threshold of space. Beneath him, as he travelled weightless over the top of X-15's ballistic arc, the soft blue curve of the Earth's horizon cut across the deep black of space. On one of the flights he lost, and had to reset, a flight control circuit critical to his ability to safely re-enter the Earth's atmosphere twenty-one times in ten minutes. Again, Engle was unperturbed, regarding it as a minor distraction. While he flew a precise profile to a height of 271,000 feet with his right hand, with his left, he calmly reset his yaw damper switch every time it tripped. After the first time it happened, he didn't think it was worth even mentioning again. At the post-flight debriefing, the engineers were again left scratching their heads.

Or marvelling.

There was no doubt about it, Jovial Joe was doing the best job in the world. But as a serving Air Force pilot he knew it would come to an end when he was handed a new assignment over which he'd have no say. And Engle wanted to go to the moon, a journey that he regarded, through the eyes of a test pilot, as *a tremendous envelope expansion*. He applied for NASA, and was accepted.

But, as Bob Crippen was about to discover, for an aspiring astronaut, NASA was not the only show in town.

FIVE

Edwards Air Force Base, 1966

BOB CRIPPEN WAS encased in a full pressure suit, his helmet secured to its torso with a locked metal ring. The zoom climb was the climax of his Edwards syllabus – a flight to as close to 90,000 feet as a Lockheed F-104A Starfighter could take him. It was a test of both the nerve and the ability of Yeager's students in an environment that was usually the sole preserve of exotica like the X-15. Crip checked his suit and switched off the cockpit pressurization. At the height he was ascending to, the air-breathing Pratt and Whitney J79 engine would be starved and, without it, the Starfighter's own system couldn't function. If the pressure suit failed, Crip's blood would boil in the thin air of the upper stratosphere.

He levelled off at 35,000 feet, glancing at another F-104 flying chase off his shoulder. Checking his heading, he advanced the throttle through the gate to full afterburner and felt 15,800lb of thrust push him from behind. He scanned his instruments – temperatures, pressures, fuel – and watched the Mach meter spin up as he accelerated. He let the height creep up from 35,000 feet as he urged the silver-skinned Starfighter past Mach 2 – twice the speed of sound. Then he pulled back gently on the stick, loading on g as he raised the nose to 45° above horizontal. Rocketing away from Earth, higher than anyone else in the sky above Mojave, Crip and the Starfighter kept climbing. The altimeter recorded his progress. 50,000 feet . . . 55 . . . 60 . . . the J79's afterburner – raw fuel sprayed into the jet pipe to augment the engine's thrust – suffocated and extinguished. Crip kept climbing, eyes checking the aeroplane was on a knife-edge angle of attack and g-load, scanning the rising engine temperature . . . 65,000 feet. He shut down the engine to prevent it overheating. With that the cabin pressure was lost and he was glad of the embrace of the pressure suit as he soared past 70,000 feet, nudging the stick forward a touch to keep her from stalling. But from here on he was travelling like an artillery shell, the jet's momentum carrying her up and forward, trading speed for height. As the Starfighter carved a ballistic arc to near 90,000 feet he felt himself rise out of the seat, a few seconds of weightlessness – and silence – before the razor-winged jet pointed her nose back towards the ground and gathered the speed he needed to get her flying again. Zero-g, the indigo blue-black of the edge of space, a hint of the curve of the Earth. He wanted more.

Both NASA and the Department of Defense were asking for astronaut applications. The forms gave him the option to choose NASA, DoD, or not express a preference. *I'll take either*, he thought. He figured the more bets you place, the better your chances. But when he graduated from Edwards, the Navy asked him to jump one way or the other. From where Crip was sitting, NASA appeared to have a lot of astronauts. By contrast, he knew the military had just eight.

Maybe, he figured, *my chances of flying are better in that direction.* DoD, he told the Navy. And so in June 1966, Bob Crippen was named in a second selection of astronauts for a classified military space programme. He was looking forward to working with his friend Dick Truly again.

Truly had been listening to the car radio when it was announced that Secretary of Defense Robert McNamara had approved an Air Force space station. At the time it hadn't registered, but he'd soon learned that his Commandant at the Aerospace Research Pilots School was eager to see the Air Force claim what he regarded as its rightful place in space. Having glimpsed the curvature of the Earth on his own high-altitude flights, Chuck Yeager now looked forward to Air Force space stations, shuttles armed with particle beam weapons and lasers. 'No bluesuiter,' believed Colonel Yeager, 'wanted to surrender space to NASA.' He wanted to make sure that his students led the charge, and to that end Yeager travelled to the Pentagon to make his case. Rather than invite applications for a military astronaut corps, he argued, it would save time and money simply to select them from the existing pool of graduates from his school. The Air Force agreed. Including Truly's Class 64A there were sixty-six pilots to choose from, minus those who had already been accepted into NASA's civilian Astronaut Office. Working on the assumption that anyone who'd graduated from ARPS wanted to be an astronaut – in any case, they had to do what the Air Force told them to do – Yeager didn't actually tell those whose records he was poring through that their hats were in the ring.

By the autumn of 1964, however, Truly and his friend and fellow US Navy student Jack Finley knew something was up. The rumour was spreading throughout the class of thirteen that some kind of selection was going on. It finally dawned on them that they were in contention, yet it still never really registered with Truly that he was in with a chance. Yeager whittled down his list of sixty-six names to a shortlist of fifteen. Then, at last, the Commandant shared the news of the selection process with those who'd made the cut. And that included Truly and Finley. They would be off to Brooks AFB again for further medicals, interviews and psychiatric assessments. None of that did anything to dampen Truly's excitement. In November, Yeager flew to Washington again to make the final selection. Six USAF pilots made the final eight. Two Navy: Truly and Finley. *I'll be damned*, thought Truly. He'd been chosen to be an astronaut.

A year later, on his birthday, 12 November 1965, Truly was introduced to the world alongside the seven other men chosen for a programme the Air Force called the Manned Orbiting Laboratory, or MOL. Clustered around a model of the Titan IIIC rocket they were expected to ride, the eight astronauts smiled for the camera. Al Crews, Mac Macleay, Mike Adams, Jim Taylor, Greg Neubeck and Dick Lawyer wore their Air Force uniforms; alongside them, Truly and Finley were in the dark blue of the Navy. It was the first and last time any of them were ever invited to talk to the press about their new assignment. They couldn't even tell their wives the detail of it. With a nod to both Yul Brynner and NASA's Mercury Seven, the group dubbed themselves 'The Magnificent Eight', but they would enjoy none of the fanfare that attended their NASA counterparts. They joked that when they finally got to fly, the Air Force might put out a statement announcing: 'An unidentified spacecraft left from an unidentified location carrying an unidentified crew on an unidentified mission.'

On 13 November, the day after their introduction, they disappeared into the black world: the highly classified shadows of the US Department of Defense that it chooses not to share with the public.

Bob Crippen joined the programme seven months later alongside Air Force pilots Gordon Fullerton, Hank Hartsfield, Karol 'Bo' Bobko and, from the US Marine Corps, Bob Overmyer. A year later, a third selection was made when Jim Abrahamson, Bob Herres, Don Peterson and Bob Lawrence joined the programme. Lawrence, who'd earned a PhD in Nuclear Chemistry two years earlier, looked set to become the first African-American in space. Ultimately, two of the MOL pilots chose to leave the programme. Mike Adams left to take up a much-coveted slot on the X-15 programme, while hard-charging Jack Finley decided that, with the country at war in Vietnam, he belonged on the frontline, leading a Navy squadron in combat. Those who remained were, as Bob Lawrence's doctorate illustrated, anything but your average collection of fighter jocks.

Growing up, Don Peterson read German science fiction in translation because he thought it more realistic than the American stuff. When the Air Force was looking for pilots to fly a nuclear-powered bomber then under development, they sent him back to college to study for a Masters in Nuclear Engineering. He would later go on to write scientific papers with titles like 'Principal Sources and Dispersal Patterns of Suspended Particulate Matter in Nearshore Surface Waters of the North East Pacific Ocean and the Hawaiian Islands' and 'Fatigue and Fracture of Ultrahigh

Strength Steel and Titanium Roll Bonded and Diffusion Bonded Laminates'.

You didn't get that from Chuck Yeager.

When he approved the MOL programme, President Lyndon B. Johnson claimed it would 'bring a new knowledge of what man is able to do in space'. But even the men who'd been chosen to acquire that knowledge were yet to discover the real nature of the job they'd been selected to do.

On the day the Air Force introduced the first group of MOL astronauts to the world they played their hand carefully. To avoid any possibility of a slip, none of the men meeting the press had yet been read into the real substance of the programme. What was made public was that the Manned Orbiting Laboratory was a two-man space station, designed to perform thirty-day missions exploring the utility of men in orbit, from which the astronauts would return to Earth in a modified version of NASA's new two-man Gemini capsule. No more was ever said.

The following day, though, the crews, all now enjoying high-level security clearances, were briefed on the true nature of their new assignment. It was, from the outset, going to be an operational mission. For public consumption, 'Laboratory' sounded wholesome enough, but as Hank Hartsfield, one of the second group of MOL pilots, pointed out, 'even though it was called a laboratory, it was different kinds of experiments'.

As the Air Force had struggled to make a case for MOL, it produced a document labelled *Candidate Experiments for Manned Orbiting Laboratory Vol. 1*. Included in the list of possibilities were 'encapsulation and recovery of space objects, changing the orbital trajectory of enemy satellites and a projectile firing RMU – Remote Maneuvering Unit – for neutralization of enemy satellites'. The space station was being pitched as a means to capture enemy satellites, knock them out of orbit, or shoot them down. This didn't cut it with Secretary McNamara. America's restraint in choosing not to complain to the Kremlin over *Sputnik* and its successors orbiting overhead the nation's capital established a precedent: satellites could fly unopposed and unprotested over whatever and wherever they wanted. Placing a weapons system in space threatened to change that, opening up space as a new front in the Cold War in exactly the way Curtis LeMay and Chuck Yeager anticipated. When, instead, the Air Force reframed MOL as a dedicated reconnaissance platform – a system that wouldn't upset the status quo – McNamara became more receptive. The key to his change of mind lay with the introduction of two systems: a high-resolution camera codenamed DORIAN and a powerful radar capable of surveying and tracking activity on the Earth's surface.

The MOL space station was a 10-by-55-foot cylinder capped with a Gemini-B capsule, from which the astronauts entered through a 26-inch hatch placed between their little spacecraft's seats. But barely a third of the space station's total length was given over to the two-man crew. The front of the 400-cubic-foot living area was devoted to the 'laboratory', the systems, controls and displays, while against the rear bulkhead was basic accommodation including sleeping bags, galley and lavatory. There was no shower and, lacking the photo-voltaic solar-panel wings carried by later craft, the Air Force space station was powered by alkaline fuel cells that generated electricity by combining hydrogen and oxygen to produce water. MOL's endurance was limited by the quantity of the hydrogen and oxygen it carried on board. There was also an exercise bike. At the time of the programme's inception, little was known about the physiological effects of long-duration spaceflight. The bike might at least prevent the possibility, as Hank Hartsfield put it, that the crew would 'come back jelly'.

Launched from Vandenberg Air Force Base, California, into sun-synchronous polar orbit at 96° to the equator, the MOL space station would, because of the Earth's rotation beneath it, overfly the whole planet passing below, meaning that in any twenty-four-hour period the crew saw everything; once in daylight, once in darkness – or dawn and dusk, depending on the time of the launch. From the laboratory, aided by the brand-new IBM 4Pi digital computer connected to cathode ray displays, they were able to reposition the space station or, using joysticks, train their camera to zoom in or out to improve the quality of the imagery. Their presence on board – and the judgement they could exercise – meant limited quantities of film wouldn't be wasted taking pictures of targets obscured by cloud. Furthermore, if targets were hidden by overcast, the crew could use the radar to see through it – a capability that was of particular interest to the Navy.

Behind the small crew compartment – which zero gravity at least allowed them to make full use of – and occupying the rear 41 feet of the space station was the MOL's *raison d'être*: the massive, caravan-sized DORIAN camera system that operated along the same principles as an astronomical telescope, but pointing down, not up. At its heart was a large curved mirror that collected the available light then focused it on the camera lens. The power of any reflecting telescope is directly related to the size of the primary mirror. The bigger the mirror, the further the telescope can see and the larger it can make the image. The DORIAN system carried a primary mirror built by Perkin-Elmer in Connecticut that was just shy of 6 feet in diameter. Larger than the mirrors inside many ground-based telescopes, the silica glass blank – the unfinished disc from which the mirror was made – would require thousands of hours of precision grinding, polishing and testing. But when it was finished its performance was expected to be extraordinary. From low Earth orbit at an altitude of between 150 and 160 miles, MOL's DORIAN camera was, on paper at least, capable of focusing on an object as small as a softball. Operationally, because of the distortion caused by moisture in the atmosphere, that figure might double. But that still meant that MOL could zero in on any single square foot of the globe. Or, rather, any single 8-by-8-inch square.

With its mission clear, MOL, although still led by the Air Force, was assigned the code number KH-10 and became part of the KEYHOLE programme that covered the activities of all America's spy satellites. And that meant it came under the wing of the organization whose very existence was classified: the National Reconnaissance Office.

SIX

IN THRILLER WRITER Alistair MacLean's 1963 bestseller, *Ice Station Zebra*, American and Soviet forces race to recover the secrets of a US film capsule, dropped from a spy satellite, that's come down in the Arctic. MacLean's book – and the film starring Rock Hudson that followed – perfectly captured the hair-trigger potential in any Cold War encounter. It was also based on a true story.

In February 1958, the Air Force announced a satellite programme called *Discoverer*. Its primary purpose, it was said, would be biomedical research into the effects of space on everything from seeds to human bone marrow. The Air Force had been working to develop a satellite capability since 1955. Designated WS-117L, the programme had lacked urgency until the launch of *Sputnik*. Four days after the Soviet satellite flew over Washington for the first time in 1957, President Eisenhower was asking the Department of Defense about America's plans for a reconnaissance satellite.

The result was *Discoverer*. But while that was the name shared with the public, it was a cover for the project's classified name: CORONA (so named because the CIA's Director for Plans happened to catch sight of a typewriter branded Smith-Corona when deciding what to call the project). Responsibility for the programme, designed as an interim project to provide the earliest possible space reconnaissance capability, was split between CIA and the Air Force. A classified DoD document listed among its aims the 'Testing of techniques for recovery of a capsule ejected from an orbiting satellite.' Transmitting images from space to the ground was beyond the state-of-the-art; the capsule would return exposed camera film to Earth for developing on the ground. CORONA was to be the nation's first spy satellite.

The programme's early years were plagued by failure. *Discoverer I* blew up on the pad after mistakes made during pre-launch checks caused all sorts of chaos. That one didn't count, it was decided, and so it became *Discoverer 0*. The second *Discoverer I* made it into orbit, but not the one intended. As the bird tumbled end over end through space, the sporadic telemetry that was picked up meant that the Air Force was only able to announce with any confidence that she was aloft four days later. *Discoverer II* did reach the right orbit but, because of a malfunctioning radio link, failed to drop the capsule over the Pacific where, as it fell beneath a parachute, it could have been plucked from mid-air by the specially equipped recovery planes of the 6594th Test Group. Instead it ejected it over Spitsbergen, an island in the Norwegian Arctic. The Air Force Colonel responsible for its recovery jumped on a C-54 cargo plane from Edwards to Thule in Greenland. From there a C-130 flew him to Spitsbergen where he directed a fruitless six-day search using Norwegian skiers, search planes and helicopters. All that was found was a set of snowshoe tracks leading from where the bucket had landed to a coalmining concession held by the Soviets since the 1930s. The Colonel persuaded the island's governor to send a telegram to the boss of the Soviet mine asking if he'd seen the capsule. In his reply, the Russian said his people 'had not seen that container . . . we got no information about this'. Four years later, Alistair MacLean spiced up his fictional version of the story with the addition of a nuclear submarine and a British secret agent.

To lend weight to CORONA's official cover story, *Discoverer III* carried four mice with radios attached to their backs. They were said to be trained. But that couldn't prevent them from being pronounced dead before launch. Four replacements were found but suffered a similarly meaningless death when the satellite's booster fired while its nose was pointing towards Earth, rather than space. It was a short flight.

The disasters continued unabated to the point where cancelling the programme was considered. But with the successful recovery of a payload from *Discoverer XIII* the tide turned. Then, on 18 August 1960, *Discoverer XIV* was launched. Six days later, just after 8.15am on 24 August, Edwin Land, the chair of the President's Board of Consultants on Foreign Intelligence Activities, unspooled a reel of developed film from the satellite across the floor of the Oval Office for the President. 'Here's your pictures, Mr President,' he said as if he were a magician pulling a rabbit out of a hat. Eisenhower was stunned. As he prepared for the National Security Council meeting that followed, it was immediately clear to him that they were living in a new world. In just seven orbits, *Discoverer XIV* had captured images of over 1,650,000 square miles of Soviet territory with a resolution of less than 50 feet. The take, revealing scores of air bases, surface-to-air missile batteries and a previously unknown ballistic missile site, offered more than the sum total of four years' worth of U-2 spyflights.

Concerned that knowledge of the satellite imagery would provoke a response from the Soviets that might limit further reconnaissance from space, Eisenhower quickly decreed that none of it was ever to be publicly released. It was the first step on a path that saw the country's spy satellite programme disappear into the highly classified 'black' world.

In an effort to claim satellite reconnaissance as its own, the Air Force had done its best to publicly associate itself with the nascent *Discoverer* programme, sanctioning the release of a nine-page document that sailed dangerously close to revealing the satellite's true purpose. In the end, the pamphlet was counter-productive. It reinforced the administration's doubts that the Air Force couldn't (or wouldn't) manage a completely covert programme. More importantly, Eisenhower was determined that satellite imagery should not be allowed to serve the interests of the Air Force, nor for that matter the CIA. It was an asset of such import-ance to the nation that it should not be controlled and directed by a single service or agency. The result was the creation of a civilian office reporting to the Secretary of Defense, which, in September 1961, became the National Reconnaissance Office, or NRO. The new spy agency had no headquarters building, but was instead found behind an unmarked metal door on the fourth floor of the Pentagon's C-ring. Beneath the umbrella of the NRO, the Air Force, CIA and Navy fought for influence.

The initial reaction to the photographs from *Discoverer XIV* was simply uncritical amazement. 'Like a dog that walks on its hind legs,' one General suggested, 'remarkable that it happens at all.' Almost immediately, though, the intelligence community wanted more. What form that might take highlighted the different requirements of the Air Force and the CIA.

CORONA – labelled KH-4 by the KEYHOLE coding system used by the NRO to identify its photographic reconnaissance assets – functioned as a 'search' satellite, its job to look for what was not known. In imaging large swathes of 'denied area' – enemy territory – it provided a strategic overview. From KH-4 imagery – classified TOP SECRET RUFF – the NRO could glean answers to the big questions like 'Is there a bomber gap?' or, in the wake of *Sputnik*, 'Is there a missile gap?' A search capability, though, which gathered exactly the kind of intelligence the CIA was after was only one part of the equation. The Air Force, by contrast, needed a surveillance capability; that is, a system that could look more closely at what was known. Once a search bird like CORONA uncovered evidence of enemy installations and equipment, the Air Force had to assess it in detail. They needed a satellite, equipped with a spotting camera, that could focus on a particular location or object. In contrast to the CORONA with its resolution of 50 feet, they wanted an agile, manoeuvrable bird with a resolution, from an altitude of 95 miles, of just 2 to 3 feet. The Air Force programme which would ultimately sit within the NRO was approved just a month after Eisenhower first studied *Discoverer XIV*'s photographs in the Oval Office. It was code-named GAMBIT, and known within the NRO simply as 'G'.

Given the KEYHOLE designation KH-7, the first GAMBIT was launched into a polar orbit from Vandenberg on 12 July 1963. The pictures from early flights showed promise, with the best of them meeting the ambition of 2- to 3-foot resolution. On the third test flight, colour pictures of a game of football were returned. They were of no value to the intelligence community, but they showed at least that GAMBIT was capable of being pointed accurately. It turned out to be a rare reassurance. Like *Discoverer*/CORONA before it, GAMBIT's gestation was painful. But it was the performance of the spacecraft, not the KH-7 camera it carried, that was at the root of it. A problem diagnosed early was that the satellite's heat-seeking control sensors couldn't tell the difference between winter in Antarctica and outer space. The confusion would send it tumbling out of control. Nagging problems, a later NRO history disclosed, 'transformed a healthy-looking GAMBIT into a zombie – a stupid creature circling the earth in unauthorized orbits, totally disinterested in attending to its assigned duties'.

For the General running the programme, the frustration of watching a bird 'go gypsy' was agonizing. On the ground at the Air Force's Sunnyvale Satellite Control Facility it was often difficult to diagnose the problem, let alone do anything about it. It lent weight to the Air Force argument that one of the purposes of placing its personnel in space was 'to correct malfunctions'.

In 1967, at a time when both CORONA and GAMBIT had overcome their teething problems, events in the Middle East suggested that, as well as being on hand to fix anything that went wrong, military astronauts might play a more critical role.

On 5 June that year, Israel launched a devastating attack on Egypt, Syria and Jordan. By noon on day one, Israel's lightning strike had destroyed Egypt's air force on the ground, her tanks smashing Egyptian and Jordanian armour, before doing the same to Syria's. Six days after hostilities began, Israel had won. The CIA had expected the war. But despite the NRO having both CORONA and GAMBIT satellites on orbit during the fighting, it was all over before either bird was able to deliver a single image.

The Six-Day War highlighted the critical failing of the KEYHOLE programme: it was incapable of delivering real-time intelligence from space. The MOL crews were briefed on what CORONA and GAMBIT could do. And they knew that their manned programme offered something more. They could rotate the DORIAN camera's reflecting mirror as they passed over a target so that pictures could be taken from two different angles. Combining the pair made stereo imagery possible. But they could also look down from low Earth orbit with their camera and sensors and tell Washington what was going on *now*.

SEVEN

1966–7

LIKE THOSE FROM GAMBIT and CORONA, the pictures taken by the Manned Orbiting Laboratory's KH-10 DORIAN camera had to be physically returned to Earth. There were either a limited number of re-entry buckets similar to those used by the National Reconnaissance Office's unmanned spy satellites or, at the end of their thirty-day mission, the two-man MOL crew would suit up, then pull themselves through the 5-foot-long tunnel linking the lab with their Gemini-B spacecraft carrying with them a data capsule of the take. After powering up the Gemini, they'd separate from the MOL, then fire their retro-rockets to initiate their return to Earth. After splashing down in the Pacific beneath parachutes, they'd be picked up by the Navy. On the face of it, this was an even more cumbersome way of gathering intelligence from space than employing the 6594th to snatch CORONA and GAMBIT buckets from Pacific skies. The difference, though, was what the MOL crew could do while on orbit. In the absence of a capability to transmit real-time imagery from space, the Air Force planned to make use of 'man's unique capabilities as a broad-band multiple sensor'. A trained flight crew could not only exercise their judgement about what to photograph to ensure that pictures were taken of 'the right target at the right time', but could also develop and interpret photographs on board then communicate their findings via encrypted radio. They offered a measure of real-time intelligence, using the most capable high-resolution photographic system available. A conversation with the ground was possible. They could double-check, look again, draw conclusions, or easily focus attention elsewhere – a process that, relying on unmanned systems, could last weeks. Air Force studies suggested that MOL crews would capture as much as three times as many photographs of 'active' targets per day than an unmanned spy satellite. In order to maximize their contribution, the MOL crews were exposed to existing TALENT and KEYHOLE imagery, the former codename applying to imagery taken from high-flying Air Force and CIA spyplanes. They pored over detailed satellite photographs of facilities in the Soviet Union and China, while a camera simulator was built to help train them to operate the DORIAN system under development. Once operational, it promised to deliver images that were better than anything they'd seen so far.

By November 1966, the Air Force was ready to flight-test MOL hardware. Of particular interest was the performance of the spacecraft's heatshield. A number of different options had been considered for moving the astronauts from their Gemini-B capsule to inside the space station. Spacewalking was a possibility, as was an inflatable tube linking the capsule hatch to that of the MOL crew quarters. Both methods seemed a little unsatisfactory. Instead, manufacturer McDonnell-Douglas proposed cutting a hatch into the capsule's crucial heatshield. No one was absolutely certain it would work, though. To test it, an unmanned capsule – a veteran of a previous unmanned test flight – was fitted to a reduced-length mock-up of the MOL space station, attached to a Titan IIIC stack and prepared for launch from Kennedy Space Center. Instead of NASA's familiar red 'United States' running down its length, the Titan bore the words 'United States Air Force' beneath the well-known insignia that identified the country's combat aircraft.

The test launch went without a hitch. Lobbed into a punishing re-entry profile after a brief sub-orbital flight, the capsule splashed down safely near Ascension Island in the mid-Atlantic. The modified heatshield proved its worth. When it was examined after the flight, not only was there no burn-through, but the heat had fused the hatch to the rest of the heatshield. And in the process of validating the design, the capsule became the first spacecraft ever to have been reused.

The MOL crews greeted the news with relief. Invited down to the Cape in August 1966 to observe an earlier launch of one of the big Titan IIIC rockets similar to the one they would be riding, the astronauts had watched as the 140-foot stack rose quickly from pad LC-41, its two Aerojet LR-87 engines in the rocket's core augmented by a pair of UA1205 strap-on boosters, each adding over a million pounds of extra thrust. On board was a clutch of communication satellites. As it accelerated away, streaming fire and smoke, the sound of the engines tearing at the air, it was hard not to feel a sense of exhilaration. Until, seventy-eight seconds into the flight, the shroud covering the payload in the rocket's nose ripped off. Five seconds later, the broken rocket was destroyed by the range safety officer.

The sucker blew up! Dick Truly said to himself as he

watched the aerial catastrophe. For the whole group, it was a sobering reminder of what could go wrong, and how quickly.

In the end, Bob Crippen, frustrated that he would be absent from the launch after his jet developed a fault en route to the Cape, was lucky to miss it. It was Crip's job, though, to try to establish whether or not they had any chance of surviving a similar disaster, or 'malfunction'. The combination of Titan IIIM booster and Gemini-B was unique to the Air Force. Like NASA's Gemini crews, the Air Force astronauts were strapped into ejection seats, but complicating their possible escape was the use of two solid fuel boosters secured to the side of their rocket. Unlike the Titan's liquid fuel engine, the solids, once fired, couldn't be throttled back or turned off. Instead, the boosters were provided with a more conclusive way of terminating their thrust. By blowing the tops off them while they still burned furiously until the fuel was expended, they were effectively neutered, turned into little more than fiery toilet rolls, the speeding air blowing straight through them.

Crip flew down to the Ling-Temco-Vought Company plant in Dallas where an abort simulation had been set up. His challenge was to see whether, in the split-seconds available following a terminal fault in the rocket, the pilot's reactions were sufficiently quick to select and abort and eject the capsule from the dying stack.

While the MOL programme headquarters was based inside the Air Force's Space and Missile Systems Organization in the unprepossessing surroundings of Los Angeles Air Force Base, the crews travelled around the country visiting the contractors working on different aspects of the programme.

While none worked in isolation, each of the MOL pilots had particular areas of responsibility. After witnessing the Titan explode after launch, Gordon Fullerton, 'the booster guy', discussed the fault with the rocket's manufacturer, Martin-Marietta, in Denver, Colorado. Hank Hartsfield worked closely with Douglas in Huntington Beach, California, on the spacecraft; Lawyer was assigned to pressure suits; Bob Crippen collaborated with IBM on the computers; Jim Abrahamson concentrated on the simulators; while Dick Truly and Mac Macleay were often found at General Electric in King of Prussia, Pennsylvania, focused on the systems controlling the space station's giant camera.

When travelling around the country visiting the programme contractors, the MOL flight crews left behind their military IDs and wore civilian clothes. Their efforts to be inconspicuous weren't always convincing.

Visiting the East Coast in winter, Don Peterson and Bob Lawrence, wrapped up in trench coats against the cold, pulled into a petrol station. The attendant looked them up and down, approached as inconspicuously as he could, and asked, 'FBI?'

He's been watching Bill Cosby and Bob Culp, thought Peterson. 'No,' he replied.

The attendant thought for a moment. Peterson could almost see the guy's mind turning it over: *crewcuts, suits and ties, trench coats . . .*

'CIA?'

Getting pegged as G-men or spooks by kids pumping gas was the least of Peterson and Lawrence's concerns. It was relatively easy for most of the MOL crews to travel incognito. Not so Lawrence. At a time when it was sometimes a struggle to find a restaurant that would serve them both, Lawrence, a black astronaut, was a novelty. He shunned the attention, but the press learned to recognize him, causing a nagging anxiety inside the programme office that his whereabouts would leave a trail of crumbs for reporters who were otherwise forced to cover MOL on scraps.

The crews were also frequent visitors to Houston where they'd compare notes and take advice from the Apollo astronauts. While they relished the opportunity to get to grips with the particular technical challenges presented by MOL's military mission, aspects of their training overlapped with their civilian counterparts' at NASA. Because, like them, the Air Force pilots had to prepare for the possibility that their capsule might not return to Earth where it was expected to.

The Panama Canal Zone was home to the US Army's Jungle Training School. Throughout the 1960s, increasing numbers of American soldiers passed through Fort Gulick to learn the skills they needed to live and fight in Vietnam. Astronauts, under the watchful eye of the USAF's Tropical Survival School, shared the forests with them. To keep them alive, they had an Air Force Manual 64-5. Its title was self-explanatory: *Survival*. It assumed nothing about its reader's ability to stay alive. 'Dangerous beasts,' it warned, 'Tigers, Rhinoceroses, Elephants, are rarely seen and best left alone.' There was desert survival training in Nevada, and even, because the MOL's high-inclination orbit created the possibility of their coming down in the polar regions, rudimentary Arctic survival. They were briefed on contingency plans for using military force to extract them from hostile territory, or capture by armed rebels, but in the end it was the water

survival training for returning in one of the three designated recovery zones – the Pacific, Atlantic and Indian
Oceans – that nearly killed one of them.

So far the water survival had been fun. Dick Truly hadn't
made it out to Buck Island in the Caribbean like some of his
colleagues, but he'd learned to scuba-dive in the Douglas
Aircraft Co. tank at Huntington Beach. Now, trussed up in a
full pressure suit, he was standing on a little deck at the aft
end of a boat bobbing on the Gulf of Mexico swell. Ahead
of him, a power boat dug into the water as it accelerated
away. Truly was straight away hauled off the deck beneath a
parasail. Above him, the flexible wing scooped in air through
its leading edge. A few hundred feet up, he disconnected
from the tow and drifted back down towards the Gulf of
Mexico. After splashing into the sea, he flopped into the life
raft.

An hour later, they came to pick him up. Truly slipped
into the sea in preparation for being winched up. He
immediately knew something was wrong. Not zipped up,
the neck of the pressure suit was letting in water, but
with the spacesuit's visor closed, Truly began to suffocate.
He waved frantically at the crew on the boat, but they just
waved back, unaware he was fighting for breath. His
struggles to get the visor open became increasingly
desperate as his lungs began to scream. *I'm going to die*, he
thought. At this time, every American astronaut to lose his
life had done so during training. Truly was lucky not to join
them.

The MOL astronaut group had so far escaped tragedy,
but that was about to change. Flying a stub-winged F-104
Starfighter with one of the Edwards instructors, Bob

Lawrence had already successfully completed two steep approaches when they poured on the coals and climbed out to a position 10 miles from the end of Runway 04. They cut the power at 25,000 feet and lowered the undercarriage, before descending in a steep 30° unpowered glide to the runway in a profile designed to simulate a returning spacecraft. No one knows for sure why they crashed into the runway nearly 3,000 feet short of their aiming point – *the instructor pilot screwed up*, Crip figured – but on impact the Starfighter gear buckled and flames burst from the jet's belly before it bounced into the air again. The instructor ejected first. Lawrence followed him out, but he was too low, the angles too severe. He died on impact less than 100 yards from where the broken jet tumbled to a dusty halt. His chest was smashed, his heart sliced open. He was still harnessed to the seat.

Thirty-two years old, Lawrence left behind a wife and child. *It was the flying game*, Crip reflected. You go to the memorial service and that's about it. You don't linger. But Bob Lawrence was one of the good guys. *It was a real shame*, he thought. And the tragedy of Lawrence's death was deepened because it followed so soon after that of another pilot the MOL crews regarded as one of their own: Mike Adams, one of the original MOL 'Magnificent Eight'.

Three weeks earlier, Adams had been launched above the Mojave in the X-15, the experimental rocket plane that had lured him away from the MOL programme. As he descended from a height of 266,000 feet the X-15 began to yaw dangerously before entering a hypersonic spin that led to its destruction.

Adams' fatal flight had taken him to a height of over 50 miles. He was awarded his astronaut wings posthumously.

Adams was the first of them to fly in space. And the first of them to die for it.

The deaths of Mike Adams and Bob Lawrence stained 1967 darker: they had not been the first astronauts to die that year. On 27 January, while conducting a launch test on Pad 34 at Cape Canaveral, the crew of *Apollo I* – Gus Grissom, Ed White and Roger Chaffee – were killed when a combination of faulty wiring and a pure oxygen environment turned their capsule into a furnace.

EIGHT

Houston, 1968

EIGHTEEN MONTHS AFTER the *Apollo I* tragedy, John Young still had to swallow hard when he thought of the deaths of Grissom, Chaffee and White. And yet, for all his anger at the mistakes that led to the unnecessary deaths of the three astronauts, fond memories of 'that old rascal' Grissom couldn't help but raise his smile. Grissom and Young had been firm friends. It was a bond formed in space. On Young's first mission, the maiden voyage of the two-man Gemini capsule, he'd flown with the veteran Mercury astronaut, even sharing with him a contraband corned beef sandwich he'd smuggled aboard the capsule in his spacesuit.

After he'd come to Houston following his record-breaking ascents aboard the Navy's Phantom jet fighter, Young was by the summer of 1968 on his way to becoming NASA's most experienced astronaut. A trim 5ft 9in, with green eyes and a flick of neatly parted brown hair, he would be described by the professor teaching lunar geology as the 'archetypal extra-terrestrial'. And there was something *other* about Young. He had an economy of movement about him that could lend him an almost reptilian stillness, as if, perhaps, he was conserving his energy for mulling over thorny engineering problems or considering humanity's place among the stars. But Young was also self-deprecating and funny; his aw-shucks tones and bone-dry sense of humour masked that sharp intelligence and thoughtful nature.

As Commander of *Gemini X* in 1966, Young's second flight saw him twice rendezvous without the aid of radar and travel further from Earth than anyone had previously ventured. In July 1968 he was named as Command Module Pilot for *Apollo X*, designed to be a dress rehearsal for the first moon landing. But he was becoming concerned that, after the American space programme had clawed its way back ahead of the Russians following the scalding shocks of *Sputnik* and Yuri Gagarin, it was in danger of losing the race to be first to the moon. It was Michael Collins, his crewmate on *Gemini X*, who helped fuel his concern. After meeting cosmonauts at the Paris Air Show, Collins reported that they'd talked of an imminent manned flight around the moon. It didn't matter, Young thought, whether they planned to land or not. In August, he wrote a memo to NASA headquarters in Washington arguing: 'If the Russian cosmonauts beat us to the moon, if only in lunar orbit . . . the Russians will win the moon race.'

A month later, a Proton-K rocket launched from Baikonur Cosmodrome in Kazakhstan carrying a *Zond 5* probe bound for the moon. On board was a pair of Horsfield tortoises which became the first earthlings to circumnavigate the moon.

By then, though, before a single manned Apollo flight had launched, an audacious decision to reorganize the Apollo flight schedule to send *Apollo VIII* swinging around the moon had already been made. Young liked to feel that he'd played his part in making that happen.

Yet even before the mission that brought home the famous Earthrise photograph was launched, thoughts inside Houston were already turning to what might follow the moon missions. And that meant that the men whose vision and leadership had got NASA to the brink of a crew flying to the moon were going to have to more or less tear up everything that had got them this far and start from scratch.

NINE

Houston, 1969

'IN MY LIFETIME,' Max Faget emphasized, 'all spacecraft will land on parachutes.' Faget, Director of Engineering and Development at NASA's Manned Spacecraft Center in Houston, was so confident of his assertion that he was prepared to bet a bottle of whisky on it being true. He had good reason to feel confident. It was Faget who, weeks after *Sputnik* flew in 1957, had argued that, as he put it, 'scrunching' a man into a blunt-bodied capsule like those he'd helped develop for the US Navy's Polaris missile warheads was the best way for a returning manned spacecraft to cope with the heat of re-entry. The simplicity of his idea – not to mention the capsule's light weight, which meant it could be launched on top of existing Air Force missiles – meant his arguments won the day. The result was the Mercury capsule, an angular teardrop-shaped craft that swung down to Earth beneath parachutes rather than gliding back to a runway.

Faget called himself 'a conceptualizer', which captured the essence of what he did. A brilliant, ideas-driven engineer, he made up for his lack of stature with an energy and vitality few could match. Born in British Honduras to American parents, his father, a doctor employed by the British government as a replacement for men sent to the World War One trenches, came up with the first practical treatment for malaria. A streak of ingenuity passed from father to son was combined with a love of air and space fuelled by model aeroplane building and reading *Astounding Science Fiction* magazine. A founder member of NASA's Space Task Group, he'd moved to Houston in 1962 when his boss, Dr Robert Gilruth, became the first Director of NASA's new Manned Spacecraft Center.

Now, after attending a presentation in early 1969 at which the Air Force had extolled the merits of a reusable winged spacecraft, Gilruth gently seeded the idea with Faget, conscious that, with the Apollo programme perched on the edge of success, NASA had to plan for what followed. Over lunch, Gilruth mentioned what he'd heard from the Air Force: 'Max, these guys are talking about a crazy thing. Why don't you look at it and see what it's about.'

Unable to resist the challenge, Faget started kicking the idea around with his colleague Caldwell Johnson, who since prior to the move to Houston had lent precision and elegance to Faget's freewheeling invention. Faget told Gilruth, 'Bob, you know it may be feasible.' Back home in his garage, away from prying eyes at MSC, he and Johnson constructed a beautiful scale model of their winged spacecraft design out of balsawood, paper and glue. The trouble was, it flew like crap. Two weeks after Gilruth had first planted the seed, and still unable to get their model to glide with any composure, Faget and Johnson took it in to show him. He immediately diagnosed the problem: a strut connected to the model's tailplane that was generating turbulence. 'You guys should have known that would give you terrible flow distribution,' he teased them. He was right, and yet it was a reminder that when it came to building aeroplanes, NASA had no experience. If they were going to build one, they first had to work out what they needed to know. And that meant gathering together experts in every field.

Engineer Dottie Lee got the call on a Friday afternoon. She was with contractors from Avco in Building 13, working on analysis of the Apollo Command Module heatshield. In May, *Apollo X* would all but land on the moon before returning home; then in July, Neil Armstrong, Buzz Aldrin and Michael Collins were due to fly the big one: *Apollo XI*. Lee's boss called her up to his office.

'Monday morning,' he told her, 'you're to report to Building 36. You don't tell anybody where you're going or what you're going to do. Read this document.' He handed her a sheaf of paper that elaborated on Gilruth and Faget's ideas and concepts for a vehicle that could reach orbit then glide back down to Earth and land horizontally.

Since Lee had witnessed Faget's theatrical introduction to the concept of a winged reusable spacecraft when he threw around his balsawood model, the Chief Engineer had added flesh to the bones. Now he was assembling the team of experts he wanted to develop it. Lee returned to the men from Avco and told them, 'I'm going to be out of pocket for a little while'; then added, conspiratorially, 'I cannot say anything of where I am . . .'

Over the weekend, Lee studied the document she'd been given. On Monday morning, she arrived at work and walked to Building 36. Working alongside her in the windowless high bay was a group of around twenty engineers, each specializing in a different aspect of the design, covering bases like structures, materials, aerodynamics and thermal protection. It became known as the 'Skunk Works' after the secret Lockheed facility that had given birth to both the country's first jet fighter and the U-2 spyplane. Inside Building 36, the group looked at different configurations and shapes. Three designers worked at drawing boards, making changes as they were suggested. A new design would be passed throughout the group in hours, annotated by different members of the team before being passed back to the draughtsmen.

They were working with a blank piece of paper. Still unknown were the size and weight of the vehicle, its launch configuration, how long it would stay in orbit, and its propulsion, but between them they established a foundation. And, as hoped, their study revealed areas in which NASA's expertise was lacking. So far, for instance, none of NASA's spacecraft had used hydraulic systems. To cover that they'd need to train or hire people who understood hydraulics. Potential problem areas were uncovered: with a cluster of heavy rocket engines mounted in the tail, it was near impossible to get the vehicle's centre of gravity far enough forward for it to be aerodynamically stable.

While they worked, Max Faget largely left them to it, but like a tooth fairy, he would visit unseen at night, leaving evidence of his presence in the form of changes and suggestions to their drawings.

Throughout, security at Building 36 was tight. A guard stood by a locked door, allowing members of the group in and out but barring entry to anyone else. If Lee needed to run numbers by contractors, she could only identify the project she was working on with a fictional job number. Returning from off-site visits, she'd have taxis drop her off at a different building to hide the location of the group.

After a decade of scepticism about returning to Earth in a winged vehicle to land at a runway, Faget was now becoming an evangelist for it. Or at least it looked like that. The design being defined in the Skunk Works had wings and wheels, but what really characterized the effort was that Faget didn't really plan to fly back through the upper atmosphere, but fall in much the same way as a Mercury or Apollo capsule. What Faget's team were really doing was designing a blunt-body re-entry vehicle in the shape of an aeroplane. It was designed, like his capsules, to present as large and unstreamlined a surface in the direction of travel as possible. His point, in 'flying' his balsawood glider at that 60° angle of attack to the direction of travel, had been to demonstrate that. Using Faget's tried and tested approach to re-entry would ensure the new spacecraft – even if it was shaped like an aeroplane – would decelerate and lose altitude more quickly, both of which meant less time subjected to the very highest temperatures.

Faget's design had short straight wings, *like the X-15*, he thought. At re-entry, and for much of the descent through thin air, wings served no purpose. In fact they were an unwelcome encumbrance, but keeping them short and straight was the next best thing to not having them if you wanted to fall like a stone from 400,000 feet. Where they came into their own, though, was low down in the thicker air. Once his spacecraft had fallen to lower altitudes, Faget's design would simply lower its nose to gather forward speed, resulting in air over the wings and, with that, lift. Down low, those straight wings would generate enough lift for

Faget's design to land like an airliner pretty much anywhere it wanted. 'Ordinary international airports,' he enthused, 'Houston Intercontinental, or Kennedy ... carrying less cargo it could even come down at such smaller fields as Houston's Hobby Airport or at LaGuardia.' His enthusiasm was infectious. The work done inside Building 36 was labelled 'The DC-3 Study' in honour of an earlier pioneering flying machine.

With its classic 1935 DC-3 design, the Douglas Aircraft Company created the world's first modern airliner, helped win World War Two, and ushered in the age of commercial air travel. Houston's livewire Chief Engineer thought his new design just might have the potential to do the same thing for space. NASA management were already referring to the new spacecraft as a Space Shuttle.

TEN

On 21 August 1968, Soviet tanks had rolled into Czechoslovakia. A first bucket drop of images taken by a KH-4 CORONA satellite launched two weeks before the invasion showed no sign of a build-up of Russian men and materiel. A second set of photographs, taken just days before, was unambiguous: the Red Army was massing on the border. But because of the time taken to recover, transport, develop and analyse the film, this imagery wasn't seen in Washington until after the event. It was a miss that cut deep inside the National Reconnaissance Office. It's little wonder, then, that in early 1969 the Secretary for the Air Force said of the Manned Orbiting Laboratory that its development was 'an urgent need'.

There had been slips in the programme. Budget constraints had seen the scheduled first flight pushed back from 1967 to 1970, then, ultimately, April 1972. The crews treated it with predictably black humour by holding an annual 'three-years-to-first-launch' party. Like the speed of light, three years to launch seemed to be a constant.

With each cut, they were forced to spend months rephasing the programme, asking contractors to soft-pedal until further funding came on tap. Ultimately, though, delays, while they might reduce the cost in the short term, only added to the total cost: even if you saved $100 million in year one, three years at $500 million per annum was always going to be more expensive than two years at $600 million per annum. At a point when costs were supposed to have peaked, they continued to grow. In 1969, MOL accounted for a quarter of the USAF's total research and development budget.

There had been costly technical challenges too. Lessons learned from the 1967 *Apollo I* fire forced McDonnell-Douglas to make changes to the Air Force Gemini-B capsule. Work on the high-resolution KH-10 DORIAN camera was proving to be more costly than anticipated. But in other areas, work was progressing well. Major difficulties with the avionics were behind them. The software development was now on track; so too the communications systems. And the crews could now touch real flight hardware. The hull of the first space station had been shipped from Douglas's Huntington Beach plant to General Electric in King of Prussia, Pennsylvania, where the vehicle's systems were being installed; what Hank Hartsfield called 'the guts of the lab'. Closer to home, the facilities needed to both launch and support MOL on orbit were nearing completion.

The command and control of Air Force satellites had come a long way since the Ballistic Missile Division first opened a field office in a motel in Palo Alto in 1958. In 1960, an interim control centre was set up inside the grounds of Lockheed's Sunnyvale facility. In order to fit in alongside Lockheed's own workforce, Air Force personnel wore civilian clothes. To avoid mistakes they were advised: 'A business suit, shirt, and tie compose the normal attire worn by businessmen. The coat may be removed within the office of work.' The facilities consisted of just two rooms, one barely larger than a closet. The larger of the two, the control room, was home to a table, folding chairs, a blackboard and a radio. Unencrypted telephones and one-hundred-words-per-minute teletype machines were used to communicate with white-coated technicians at the Vandenberg launch site and tracking stations in Hawaii and Alaska. The latter consisted of a van on a hill dubbed Readout Ridge. After a test, controllers there parcelled up their magnetic tapes, strip-tape and handwritten notes and posted the data back to California.

By mid-1965, things had changed out of all recognition. Building 101, the Air Force Satellite Control Facility's purpose-built site at Sunnyvale, had a control room at its heart containing over thirty computer terminals and employed nearly 750 military and civilian personnel. But in 1966, with the decision that Sunnyvale rather than Houston would handle MOL mission control, the workforce needed to near double. To house them, an ambitious construction programme was also funded. Once complete, the new four-floor building, a large windowless box, would contain twelve new mission control centres. To accommodate it, traffic was re-routed, streets and sidewalks demolished and a new powerplant built. Officially, the striking-looking sky-blue Satellite Control Facility was labelled Building 1003. Surrounded by a jumble of skyward-facing dish antennas, it was more often known simply as the Blue Cube. The Air Force was no longer relying on a van on a hill in Kodiak to track its spacecraft.

On 1 July 1963, a ship reached the island of Mahe in the Seychelles after sailing from the United States. On board was everything needed to build a satellite tracking station. The cargo took five days to unload, but by the end of August the new site was declared ready for business. For over a decade it was supplied by a single Grumman HU-16B Albatross amphibian operated under the auspices of Pan-Am International. Apparently unhindered by isolation, illness – on one occasion, polio – and coups, the Indian Ocean tracking station provided telemetry in support of the Air Force space programme as an integral component of the satellite control network. There were similar stations in New Hampshire, Guam, Hawaii, Alaska and Thule in Greenland. Each, like Mahe, faced its own set of environmental challenges – except, perhaps, the one in New Hampshire. The final piece of the jigsaw was Vandenberg.

Since the pioneering early days of the space race, rockets flying from Vandenberg Air Force Base in California had placed over 300 satellites in orbit, including every one of the NRO's KEYHOLE birds. Apart from the inconvenience of having the Southern Pacific Railroad running right through the middle, it was the perfect location from which to do so because it offered a trajectory south into polar orbit that didn't make landfall again until it hit Antarctica. It wasn't just the safety of those beneath the flightpath that made this so attractive, but also security. The NRO could be certain that a launch malfunction would not gift their most precious secrets into someone else's back yard.

By spring of 1969, the construction of a new launch pad at Vandenberg, SLC-6 – 'Slick Six' – was substantially complete. Located in the south-west corner of the base, the twin-pad complex was custom-built to support the Air Force's manned missions. MOL crews had become increasingly frequent visitors to Vandenberg looking for base housing and familiarizing themselves with the new crew facilities. The training building was finished and in the process of being kitted out, while fourteen pale blue space-suits, each bearing the name of one of the MOL pilots on the left side of the chest, were manufactured. For Crip, Dick Truly, Gordo Fullerton, Hank Hartsfield, Don Peterson and their comrades it was tangible evidence that the programme was real. They were yet to be given flight assignments, but Crip was looking forward to moving to Vandenberg from LA. *Just a few months away*, he thought.

On the afternoon of Saturday, 7 June 1969, the Secretary of the Air Force was invited to the White House to brief President Nixon, Secretary of State Henry Kissinger and the Secretary of Defense on the status of the MOL programme. In the Oval Office, he outlined the progress that had been made and the vital contribution to the nation's security that he was certain MOL and its KH-10 DORIAN camera was going to make. On Monday, Kissinger phoned to tell him that his presentation had been excellent. The future seemed assured.

And space was a boom industry. NASA had sent men around the moon on Christmas Day, 1968, shown us photo-graphs of Earth from space, and prompted *Time* magazine to describe a Utopia composed of 'social justice, peace, an end to hypocrisy'. They went on: 'the moon flight of *Apollo VIII* shows how that Utopian tomorrow could come about'. To the public, NASA was as glamorous, capable and sure-footed an organization as it was possible to imagine; yet on the verge of achieving the thrillingly ambitious goal laid out by Jack Kennedy in 1961 of 'landing a man on the moon and returning him safely to Earth', this was deflecting attention from what lay beneath. The race to put a man on the moon was essentially a stunt; a single, narrowly focused effort leading nowhere except to success or failure. The Apollo programme had a ravenous appetite for money, expertise, manpower and attention. There was little opportunity given to what might follow the programme's conclusion. The only thing on the books at this time was a plan to use left-over Saturn V components as the basis for a space station called *Skylab*.

In early 1969, President Nixon had set up a Space Task Group chaired by the Vice-President, Spiro Agnew, to consider 'the direction which the US space program should take in the post-Apollo period' and make recommendations. NASA proposed large space stations, space tugs and nuclear rockets and, underpinning it all, a reusable space transportation system – a Space Shuttle. The latter, it struck the military representative on the task group, offered considerable military potential.

Perhaps caught up in the euphoria surrounding the Apollo moon missions, the Vice-President began to bang the drum for a mission to Mars. And he didn't stop. 'Would we want,' he asked, 'to answer through eternity for turning back a Columbus or a Magellan?' Agnew envisioned a Mars mission as nothing less than an 'overture to a new civilization'. Clearly the Vice-President was a subscriber to *Time* magazine.

NASA's representative jumped on it, claiming that it had been the plan all along. Unsurprisingly, Agnew's enthusiasm took hold at NASA. Wernher von Braun, the visionary German rocketeer behind both the V-2 missile and, once he'd settled in the US after the war, the Saturn V rocket used for the voyages to the moon, was invited to a meeting of the

Space Task Group to explain how a Mars mission could work. He outlined a plan for twelve astronauts to set a course for Mars on 12 November 1981. After two months exploring Mars they would swing past Venus on their journey home to return to Earth orbit on 14 August 1983.

It was heady stuff. And it was all pie in the sky. But while, as the Space Task Group's report was drawn up, NASA still had a few months to savour the prospect of their dreams becoming real, the men working in secrecy on the Air Force space station programme were about to have theirs shattered.

ELEVEN

THE 10TH OF June, 1969. They called it Black Tuesday.

Bob Crippen was getting ready for work at his house near the Long Beach naval yard. It was just a half-hour drive into the programme office in El Segundo, but his secretary called him at home because the news couldn't wait.

On the other side of the country, in King of Prussia, Dick Truly was working inside one of the classified areas of the General Electric plant. He was arguing with one of the contractor's software engineers over a portion of code. It was small but important. Worth getting right. Truly felt a tap on his shoulder. It was Mac Macleay. Absorbed in the discussion, he shrugged it off. 'Go away,' he told his friend. Insistent, Macleay pulled him up and, out of earshot of the contractor, told him quietly, 'The programme's cancelled.' Macleay himself had the news straight from the GE president. It was solid. The company boss had had his secretary with him, and it was clear to Macleay that she'd been crying her eyes out. Truly looked back at the computer tech with whom he'd been haggling, his mind blank, purged by the shock of what Mac had just told him. He couldn't even remember what they'd been arguing over. It didn't matter any more.

The programme's head, General Bleymaier, was incensed. No one gave a damn. He'd fought through the night to get the announcement pushed back from 10.30 Eastern Standard Time so news wouldn't break early in LA. 'Let me tell my people first,' he'd asked.

'Well, [Deputy] Secretary Packard's got a busy day,' he was told by the Department of Defense in Washington, 'and that's the only time he can do.'

Dick Truly's wife, Cody, called Mrs Bleymaier at home to see if it was true. Cody had a chance to speak to her husband on the phone later in the day. Suddenly, Dick's dreams were nothing. He flew home to LA the next day. By the time he'd managed to get through to the programme office on the day of the cancellation, he and Mac had already had too much to drink to fly themselves home.

Throughout the first six months of 1969 there had been a battle waging within the National Reconnaissance Office. And for much of the time, the Manned Orbiting Laboratory and the Air Force were poised for victory.

Since 1966, the CIA had been fostering the development of a replacement for its KH-4 CORONA satellites. The new spacecraft, labelled the KH-9 HEXAGON, or often simply 'H', was going to be a monster. Fifty-four feet long and weighing nearly 25,000lb, it was a robotic orbital spy the size of a Greyhound bus. It soon acquired the nickname 'Big Bird'. Instead of the single film recovery capsule carried by CORONA, the new design carried four. But any comparison with the earlier bird stopped there. At its heart was a KH-9 camera system capable of scanning millions of square miles of territory in minutes. In two minutes, HEXAGON could give you Syria, Lebanon, Israel, Jordan, the Sinai Peninsula and Cyprus as a single contiguous area. In a crisis, by altering HEXAGON's orbit the NRO could have that daily. Capable of staying aloft for six months, HEXAGON carried enough film to provide colour pictures of the world's entire land mass within just two missions. Although Big Bird was designed as a search satellite, the KH-9 cameras would be able to deliver images with a resolution of less than 3 feet – a figure approaching that of the high-resolution GAMBIT surveillance satellite. That the system also employed two cameras was key to the dramatic leap forward in capability. HEXAGON would take pictures in stereo. Small differences in perspective between the two images meant that, by combining them, photo interpreters would be able to view the take in 3D.

None of this came cheap, though. And, in addition, *all* the contractors working on the programme were behind schedule. Richard Nixon's new administration concluded that the country couldn't afford to pursue two such enormously expensive space reconnaissance systems as MOL and HEXAGON together. And on 9 April 1969, the President jumped in the Air Force's direction and cancelled the CIA's KH-9 programme. But it wasn't as simple as that.

From the earliest days of the spy satellite programme, the value of a system that could deliver real-time intelligence was obvious. Equally clear was that it was not technically feasible. A single test of a system that would have developed its film in space and photocopied the result before converting it into an analogue signal and transmitting it to Earth produced little more than a blur. It was like trying to convey the sound of an orchestra using walkie-talkies. All that was about to change though.

By day, Dr William Baker was head of the Bell Laboratories Research Division. Set up in 1925 by the telecommunications company AT&T as a standalone research and development facility, the scope of Bell Labs' work was kaleidoscopic. Within it, Baker's own department covered 'the fields of physical and organic chemistry, of metallurgy, of magnetism, of electrical conduction, of radiation, of electronics, of acoustics, of phonetics, of optics, of mathematics, of mechanics, and even of physiology, of psychology, and meteorology'. But alongside his work at the place Arthur C. Clarke had described as a 'factory for ideas', Baker was also embedded deep within Washington's intelligence elite. Invited, in 1956, to examine how the United States might stay ahead of the Soviet Union's efforts to confound its intelligence gathering, the influential Baker Report he produced the following year urged an overwhelming concentration on emerging digital technology. Four years later, as a member of the President's Foreign Intelligence Advisory Board, Baker helped bring the National Reconnaissance Office into existence. Before long, his day job presented him with an opportunity to bring these two strands of his working life together.

When in 1969 Bell Labs announced that two of its researchers, while studying computer memory, had discovered a sensor capable of converting light into electronic signals, it was already two or three years since Baker had shared the breakthrough with the NRO. The invention, known as the charge-coupled device, or CCD, was in essence an electronic eye and was to become the foundation upon which the digital camera revolution was built. Eventually.

By the time Bell Labs went public with the new technology, the NRO's head start had already begun to bear fruit. It was appreciated that a new KEYHOLE satellite using CCDs would require a very large mirror to compensate for the loss of image sharpness offered by film, and that huge bandwidth would also need to be developed, but the prize was irresistible.

The prospect of building an operational digital imaging satellite was years off still, but while it might have been beyond the state-of-the-art in 1969, when the CIA formally objected to the cancellation of their KH-9 HEXAGON Big Bird, they had, as part of their arsenal, the tantalizing prospect of a revolutionary new technology in the not too distant future.

Forced to consider which of the two rival KEYHOLE systems, the Manned Orbiting Laboratory DORIAN or HEXAGON, would offer the best tool for monitoring arms control, the President reversed his decision and cancelled the Manned Orbiting Laboratory.

Ironically, MOL's own 70-inch mirrors, ground and polished by Perkin-Elmer for the DORIAN camera's huge telescope, scored an own goal, helping to put the Air Force's manned space station built to carry them out of business. By demonstrating the feasibility of fabricating mirrors of a sufficiently large diameter for a *digital* spy satellite to be effective, they'd shown that MOL itself, five years late, was being overtaken by technology.

MOL's own lavishly produced mirrors, like healthy organs taken from road-crash victims, were donated to science, and transplanted into ground-based telescopes. And with the programme's cancellation the Air Force, after trying to get its men into space for over a decade, had finally to face the reality that it had been pushing water uphill. It was over.

The shock announcement broke hearts in El Segundo. Many of the MOL pilots, faced with a choice between applying for NASA or the Air Force astronaut programme, had jumped the wrong way. *Boy*, thought Bob Crippen, *I really screwed up with* that *decision*.

For nearly a month, the redundant astronauts cried into their beer. During a period they called the MOL Wake they talked over their options. Two of them, Bo Bobko and Don Peterson, decided to go back to college to study for Masters degrees. Dick Truly and Bob Crippen, the two naval aviators on the programme, flew to the Navy Annex in Washington to see what they could offer. The two grounded astronauts were warmly received and, despite an absence from the frontline of five years or more, the detailers told them they could have their pick of assignments – whatever they wanted to fly. *That*, Truly knew, *just doesn't happen*. And, it turned out, there were strings attached. Both had their eye on combat in South-East Asia. Despite deploying with frontline squadrons during the Cuban Missile Crisis, choosing test-flying meant both had suffered from a nagging sense of guilt that the fighting in Vietnam had been the preserve of friends. But so sensitive was the classified information they'd enjoyed access to on MOL, they were subject to a two-year duty and travel restriction. There was no way the Department of Defense was going to countenance the risk of having them shot down over enemy territory, captured and interrogated. Neither Crip nor Truly was going to be allowed anywhere near the Yankee Station. The Air Force pilots in the group faced the same problem. Informed he could deploy to South-East Asia to work in a Saigon command post, Hank Hartsfield reacted angrily. 'Baloney,

I'm a pilot,' he told them, 'I'm not going to sit in a darn command post and ride a chair.' He figured he might as well study for a Masters too. By the time he was done, at least they'd let him fly.

Running out of ideas, Crippen's thoughts turned to Patuxent River, the Navy's flight test centre, and a return to work as a test pilot.

Then Bo Bobko raised the idea of approaching NASA. 'Why don't we call?' he asked during a morning pilots' meeting. 'See if they could use us . . .' Initially, the others dismissed it out of hand. 'NASA doesn't need more astronauts,' they told him. 'They've got lots of astronauts!' It was a potential lifeline, though.

And, when they asked the question, all fourteen were invited to Houston.

After flying down to Houston from LA, they were shown round by Colonel Tom Stafford, Commander of the recently returned *Apollo X* mission that had just a few months earlier all but landed on the moon. NASA's Manned Spacecraft Center was modern and purpose-built, laid out like a college campus on land donated by Rice University seven years earlier: three-storey concrete and glass office buildings set in landscaped gardens, lined with paths, shrubs, man-made lakes and lawns surrounded by parking lots. The tallest office, Building 1, housed administration and senior management, Building 2, the auditorium. But Building 4 was the prize: the Astronaut Office.

'Well, you know, if you do get here . . .' Stafford began, offering a glimmer of hope as he talked through plans for the Apollo mission through *XIII* to *XIX* and *Skylab* beyond that. Stafford's words were powerfully reinforced by action.

As the MOL group pressed their case in interviews with Deke Slayton, the Director of Flight Crew Operations, in July 1969, Neil Armstrong, Mike Collins and Buzz Aldrin were already on their way to the moon aboard *Apollo XI*.

TWELVE

Houston, 1969

WITH NEIL ARMSTRONG'S descent to the lunar surface on 20 July 1969, NASA, after an extraordinary eight-year campaign, had succeeded in 'putting a man on the moon'.

Inside Mission Control, John Young wasn't sure anyone was following the journey with deeper interest than he and his *Apollo X* comrades. *Been there, done (almost) that*, he thought as he layered his own insights and feelings over what was unfolding. He regarded the *Apollo XI* crew as very atypical. The Command Module Pilot, his good friend Mike Collins whom he'd flown with on *Gemini X*, he regarded as one of the Astronaut Office's outstanding characters. Lunar Module Pilot Buzz Aldrin, who never missed an opportunity to discuss the subject of his MIT doctoral thesis in orbital rendezvous, was not, Young reckoned, as smart as he thought he was. Then there was Neil Armstrong, the measured, remote *Apollo XI* Commander who, by his own admission, was more interested in ideas and things than people. In Armstrong, though, Young saw something of himself. *Test pilots and astronauts*, he maintained, *tend to be doers and not thinkers, but that wasn't so true of Armstrong.*

And maybe not so true of me, he thought, as he considered the crew's return.

After a little over twenty-one hours on the lunar surface, Armstrong and Aldrin blasted off from the surface of the moon to rendezvous with Collins in the Command Module before returning home. Landing men on the moon, of course, was only half the challenge set by Jack Kennedy in 1961. They also had to be returned safely to Earth. And, understated as ever, John Young described the job of safely returning a crew from the moon as *no simple matter*. On his journey back from orbiting the moon in May with his crewmates Tom Stafford and Gene Cernan, Young had added to the clutch of world records he'd won flying with the Navy. At the point of re-entry into the Earth's atmosphere, the *Apollo X* crew's capsule was travelling at 24,760 mph. At over thirty-seven times the speed of sound, it was the greatest speed ever reached by human beings. Young was acutely aware that the journey home always carried the potential for surprise. And so it was proving with *Apollo XI*'s return, which was more complicated than had been expected.

A few days earlier in Hawaii, a young Air Force meteorologist, Captain Hank Brandii, called up Fleet Weather Center at Pearl Harbor and spoke to the Department of Defense's chief weather officer, Captain Sam Houston, USN. 'There's going to be a real problem,' Brandii told him. 'I want you to meet me in the parking lot of the 6594th Test Group hangar at Hickam Air Force Base.' As soon as he'd got out of his car, Brandii pulled Houston through the door and into the 'vault' to look at a set of classified pictures showing a major tropical storm building over the splashdown site. Because of the TALENT/KEYHOLE classification that applied to the pictures, Houston was the only person Brandii could tell.

Brandii's job was to use the National Reconnaissance Office's 417 satellite system to track and forecast the weather in support of the CORONA spy birds. But on this occasion, he realized his images revealed that the Apollo capsule was on course for disaster.

The first indications a storm was forming were high-level vortices tracing patterns in the sky that had an almost bird-like appearance, but Brandii knew they heralded the onset of a violent thunderstorm generating vicious high-altitude winds rising to 50,000 feet. Such was the intensity of the storms that he called them 'Screaming Eagles'. And on this occasion, the *Apollo XI* capsule was heading right for one. He knew it had the power to shred the capsule's parachutes on which the astronauts' lives depended.

To an extent Sam Houston's hands were tied too. He couldn't share the source of his information, yet he had to persuade NASA and the US Navy of the need to re-route both the spacecraft and the recovery fleet. Early in the morning on 23 July, Rear Admiral Davis, commanding Task Force 30, ordered his prime recovery ship, USS *Hornet*, 250 miles to the north-west. 'You'd better be right, young man . . .' he told Houston.

In Mission Control, John Young wasn't cleared to know the provenance of the information that had forced them to change the capsule's splashdown point, but he was impressed to see that, even with the late correction to the capsule's trajectory, the spacecraft had descended into the Pacific just 13 miles from *Hornet*'s new position. After sweating anxiously through the mission, Young was just relieved to see them home. *Safe and sound.*

* * *

Also heading home were the refugees from the MOL programme. It looked as if their trip to Houston – and the
opportunity to experience the achievement of *Apollo XI*
from inside the tent – was going to be the end of the line.
Sadly, Crippen, Truly and the rest of the MOL crews had
been right with their initial reaction to Bo Bobko's suggestion.
When it came to the crunch, Deke Slayton hadn't varnished
it. 'Hey, guys,' he told them, 'I got more astronauts than I
know what to do with. I really don't need you.' Slayton
didn't even have flights for all the astronauts he *already*
had.

THIRTEEN

Houston, 1969

W<small>HEN</small> D<small>EKE</small> S<small>LAYTON</small> told George Mueller, the man with responsibility for manned spaceflight at NASA headquarters in Washington, that he'd turned away the redundant MOL astronauts, Mueller asked him to think again. 'I know and you know that we don't really need them,' Mueller told him, 'but we're gonna need the Air Force one of these days, and it won't hurt to make them happy just this once.'

When Mueller took over as Administrator in Washington in 1963, he'd immediately drafted in a brilliant Air Force Programme Manager called Major General Sam Phillips who'd previously earned his spurs successfully ushering into service the Minuteman ballistic missile. Phillips had proved to be a crucial hire. Mueller was quick to see that he was going to need Air Force help again. 'One of these days' meant *now*. In achieving its goal of putting men on the moon, NASA appeared to the outside world to be at the peak of its powers; but, cruelly, success meant that at the same time it was in danger of putting itself out of a job. Mueller knew that if he couldn't win Air Force support for the embryonic Space Shuttle programme it was entirely possible that Apollo, or at least *Skylab*, the space station being developed using left-over Apollo hardware, would mark the end of NASA's participation in manned spaceflight.

Even before the Space Task Group presented their report in September, they were warned by one of Richard Nixon's closest advisers not to make a recommendation that the President couldn't sign up to. Reluctantly, they were persuaded to row back from actually *recommending* a manned mission to Mars. And yet for a document that was trying not to do so, it made a pretty good job of it. The possibility of a manned mission to Mars is mentioned repeatedly, and suggested as a 'long range goal'. Alongside it, they recommend continued lunar exploration, a space station for fifty to a hundred occupants, and a reusable, nuclear-powered spacecraft to transport 'men, spacecraft and supplies' to and from the moon and beyond. All of this exotic kit was, once *in* space, supposed to stay there. And so also included was a Space Shuttle designed to travel to and from Earth orbit to get to it all. While the country fought a major war in South-East Asia and faced civil unrest at home, the naive, ambitious recommendations of the Space Task Group were completely misjudged and were rejected out of hand. All that survived was a trickle of funding to allow *studies* into both a much smaller ten- to twelve-man space station and the Space Shuttle.

With so little prospect of spaceflight, Deke Slayton thought bringing the MOL crews into the NASA Astronaut Office was disingenuous. He certainly had no shortage of work for them, there was just no chance of them flying for about a decade. Mueller suggested Slayton just level with them. If any of them weren't happy with that, then no one was forcing them to come to Houston. He tried that, but only Bob Herres took him up on it. And one departure didn't do enough to trim the numbers. Instead, Slayton looked at their birthdays. Thirty-five was the upper age limit for a normal astronaut selection. If he applied that to the MOL group, then just seven made the cut. That meant jobs for Bob Crippen, Dick Truly, Bo Bobko, Gordo Fullerton, Hank Hartsfield, Bob Overmyer and Don Peterson. For the rest it was the end of the road.

Dick Truly was the only member of the original MOL selection to make the cut. He'd been chosen for the USAF's astronaut programme without even knowing there'd been one to apply for. Now he'd made it into NASA through the back door. He was sure he was the only astronaut never to have actually applied to become one.

In the stifling mid-summer heat, Truly and his family once again squeezed themselves into the car to relocate. The long drive south-east turned out to be a test of endurance that not all of them were to emerge from intact. As they crossed Arizona their much-loved pet dog died in the car. And all along, nagging away in the back of Truly's mind, was the thought that he was only on his way to NASA on the back of a failed programme.

As he had in LA, Bob Crippen found a house near the sea. He, his wife, Ginny, and their three girls Ellen, Susie and Linda moved into a new development in Shore Acres near the yacht club. Although his career in the US Navy hadn't so far reflected it, Crippen had a passion for the sea, devouring books which brought to life the golden age of sail and the square-rigged naval battles of the early nineteenth century. When time allowed, Hornblower had offered a welcome break from the cactus and dust of the high desert at Edwards. As well as somewhere to live, Crip also needed to sort out his wardrobe. Since leaving Porter, Texas, for the Navy he'd spent his life in uniform. Now, as part of a civilian space agency, he had to actually think about the clothes he wore to work.

The MOL guys started work at Houston in August, and Deke Slayton was quick to remind them of the reality they faced. 'Guys,' he said, 'I've got lots of work for you to do, but I don't have any flights until they've built this thing called the Shuttle.' That, he told them, was unlikely to be before around 1980. Nor was it, at this point, even an approved programme. While there were briefings on NASA hardware, there was no training programme, nor any kind of indoctrination. Instead they were simply given a desk and an office and told: 'you just find out what you want to do, or make suggestions'.

Dick Truly was offered a slot on the support crew for *Apollo XV*, but as tempting as it was he just didn't think it was worth trying to hold on to what was left of the lunar programme. At the same time, though, he saw the *Skylab* team developing a space station being more or less ignored next to the glare of the moon missions. Truly had done so much work on the Air Force space station before its cancellation that he felt certain that *Skylab* was the programme to which he could best contribute. For similar reasons, Bob Crippen also opted to join the *Skylab* programme. And while there was the prospect of more technical work to look forward to, among his first assignments was one to test the *Skylab* potty in zero-gravity conditions, flying parabolas aboard NASA's KC-135 Vomit Comet. It wasn't entirely unfamiliar as it was a device designed for MOL by helicopter manufacturer Fairchild-Hiller. Asked to perform on cue, Crip and the other test subjects were able, over a couple of days, to deliver nine good 'data points'. It was a pretty long way from Corvette Stingrays and the cover of *Life* magazine.

At least what they were doing was real and useful. That was more than it appeared could be said for the Shuttle. Soon after arriving at Houston, the MOL guys (as they got labelled) were sat down in front of an overhead projector and briefed on the emerging Space Shuttle project. Since Max Faget's Skunk Works began work on it earlier in the year, the configuration of the whole stack had become more certain. In pursuit of lower operating costs, NASA hoped to launch the Shuttle from the back of a fully reusable winged booster – in essence a giant rocket-powered aeroplane. Unlike the stages of the current Saturn V rocket, which, once their fuel was spent, fell back to an ocean grave, Faget was proposing a vertically launched, piloted rocket with wings that, after carrying the Shuttle high enough and fast enough for the spaceplane to reach orbit using its own engines, then flew back horizontally to land on a runway to be serviced and prepared for its next flight. It was a machine that was, broadly speaking, a scaled-up version of the orbiting Shuttle itself; a father carrying his son on his back. And it was a behemoth. At 204 feet long, the machine Mueller and Faget were suggesting would have the hypersonic performance of the little X-15 rocket plane but be built on the scale of Boeing's new 747 Jumbo Jet – the world's largest airliner.

As if that wasn't enough, NASA even commissioned a study to explore the possibility of using Faget's winged

Space Shuttle design for ferrying payloads to and from the moon. To do so, a report concluded in October, would require twenty-nine Shuttle launches. That seemingly took it into the realms of the ridiculous, but frequent flying *was* key to the case being made for the Shuttle. If it flew as often as NASA suggested it would, NASA's Shuttle fleet would be capable of notching up those twenty-nine launches in just six months. By not throwing anything away and by flying frequently – which depended on the Air Force committing to use it – the stated aim of George Mueller, the boss of manned spaceflight, was for the Space Shuttle to bring the cost of launching a payload down from the $10,000-per-pound figure of a Saturn V to 'between $20 and $50 a pound'.

Dick Truly viewed the wilder schemes, ambitious designs and completely unrealistic cost projections with a sense of disbelief. *It was all*, he thought, *a bunch of crap.*

NASA's real problem was that the Air Force just didn't want or need a Space Shuttle. While they might have joined in in outlining what they would in theory require from a reusable spaceplane, they were in fact quite happy to continue as they were, using reliable, relatively affordable, expendable rockets launched from Vandenberg Air Force Base in California. As one bluesuiter made clear: 'Sure, NASA needs the Shuttle for a space station. But for the next ten years, expendables can handle the Air Force job.' What's more, he said, they didn't think it was important enough to them to pay for. NASA's position couldn't have been more different. With the door firmly shut on ambitious new space stations, a lunar programme and deep space missions, NASA needed Air Force support in order to present the Shuttle as a vehicle with a purpose *and* as a national asset, rather than simply a Trojan horse for the agency's interplanetary ambitions. If a Space Shuttle were to shoulder responsibility for carrying Air Force payloads into space, it would greatly enhance its chances of gaining political support.

The future of US manned spaceflight seemed to depend on keeping the Air Force on the team. Yet when it came to spaceplanes, the Air Force had history. It still bore the scars.

FOURTEEN

On 22 June 1970, a British naval test pilot called Eric Brown was conducting blind landing trials in a Westland Whirlwind helicopter at the Royal Aircraft Establishment in Bedford. It was the weekend, so the airfield was quiet. Only he and one other pilot were flying. After finishing for the morning, Brown shut down the aircraft and was picked up in a van to return to the mess for lunch. On the way they stopped to collect the other flyer. Wearing a flightsuit and carrying his helmet, the pilot climbed in, leaned over and offered his hand. 'Hello,' he said, 'I'm Neil Armstrong.'

Over lunch, Brown was surprised to learn that he was already well known to the first man on the moon. Throughout the forties and fifties the Fleet Air Arm flyer was unarguably one of the pre-eminent test pilots in the world, establishing records for number of deck landings and sheer variety of types flown that would never be beaten. And in taking up the German Messerschmitt Me-163 Komet he was also the first Allied pilot ever to fly a rocket plane. They had never met, but Armstrong didn't mind admitting that Brown had been something of a role model to him as a young pilot. So when, in 1960, Brown – at the suggestion of his superiors at the RAE – approached the Americans about

– 96 –

the possibility of securing a place on the X-15 programme, Armstrong, already chosen to fly the hypersonic rocket plane, had supported the application. Such was Brown's reputation in the test-flying community that his request was looked on favourably. At the controls of the X-15 Brown would have been in a position to become the first Briton to earn his astronaut's wings. But there was a catch: in order to take up the proffered slot on the programme Brown was required to become a US citizen – something he was not prepared to do. So, instead of punching holes in the skies above Edwards AFB together in the early sixties, Armstrong and Brown's mutual admiration remained remote – until their chance encounter at an otherwise deserted British airfield.

The Commander of *Apollo XI* was in the UK to fly a research aircraft called the Handley-Page HP.115. Armstrong had first hoped to fly it back in 1962. The HP.115's highly swept delta wing, he'd realized, flown at very high angles of attack, might offer useful insights into the flying character-istics of the spaceplane then being developed by the USAF. Armstrong had been one of the first pilots assigned to the programme the Air Force christened Dyna-Soar.

When, in the weeks that followed the launch of *Sputnik* in 1957, Max Faget had successfully argued that capsules were the quickest, simplest and cheapest way to get a man into space, the Air Force jumped in a different direction. They pursued a single-seat, reusable, delta-winged space-plane designed to launch on a rocket then glide back to Earth to land on a runway. Its name was a contraction of 'dynamic soaring', the term being used to describe its hypersonic, unpowered return to Earth from orbit. In June 1959, Boeing was awarded a contract to build it, and a series of twenty test flights conducted out of Mayguana in the Bahamas, Fortaleza, Brazil, and Saint Lucia in the Windward Islands was scheduled for the second half of 1963. When a full-scale mock-up of Dyna-Soar was displayed for the public in Las Vegas during the autumn of 1962, it offered a glimpse of the sort of future promised by science fiction. *Reader's Digest* described the black-painted Air Force spaceplane as 'looking like a cross between a porpoise and a manta ray'. In a good way. It would be a triumph, they went on to say, that would rank alongside the Wright Brothers'. Kennedy's Secretary of Defense, Robert McNamara, didn't agree. At the root of it was a belief that the Air Force was more concerned about *how* it travelled to and from space than what it did when it got there, and in 1963 he cancelled it in favour of the similarly ill-fated MOL programme.

Six years later, as Armstrong, one of the original Dyna-Soar pilots, was flying the experimental British jet he'd thought might have contributed to the spaceplane's development, the Air Force held the cards over plans for a new winged spacecraft. For there to be any possibility of them lending their support to NASA's Space Shuttle, they required a capability far in excess of what Faget's emerging design could offer. The Air Force wanted a machine that, after flying a single polar orbit out of Vandenberg, California, could land back where it had taken off. To achieve this it had to be capable of covering a distance of over 1,000 miles to the right or left of the point it re-entered the Earth's atmosphere because, as it circled the poles, the Earth was rotating beneath it. By the time the Space Shuttle had completed its single orbit to return to a point overhead the launch site, that site would no longer be directly below but would have moved over 1,000 miles to the east.

Because Faget's straight-winged Shuttle was designed to fall from the sky until a point below 40,000 feet where it dipped its nose to pick up flying speed, it only had a cross-range capability of a couple of hundred miles or so. Small, straight wings were useless in the thin upper atmosphere. By contrast, the large expanse of a triangular delta-wing would mean a returning Shuttle could start producing lift in higher, thinner air and therefore fly a greater distance from the point of re-entry. The disadvantage of the delta was that it demanded a much higher landing speed, but the Air Force didn't care about that. If they knew where the Shuttle was coming down, they could just make sure the waiting runway was long enough. Faget countered that his Shuttle could land *anywhere*; it didn't need a special super-size airfield. But ultimately it wasn't an argument Faget could win. Physics saw to that. The only safe, secure and feasible site in the continental USA from which the Air Force could launch heavy spy satellites into a polar orbit was Vandenberg. And on completion of a single ninety-minute orbit after launching from the California pad, a spacecraft – *any* spacecraft launching to the south – was always going to find itself out over the middle of the Pacific Ocean. If, from that point, your Shuttle only had a cross-range capability of a couple of hundred miles, it was going to get wet.

But cross-range was not the Air Force's only demand. The other was payload. Faget's initial designs were capable of carrying a 15,000lb payload in a small cargo bay measuring 15 by 30 feet. Alongside the work being done by Faget's team at Houston, NASA had also commissioned studies from private industry. From the country's leading aeroplane makers it required designs for a Space Shuttle capable of carrying a 50,000lb payload into a low-inclination, equatorial orbit. By the summer of 1970, though, it was clear that even this wouldn't be enough to accommodate the huge new KH-9 HEXAGON and its projected successors.

The Air Force said it needed a Space Shuttle capable of lifting 40,000lb into a polar orbit. And this was quite different from lofting 50,000lb in an easterly direction from the Cape. Launching east out of Florida, a rocket was gifted a near 1,000 mph running start thanks to the speed of the Earth's rotation. Instead of having to generate its entire orbital speed of 17,500 mph it only had to make 16,500 mph; Earth made up the difference. The closer you flew to an equatorial orbit, the more speed you gained and the more weight you could carry for the same level of thrust. A 50,000lb payload from the Cape was equivalent to just 30,000lb launched south out of Vandenberg. And while that might more or less accommodate HEXAGON – and even that was uncertain as the first Big Bird was yet to fly – it wouldn't, necessarily, be enough for its successors. The Air Force requirement for a machine capable of launching 40,000lb into a polar orbit from Vandenberg equated to 65,000lb from Florida – an extra 7½ tons. It was a massive additional burden which, if NASA were going to enjoy Air Force backing, had somehow to be achieved. In 1970, there wasn't a single design from Faget or the aerospace industry that came close; a problem compounded when NASA's management found something upon which they and the Air Force *could* agree: the size of the payload bay.

While NASA had scaled down its space station plans after the rejection of the Space Task Group report, there were still hopes of a twelve-man, four-deck space station launched on a Saturn V rocket. By July 1970, it looked unlikely that both this and the Shuttle could survive. It was estimated that, over ten years, servicing and expanding the space station would require over ninety Shuttle flights. If only one project was to survive, then it could *only* be the Shuttle. That meant that future space stations would have to rely on being assembled from modules carried inside the Shuttle's cargo bay. NASA now needed the 60-by-15-foot payload bay the Air Force required just to hold on to the possibility of one day building any kind of space station at all.

FIFTEEN

FRED HAISE SHOULD have been the first man to walk twice on the moon.

After a military flying career in which he'd flown for the Air Force, Navy and Marine Corps – *and washed out of each*, ragged his colleagues – Haise became a civilian research pilot at NASA's Lewis Flight Propulsion Laboratory in Cleveland, Ohio, eventually graduating from the Aerospace Research Pilots School as one of Chuck Yeager's students in the same class as Dick Truly. After that, the X-15 was a possibility, and he flew chase as part of that programme, but he could see he was too late to the party to have a really good shot at winning a slot. Instead, he applied to be an astronaut and was rated number one in the nineteen-strong 1966 selection that also included Joe Engle.

Dark-haired and lantern-jawed, Freddo had the appearance of the archetypal all-American hero. But while his track record, which included dogfighting gliders over the High Sierras and flying a wooden, wingless glider towed behind a souped-up Pontiac Catalina V8 convertible, saw him easily measure up against a group once described as 'hyperthyroid, super-achieving first sons of super achievers', there was no arrogance or edge to Haise. Combining outstanding technical ability with an easy-going, genial nature, Freddo was a dependable, popular figure inside the Astronaut Office.

To his delight, he was assigned the job of Lunar Module Pilot on *Apollo XIII*.

The launch on 11 April 1970 wasn't entirely uneventful. The second-stage centre engine experienced an unstable fuel flow which caused it to pogo, sending a powerful, persistent vertical chug resonating through the stack. It was enough to trip out an accelerometer that shut down the engine before the vibration could do any more damage. But soon that was behind them. Still able to make orbit, *Apollo XIII* was on its way to the moon.

Haise was alone in the Lunar Module, *Aquarius*, packing away equipment he, Jim Lovell (the mission's Commander) and Jack Swigert (the Command Module Pilot) had used to film a TV broadcast back to Earth. When he'd finished, Haise pushed off towards the tunnel connecting the LM to *Odyssey*, the Command Module, to join his colleagues. It was time for them all to get some sleep. As he floated back towards Lovell and Swigert, a heavy *whump* thudded through the spacecraft. Haise saw the hull move around him. He pulled himself through the tunnel, drifting into his position on the right-hand side of the capsule. He was confronted by an inexplicable display of warning lights. The instrument panel – electrics, cryogenics, fuel cells, comms and environmental systems – swarmed angry red and yellow. The master alarm provided a soundtrack through his headset. The computer restart light was on. None of it made sense. No plausible scenario they'd trained for in the sim had caused anything like the terrifying display he was looking at.

Haise forced himself to focus. Temperature, pressure and

quantity on one of the oxygen tanks were shot, the needles showing empty. *These are different sensors*, he thought. It was unlikely to be a false reading. And outside the window, he could see debris.

'OK, Houston,' reported Swigert, when he realized half *Odyssey*'s power – from a fuel cell dependent on a supply of oxygen – was gone, 'we've had a problem here.'

There was still a second oxygen tank, though. *Still not life-threatening*, Haise told himself. But he felt sick to his stomach. Haise enjoyed an uncanny ability to ingest and memorize checklists and procedures. He knew, without having to check, that Mission Rules dictated this was the end of their moon landing. Then the crew discovered that the second oxygen tank was leaking.

In the glassed-off VIP area at the back of Mission Control, John Young had been watching the television footage beamed back from the spacecraft with the *Apollo XIII* wives. Young was Commander of the mission's back-up crew. Watching Jack Swigert had not been part of the plan. Until just three days earlier, Swigert had been Young's own Command Module Pilot until a fear that the *Apollo XIII* pri-mary CM Pilot, T. K. Mattingly, might have contracted German measles had, despite Young's resistance, seen the two crewmen swapped. At the end of the broadcast, Young had said goodbye to Jim Lovell's wife, Marilyn. But while she drove home to Timber Cove, ten minutes away to the north-east of the space centre, Young was still in Building 30 when her husband confirmed Swigert's message: 'Houston, we've had a problem.'

The back-up Commander made his way straight down to the second from front of the four rows of consoles ahead of him to where controller Sy Liebergot was monitoring the crew's environmental and electrical systems. The picture presented by the readouts would have been rejected in the simulator as far-fetched. Liebergot felt nauseous. 'It could be the instrumentation,' he offered hopefully. At his shoulder, as he watched the oxygen bleed away from the second tank, Young told him, 'This doesn't look like instrumentation to me.'

Lovell, Swigert and Haise were already 200,000 miles from Earth. That distance increased by 400 miles with every passing minute. And their last manoeuvre, while allowing them to reach their lunar destination in the required lighting conditions, had also removed them from a course that would provide a free return to Earth after a single slingshot around the moon. If the crew of *Apollo XIII* continued on their current trajectory they would never again come any closer than 45,000 miles from home. But out of oxygen and electricity, the three astronauts would be dead long before then unless they abandoned the dying Command Module for the safety of the undamaged Lunar Module, a vehicle equipped with its own oxygen, electricity and water supply. Racing through a checklist that normally required two hours to complete in less than an hour, Fred Haise powered up the LM and, with Lovell, began transferring control of the stack from the CM to inside the little landing craft. It was a procedure neither had previously completed in the simulator with any great confidence.

With the crew's retreat to the LM, their short-term survival was more secure. The priority in Mission Control turned to how to bring them home. Using the Service Module's big

Aerojet AJ10 rocket engine to perform an immediate U-turn and return was quickly dismissed. So too was the possibility of using it to bring them back on to a free return trajectory. The effect of 20,500lb of thrust kicking into the damaged spacecraft was unknowable; but with the fuel cells starved of oxygen there was insufficient electrical power to operate it in any case. That left the crew reliant on the less powerful descent engine used by the LM which, fuelled by hypergolic propellants that ignited on contact, could be used without demanding so much of what remained of the spacecraft's electrical supply. But before any burn of the descent engine, the crew would have to be certain they were pointing in the right direction – and that was providing problems of its own.

Inside the Lunar Module, Jim Lovell was struggling to stabilize the stack, the LM's carefully calibrated controls completely thrown by its attachment to 30 tons' worth of Command and Service Module. As he wrestled with the hand-controller, Lovell pressed Haise for advice. 'Is there any way I can control this thing, Freddo?'

Back in Houston, his back-up Commander was trying to answer the same question.

During geology training for the mission in Hawaii, John Young had watched as Lovell and Haise ran from an erupting volcano. He'd joked that their fiery demise might have seen him promoted to prime crew. Now his focus was on saving their lives. Young had been working in the simulator since soon after *Apollo XIII* had got into trouble. And while Lovell was slowly getting a better feel for piloting the unfamiliar stack, he and Haise, surrounded by a cloud of twinkling sunlit debris from the blast that had damaged their ship, were unable to get a fix on the stars they needed to align their navigation computer. In the sim, Young and his own LM Pilot, Charlie Duke, tried unsuccessfully to position the windows of the LM in the shadow of the Command and Service Module to stop the jetsam from reflecting sunlight. In the end, Lovell, Haise and Swigert were just going to have to fire the engine on the assumption that guidance systems were still aligned. Young and Duke's work at least suggested that the *Apollo XIII* autopilot would keep them stable during the thirty-second burn.

Keenly aware of the demands on the crew of preparing for the course-correction manoeuvres, Young and Duke worked to find a way of reducing the number required from five down to three. *There was just no way*, Young thought, *that the crew would have time for five manoeuvres before entry back into the atmosphere.*

Jim Lovell pressed the PROCEED button, and at 08.42:43 GMT, a little over five and a half hours since they'd first radioed in news of the problem, the computer fired the descent engine to bring the crew back on course for a return to Earth. There was a pause from Mission Control after Lovell confirmed the end of the burn; then CapCom came through static to tell them, 'You're go, *Aquarius*. No trim required.' The burn had been perfect.

Now Fred Haise turned his attention to whether or not their lifeboat, the fragile, lightweight Lunar Module, which had on board the resources to support two men on the lunar surface for two days, was capable of keeping alive three men for the duration of the four-day journey home. The LM Pilot pulled out a checklist and began to scribble down some back-of-the-envelope calculations. Oxygen was fine. And, if they switched off all but the essential systems, the batteries would last. But the water needed to prevent the computer, navigation and guidance systems overheating was, he figured, going to dry up five hours before re-entry. Then Haise remembered they were carrying data from *Apollo XI*. Before leaving the LM following their return to the Command Module from the surface of the moon, Neil Armstrong and Buzz Aldrin had turned off the water just to see how long the spacecraft would last without water before the first system failure. It was eight hours. On that basis, even if they ran out of water five hours before they re-entered, Haise and his crewmates were going to make it home. An exhausted Haise was told by Lovell to try to get some sleep.

But *Apollo XIII* was hardly out of the woods. Ahead lay two further engine burns, course corrections, half an hour in a communications blackout on the far side of the moon, near freezing temperatures and, critically, the construction of ad hoc air filters to scrub poisonous carbon dioxide from the cabin – a job not finished before CO_2 levels had risen high enough to trigger a master alarm. And yet the crew of *Apollo XIII* eventually splashed down safely around one thousand miles south-east of Fiji on 17 April. In surviving their eight-day mission, Lovell, Haise and Swigert travelled further from Earth than anyone before or since. At no point had their return been guaranteed.

When John Young had first stood at the console in Mission Control looking at the readouts coming back from the crippled spacecraft, as the magnitude of the situation in which the crew found themselves became apparent, *for sure*, he'd thought, *we had lost them*. But the safe return of the crew of *Apollo XIII* using the Lunar Module as an improvised lifeboat joined the first lunar landing as a high-water mark of the entire Apollo programme. The encyclopaedic knowledge of the LM's systems possessed by Fred Haise was a crucial factor in the crew's survival. Undeterred by the drama of *Apollo XIII*, Haise jumped right back into training for a return to the moon. As Lunar Module Pilot on *XIII*, Deke Slayton's crew selection system would cycle Haise through *Apollo XVI* as back-up Commander, then give him command of *Apollo XIX*.

But by the time Lovell, Haise and Swigert blasted off from the Cape, the Apollo programme had, in landing men on the moon, already done exactly what it was intended to do. *Apollo XIII* was a sobering reminder of how fine the margins were. Apollo was never a system designed for persistence and as a result each mission to the moon carried a significant risk of losing a crew. Some inside NASA were questioning whether it remained a risk worth taking. The knife-edge of *Apollo XIII* made the final decision easier. On 2 September 1970, NASA, facing the budget cuts that forced them to choose between space station and Space Shuttle, pulled the plug on the last two lunar missions. *Apollo XVII* would be the last flight to the moon.

Instead of twice walking on the surface, the closest Freddo was going to get was that single, perilous slingshot around the moon that gifted his wounded spacecraft the energy it needed to return home. A good soldier, Haise agreed to stay on as back-up Commander to John Young for the *Apollo XVI* mission, but as far as flying was concerned, it was a dead end. After that, he'd have to pin his hopes on the Shuttle getting built.

Nor was Freddo alone. Another member of his 1966 astronaut intake was about to taste even more bitter disappointment.

Since he'd arrived in Houston off the back of the X-15 rocket plane programme as the first new recruit to have already earned his astronaut wings, Joe Engle had trained as a Lunar Module Pilot. He'd been on the back-up crew for *Apollo XIV*, and in August 1971, despite pressure to replace him on the crew of *Apollo XVII* with geologist Jack Schmitt, Deke Slayton included him alongside Gene Cernan and Ron Evans in the crew selection he sent to Washington. *It would be nice*, Slayton agreed with headquarters, *to send a geologist to the moon*, but that wasn't the way the straight-as-an-arrow Flight Operations boss made his decisions. Engle was more qualified.

The crew's Commander, Gene Cernan, was on holiday in Acapulco when he took the call from Slayton: 'Congratulations, Geno, *Apollo XVII*'s yours.' Cernan was elated, thanking Slayton excitedly, before catching himself.

'Does this include Ron and Joe?' he asked.

'Well, not exactly . . .'

To make sure that on NASA's last manned mission to the moon there would be a scientist on board, Slayton's crew selection had been overruled for the first time. To a bitterly disappointed Engle it wasn't a complete surprise. There'd been plenty of internal speculation that Schmitt would get the slot, but telling himself *it makes perfect sense* or *it's the best thing for the programme* wasn't going to make it any easier to tell his kids that he wasn't going to the moon.

Deke Slayton invited him into his office to talk things

through. Engle could tell Slayton wasn't enjoying the conversation much more than he was. The Flight Operations boss was sympathetic, as helpful as he could be, but he didn't have many places to turn. Slayton already had nearly twenty guys working on *Skylab*. He couldn't promise Engle a flight on that, but if it interested him, he made it clear he'd give Engle's claim proper consideration. There was also an as yet unconfirmed prospect, agreed earlier in the year, of an orbital rendezvous with the Soviets, but again, he wouldn't commit to including Engle on any crew for that. The only other option was the Shuttle. So far it had been little more than a study, but Slayton told Engle that it looked like it was going to happen. That was the one. 'If it's got a stick and rudder and wings,' Engle told him, then that was the programme which he could bring the most to. If a PhD in Geology was going to get Jack Schmitt to the moon, then it made sense to assign an X-15 pilot to the Space Shuttle.

Slayton looked pleased, or perhaps just relieved. 'That was my opinion,' he said, 'but if you'd wanted to fly sooner than that, I was ready to help out.'

He took it better than I would have, thought Slayton, as Engle left his office. For both Engle and Haise, any future hopes of spaceflight now rested with the Space Shuttle. And while the prospect of that was years away, there were increasing signs that it would at least become real.

SIXTEEN

Houston, 1971

Since the first DC-3 design for the Shuttle had emerged from Building 36 in the spring of 1969, at least seventy-five further variations on a theme had been drafted inside Max Faget's 'garage'. There were orbiters with straight wings, swept wings, delta wings and double delta wings. Designs with one, two, three or four main engines in the tail. Or none. Early designs carried all their fuel on board – which increased their size at the expense of the payload they could carry. There were big ones and small ones. Some were not at all easy on the eye while one, a design labelled MSC-036A, was notably rakish-looking. There were twin-tailed shuttles and orbiters with canards projecting from either side of the nose. Some had jet engines (in the wings or in the nose) for landing, others were given swing-down rocket engines for launch, and one or two had solid-rocket engines for use in an abort. Then, on 10 August 1971, a design labelled MSC-040 was produced by Faget's design and engineering team. And with this one, his Skunk Works nailed it. A few nips and tucks aside, with MSC-040 NASA now had their template for the Space Shuttle. But even before the successful design emerged from Building 36, it was already clear it was going to need more power. Quite a lot more.

From the earliest days of the Space Shuttle programme it was understood that there were particular aspects of it that would dictate the development schedule of a new spacecraft. The main engines, three of which were to be mounted in a triangle below the tail, were top of this list. To try to get a head start, NASA commissioned detailed proposals from engine manufacturers before it had even secured funding to build the Shuttle itself. Then, in January 1971, they agreed to give the Air Force everything it wanted. As well as those big delta wings Max Faget had so resisted, the agency was also committed to building a Space Shuttle capable of heaving 65,000lb into orbit inside a payload bay large enough to carry a house trailer. And so, in the same month, the two companies competing for a contract were told by NASA that the goalposts had moved. They needed to increase the thrust delivered by each engine by over a third. It was an extraordinary jump. Even before demanding more power, NASA had been asking for an engine stretching the boundaries of what was possible. They wanted the highest-performance rocket engine ever built; something little short of an engineering masterpiece.

At the Rocketdyne plant in California's San Fernando Valley, company vice-president Paul Castenholz knew he was at a disadvantage. In its heyday, building engines for Wernher von Braun's Saturn V rockets, Rocketdyne had employed over 23,000 people. With the Apollo hardware now built, that number had dropped to under 2,000. On top of that, rivals Pratt and Whitney had been working on an engine for the Air Force of exactly the type NASA had specified, while his own firm had devoted its energy to developing a radical new kind of rocket engine, the Aerospike, which NASA had now ruled out of contention.

NASA had specified an engine burning liquid hydrogen and liquid oxygen capable of delivering 415,000lb of thrust, 14,670 feet per second in exhaust velocity, and 3,000lb per square inch of pressure. And, unlike any other rocket bar the 57,000lb thrust XLR-99 used in the X-15 rocket plane, they wanted it to be reusable. *It would be like*, thought one Rocketdyne engineer, *inventing an Indy 500 car with a 50,000-mile warranty*.

Pratt and Whitney already had an engine called the XLR-129. Developed for an abandoned classified Air Force programme to build a hypersonic rocket glider codenamed ISINGLASS, the Pratt and Whitney engine was capable of delivering 250,000lb of thrust at the required pressure. They had something upon which they could build. But with Aerospike out of the race, Rocketdyne had nothing. Castenholz was quick to appreciate that drawings and promises had no chance of winning the prize. To leapfrog their rivals, Rocketdyne were going to have to *show* NASA that they could build them the engine they wanted. And he managed to persuade the company to stake $3 million of corporate funds to do it. Rocketdyne's engineers worked round the clock, sleeping in the factory's on-site hospital night after night while they built a full-scale test bed. Using what knowledge they could bring from the J-2 they'd built for the Saturn rocket, by late 1970 they were able to light their engine in their Nevada field lab near Reno. Unable, at this point, to cool it, their test rig could run for no more than five seconds before it would self-immolate and destroy itself. But by 1971, the Rocketdyne Shuttle's main engine thrust chamber, in its final, brief full-power burn, delivered 505,700lb of thrust, an exhaust velocity of nearly 15,000 feet per second, and a chamber pressure over 3,172 psi. In just months, they'd built an engine that smashed NASA's requirements.

And when, after agreeing to build the Shuttle the Air Force said it wanted, NASA raised the bar from 415,000lb of thrust to 550,000, Pratt and Whitney were nowhere near. Armed with film of the Nevada test runs and over ninety volumes of documentation, in March 1971, Castenholz delivered the Rocketdyne proposal to managers at NASA's Marshall Space Flight Center in Huntsville, Alabama, the centre responsible for the Shuttle's engine development. He also took with him the thrust chamber itself. He wanted everyone on the evaluation board to see and touch it. Four months later, Rocketdyne was awarded the contract.

There was just one critical component missing from their demonstrator, however. Castenholz's team had simply had no time to develop and build the turbopump the engine needed to generate the high propellant pressures required. For the purposes of their demonstrator, Rocketdyne had fed liquid hydrogen and oxygen directly into the engine using a powerful external pump. Rocketdyne had only shown their engine could deliver *if* they could build a working turbopump. It was a big *if*. Not much bigger than a healthy-sized lawnmower engine, the turbopump had to spin at a rate of 36,000 revolutions per minute, at pressures of 8,000lb per square inch, and at power levels of 75,000 horsepower. By comparison, the engines that drove the *Titanic* generated just 55,000 horsepower and were spread over an area the size of around fifteen tennis courts. Justifiable elation at beating off the competition from Pratt and Whitney had to be tempered by the reality of the task that now lay ahead of them.

As Rocketdyne contemplated the chasm that existed between their demonstrator and the production of a reliable, reusable rocket engine certified for human spaceflight, NASA was still waiting for the President to greenlight the development of the flying machine the new engine was designed to propel into space. Yet, while a decision on NASA's Space Shuttle remained beyond the horizon, Richard Nixon had shown he was quite prepared to commit large sums of money to America's future in space. But when, in September, National Security Advisor Dr Henry Kissinger gave word that the President had authorized development of an ambitious, revolutionary new spacecraft, it was not the country's civilian space agency who would be the beneficiaries. Instead, it was the black world of the National Reconnaissance Office.

The charge-coupled device technology discovered by Bell Laboratories in the late sixties was now sufficiently mature to allow the NRO to pursue their long-held ambition to build a digital spy satellite capable of providing real-time photographic intelligence from space. To be built in complete secrecy by Lockheed in California, the new spacecraft was known, initially, as Program 1010. It soon acquired the codename KH-11 KENNEN and was projected to be operational by the end of 1976. Not many people working for NASA enjoyed the high-level security clearance that knowledge of the new NRO programme required, but MOL veterans like Bob Crippen and Dick Truly, taken in by Houston after the promise of digital spy satellites had put their military space station programme out of business, were among them.

So too was the Director of NASA's Ames Research Center, Dr Hans Mark. And, over the next decade, it would be Mark more than anyone else who provided the bridge between the country's military and civilian space programmes. And at the heart of it all would be the Space Shuttle.

SEVENTEEN

Ames Research Center, 1971

WHEN ASKED HOW he was getting along with Hans Mark, a former colleague once replied, 'It's mindboggling having to deal with a genius.'

Mark's interest in space was inspired on his twelfth birthday in 1941 when his parents gave him a copy of a book by P. E. Cleator, the first head of the British Interplanetary Society, called *Rockets Through Space*. In it, his father had inscribed the message 'The phantasy of today is the reality of tomorrow. For Hans, from Herman.' Mark determined that he would like to live his life on that basis.

His father was well qualified to consider the advance of science. In 1927, he had taught Edward Teller quantum mechanics at Karlsruhe University. The Mark family escaped from Austria when the Nazis invaded in 1938. After an arduous journey west, they settled in New York. Hans shared his father's aptitude for science, but the atom bombs that brought an end to World War Two suggested that it was nuclear physics, not space and rocketry, that offered him the brightest future. By 1959, as an assistant research physicist at Berkeley in California, Hans was helping Teller teach his advanced quantum mechanics course to graduate students. In 1956, Teller, the self-proclaimed 'Father of the Hydrogen Bomb', employed Mark as a researcher at the university's Livermore Laboratory. Rooted in their experience in pre-war Europe, both men shared a common interest in maintaining their adopted country's military strength. Through the early sixties, Mark ran a division of 150 people in support of the design and development of nuclear weapons. Central to his work strengthening America's atom bomb arsenal was an effort to monitor and detect advances in Soviet weaponry and, as a result, since his arrival at Livermore, Mark had enjoyed the highest TALENT/KEYHOLE security clearances. When, after deciding that there were limits to how much he could do to make nuclear bombs any more militarily useful, he moved to Mountain View to take up a new post as Director of NASA's Ames Research Center, he retained them.

Tall and urbane, his face framed by a steel-grey crewcut and thick dark eyebrows, Mark had just the faintest trace of the Austrian accent he'd brought with him to America as a boy. He started work at Ames in February 1969, a month before Richard Nixon set up his Space Task Group.

In a parallel effort it hoped would inform the Space Task Group Report, NASA set up a committee made up of the directors of the agency's different centres. At the first meeting, Hans Mark found himself sitting next to the Director of Marshall Space Flight Center, Wernher von Braun. The charismatic rocketeer had come to America from Germany in very different circumstances to the Mark family. After leading the development of the V-2 ballistic missile that assaulted London in the latter days of World War Two, von Braun's expertise and passion had been spirited to the

US at the war's end where they had since provided both the foundation and vision for America's space programme. And it was von Braun, the country's greatest advocate of manned spaceflight, who proved crucial in rekindling Mark's childhood enthusiasm for human space exploration.

At the time he joined NASA, Mark was no cheerleader for the Space Shuttle. In fact, after his time at Livermore had opened his eyes to the capabilities of unmanned spacecraft like CORONA and GAMBIT, he couldn't help feeling that perhaps adding human passengers to the equation was only piling on unnecessary complication and expense. Von Braun argued persuasively that only the on-the-spot judgement, creativity and resourcefulness provided by a manned crew allowed for genuinely sophisticated and flexible space missions; but it was the invitation to Mark to join the *Apollo XIII* Accident Board that really proved it to him. While on that occasion the astronauts' intelligence, knowledge and ingenuity had been applied in the service of saving themselves, it was evident that the same qualities could be brought to bear on any mission that demanded it. Mark's conversion was sealed on board a yacht, during the downwind leg of a race in San Francisco Bay.

Mark and von Braun shared a love of boating. Whenever the Marshall Director visited Ames, Mark took care to ensure a sailing trip was on the itinerary. With the yacht trimmed and the wind behind them, the crew were able to relax while von Braun, dressed in a white cricket jumper, his hair tousled by the sea breeze, held court. Asked whether or not there would be a return to the moon following Apollo, von Braun compared the journeys of astronauts to those of Scott and Amundsen to the Antarctic. After the explorers' 1911 heroics in reaching the South Pole, little had been done to capitalize on their pioneering efforts until the availability of what von Braun called 'enabling technology'. Just as more substantial exploration of the Antarctic was dependent on the aeroplane, he explained, our proper understanding of the moon would follow the construction of a space station. And that, von Braun was certain, would first need a Space Shuttle. By the end of 1971, Mark was persuaded. Soon after that he would be seen proudly wearing a tie patterned with the Orbiter's silhouette. He was a believer, and belief, in the end, was what drove NASA's pursuit of the Shuttle. Mark recognized that. 'I do not know,' he wrote, 'how to quantify the value of putting people in space.'

While NASA continued its wait for the President to make a decision on whether or not to build the Shuttle, it employed the services of an organization it hoped could do just that.

Consulting firm Mathematica Inc. was the creation of economist Oskar Morgenstern. As with Hans Mark, Morgenstern was resident in the United States as a consequence of Hitler's annexation of Austria in 1938. A Professor of Economics at the University of Vienna, Morgenstern was visiting Princeton when the Nazis seized Vienna. He remained at the American university, where he met the Hungarian-born mathematical genius John von Neumann. A prodigy who as a child could memorize and recite the phone book, von Neumann had earned his PhD in Mathematics at just twenty-two. Still in his twenties he took up, alongside Albert Einstein, one of five professorships at Princeton's Institute for Advanced Study. Von Neumann's polymathic brilliance ranged from quantum mechanics to the hydrogen bomb. In 1944, along with fellow émigré Morgenstern, with the publication of the landmark *Theory of Games and Economic Behaviour* he created game theory, a mathematical model for the prediction and analysis of rational decision-making. When von Neumann went on to invent the architecture of the modern computer, his co-author founded Mathematica, a company that, given its family tree, seemed plenty smart enough to perform the kind of voodoo economics necessary to make the Space Shuttle look like financial good sense.

After NASA hired them in 1970, they did. Over twelve years, between 1978 and 1990, Mathematica concluded, the Space Shuttle would save the country over $14 billion compared to simply continuing with the then current generation of expendable rockets. There were plenty of beautifully plotted graphs whose lines illustrated the numbers. It was impressive stuff. The problem was that the Office of Management and Budget thought the Mathematica report was pure fantasy. One member of a committee set up by Nixon in the summer of 1971 to assess the Shuttle programme thought the figures showed evidence that NASA and Mathematica must be smoking marijuana. The accountants' disbelief was not without justification. Mathematica's promised $14 billion saving was based on an assumption that the Shuttle would fly 736 times in twelve years. Sixty flights a year; more than one a week. Just to *match* the cost of the expendable rockets would require thirty-nine flights every year. But in the end the cold cost/benefit analysis didn't matter. Or rather, as Hans Mark had decided, 'it was almost certainly beside the point'.

Fortunately, there were powerful allies in Washington who understood this.

In August 1971, Caspar Weinberger, the Deputy Director

of the Office of Management and Budget, learned that proposed cuts to NASA's 1972 budget would signal the end of America's participation in manned spaceflight. Faced with this stark reality, Weinberger wrote to the President. 'I believe,' he told Nixon, 'this would be a mistake.' Going on to address the further possibility of cancelling the last two remaining Apollo moon missions, Weinberger explained his position more clearly. Stepping away from manned spaceflight would, he wrote, 'be confirming, in some respects, a belief that I fear is gaining credence at home and abroad: that our best years are behind us'. Nixon's response, handwritten in the margin of Weinberger's original memo, was succinct: 'I agree with Cap.'

If the President had any lingering doubts about approving the development of the Shuttle they were perhaps put to bed in December 1971 when he flew to the Azores for a two-day summit with the French President, Georges Pompidou. Nixon flew to the Portuguese mid-Atlantic outpost aboard a Boeing VC-137, a specially appointed Presidential version of the 707 airliner known as Air Force One. Painted in an elegant sky blue and white livery, the jet usually bestowed bragging rights upon the US President in any company. But that Sunday, 12 December, Pompidou arrived in Lajes aboard Concorde 001, the first prototype of the needle-nosed Anglo-French supersonic airliner. The appearance that America was lagging behind the Europeans embarrassed the US President, who had been disappointed a year earlier when Congress had strangled the country's own supersonic transport project. At dinner on Monday night Nixon observed that Pompidou's journey to the Azores had been three times quicker than his own. 'I do not speak in envy,' he claimed, 'I only wish we had made the plane ourselves.' When he returned to the US, a decision on the Space Shuttle was imminent.

NASA had tried, without ever being entirely convincing, to justify the Shuttle on the grounds of cost, national security and jobs, but in the end perhaps it was simply that Nixon wasn't prepared to preside over either the end of US manned spaceflight or, following the abandonment of Boeing's supersonic airliner, a further retreat from the country's world-leading position in aerospace technology. America would have a Space Shuttle because America *should* have one. And on 3 January 1972, NASA were told to draft a statement for the President. The Space Shuttle had been approved.

Two days later, at the Western White House in California, NASA's new Administrator James Fletcher met with Richard Nixon. As the two men sat in armchairs flanking a low table carrying a large lamp, they passed a model of the Space Shuttle between them while Fletcher explained how it worked. A fifteen-minute meeting quickly turned into a half-hour conversation with Nixon's advisers looking on, wondering whether or not Fletcher was going to get his model back.

'I have decided today,' Nixon began the announcement that followed, 'that the United States should proceed at once with the development of an entirely new type of space transportation system, designed to help transform the space frontier of the 1970s into familiar territory, easily accessible for human endeavour in the 1980s and 90s ... It will revolutionize transportation into near space by routinizing it. It will take the astronomical cost out of astronautics.' Prior to the statement, Nixon had admitted to Fletcher that 'even if it were not a good investment, we would have to do it anyway, because spaceflight is here to stay . . . and we'd best be part of it'. The President finished his public announcement with a flourish that perhaps better reflected his real reason for approving the programme: '"We must sail sometimes with the wind and sometimes against it," said Oliver Wendell Holmes, "but we must sail, and not drift, nor lie at anchor." So with man's epic journey into space – a voyage the United States of America has led and still shall lead.'

Earlier drafts of the President's statement had referred to the new vehicle as the Space Clipper. Of a number of suggestions provided by NASA, including Astroplane, Hermes, Pegasus and Skylark, it was the one favoured by White House staff. Nixon himself dismissed them all. Instead of gracing the new spaceplane with any kind of romance or style as earlier spacecraft like Mercury, Gemini or Apollo had enjoyed, the President decided NASA's new bird would be known by the name by which it had always been known: the Space Shuttle.

For all the excitement that greeted the Presidential announcement, not everyone was so happy. Two weeks later, the National Reconnaissance Office's second KH-9 HEXAGON spy satellite was launched from Vandenberg Air Force Base on top of a USAF Titan IIID rocket. Given the reliability and, more importantly, self-sufficiency offered by the Titan, the President's words were not particularly welcome within either the Air Force or the CIA. The Generals and spooks who had until this point offered no more than a reluctant commitment to using the Shuttle as a launch vehicle now had to contemplate the reality of having to do so. Designed to be launched on top of a rocket, none of their satellites was compatible with the Shuttle.

Weeks after Nixon's statement, the NRO completed a study into 'the minimum modifications required' to make HEXAGON and its ground systems compatible with the Shuttle. The document explored the possibility of the Shuttle retrieving the hugely expensive satellite from orbit and returning it to Earth for refurbishment. It afforded a glimmer of what manned spaceflight might yet offer the military. For the time being, though, NRO's concerns were more immediate and practical.

Hans Mark's old friend John Foster, Director, Defense and Engineering inside the Pentagon, summed up the feeling within the NRO succinctly. With the Shuttle decision confirmed, he said to Mark, slightly incredulously, 'We're going to have to put HEXAGON in that thing . . .'

EIGHTEEN

Descartes Highlands, the Moon, 1972

11.57AM HOUSTON TIME, 21 April 1972. Urged on from inside by his crewmate Charlie Duke, *Apollo XVI* Commander John Young stepped down the ladder of the Lunar Module *Orion* towards the surface of the moon. After three previous spaceflights, including *Apollo X*, which had taken him to within 61 miles of the moon, it was where, he felt, he was supposed to be. 'I'm sure glad they got ol' Brer Rabbit here,' he drawled, in a reference to the Joel Chandler Harris stories he'd enjoyed as a boy, 'back in the briar patch where he belongs.'

After completing four final items on the checklist, Duke was ready to leave the Lunar Module to join Young. 'Here I come, babe!' he said.

The crew's almost childlike excitement was as clear as it was understandable. It was also a pale echo of the exuberance they'd shared inside the Lunar Module, in private, after touching down. A fault with the gimbal motors of the Command Module rocket nozzle had threatened to force a mission abort. Analysed on the ground at Houston before a decision was made to proceed, it had still delayed Young and Duke's descent to the Descartes Highlands by six hours. As a result, by the time the Commander had safely brought the Lunar Module in to land among the craters, he'd been working at capacity for thirteen hours. An immediate moon-walk, as scheduled, would have raised that to near thirty. As eager as the crew were to leave their spacecraft, they had no complaints when Mission Control decided to postpone the EVA (extra-vehicular activity) until after Young and Duke had got some sleep. Young, seemingly entirely at home on another world, slept 'like a brick' on a hammock hung at head height inside the Lunar Module. With assistance from a Seconal barbiturate, so too did Duke, lying in a hammock strung beneath his Commander's.

'Hey, John, hurry up!' Duke said as Young climbed slowly down the ladder.

'I'm hurrying!' replied the Commander, amused by his crewmate's impatience.

With both astronauts finally outside *Orion*'s claustro-phobic interior, they deployed the Lunar Rover buggy and offloaded the experiments they'd conduct during nearly three days on the moon, before Young set up the flag. 'Wait a minute,' Duke said as he bounded back to the Lunar Module to grab the camera. 'You're not getting away from there without me getting your picture.' With the Lunar Module, the Rover and Stony Mountain in the background, Duke lined up the perfect shot. 'Come on out here and give me a salute, a big Navy salute,' he said. Still displaying an ebullience for which he was not well known, Young crouched down as far as the thick white spacesuit would allow, then, in one-sixth lunar gravity, sprang one and a half feet into the air and snapped off a sharp salute. Nearly a second and a half later, he landed again.

After Young returned the favour, Mission Control told them, 'This looks like a good time for some good news here . . .'

'OK,' Young encouraged.

'The House passed the space budget yesterday, 227 to 60, which includes the vote for the Shuttle.' With Congress supporting the President's commitment to the Shuttle, it was on its way.

'Beautiful,' purred the two moonwalkers, 'wonderful, beautiful.' As Duke skipped ahead of him, back towards the Lunar Module, Young added, 'The country needs that Shuttle mighty bad, you'll see.'

John Young might have been a committed supporter of the Shuttle, yet it was not the machine he had hoped for. Like Max Faget, he'd wanted a fully reusable Space Shuttle made up of an Orbiter and winged booster that could fly back to land on a runway, before being prepared for their next flight. Instead, primarily because of the need to save on development costs at the expense of higher costs per flight, NASA was building the machine represented by the model that had so captivated Richard Nixon when he'd first announced the programme. It was an awkward-looking cluster of spaceplane, fuel tank and strap-on booster rockets that looked unlike any other rocket ever built. There were good reasons for this, the physics of which had been understood for hundreds of years.

In July 1687, sixty copies of a freshly printed new book were dispatched from London to Cambridge by horse and cart. The volume's title was *Philosophiae Naturalis Principia Mathematica*, and its author, Sir Isaac Newton, one of the greatest scientific minds of all time. A fellow of Trinity

College, Newton was Lucasian Professor of Mathematics at the University of Cambridge, and *Principia* was his masterpiece, establishing laws of motion and gravity that have both underpinned classical physics and provided the foundation for all rocketry and spaceflight ever since. His three-volume *Principia* was deliberately not presented as a work of popular science. In fact it quickly gained a reputation for its inaccessibility, but in writing a book for a narrow, specialized readership, Newton hoped 'to avoid being baited by little smatterers in mathematicks'. Despite this approach, *Principia* included a brilliantly simple thought experiment, which became known as Newton's Cannon, to illustrate the requirement for sufficient speed in achieving orbit.

First, Newton suggested, one should imagine a cannon placed upon the peak of a mountain so high that it projected beyond the Earth's atmosphere. Then imagine firing cannonballs horizontally from the barrel. With every increase in the amount of explosive charge used, the speed of a cannonball fired from the gun grows. With each increase in muzzle velocity, the cannonball falls to the ground further from where it was fired. But the Earth is round, so its surface curves away from the cannonball as it falls. If the cannonball is fired fast enough it will take longer to arc down towards the ground than the Earth's surface does to curve away from it. Launched from the mountain peak above the atmosphere, not subject to drag from air resistance, the cannonball will simply continue to fall at a constant speed without ever getting any closer to the Earth's surface beneath it. Orbit, rather than being an escape from the effect of gravity, is dependent on it.

Of course, Newton didn't know about swingball, which

illustrates the whole thing even more clearly. Hit the ball with insufficient force and it won't go round and round, instead falling back towards the pole in the centre. Enough of a wallop will send it circling round and round the pole attached by its string (gravity). But hit it too hard and the string will snap, allowing the ball to fly off into the bushes (outer space).

The principle was straightforward, and yet constructing a vehicle capable of travelling fast enough to achieve orbital velocity was not as easy as it would seem. The challenge was in building a machine that could carry sufficient fuel to keep the engines burning long enough to propel it to the 17,500 mph needed to reach orbit. The first person to describe this was another pioneering scientist, working two centuries later and half a world away, who, using the physical laws established by Newton, worked out how it might be done.

Born in a remote Russian village in 1857, Konstantin Tsiolkovsky was largely self-taught. After contracting scarlet fever at the age of ten he became hard of hearing; then, after the death of his mother when he was thirteen, almost completely deaf. Isolated and humiliated by his affliction, he came to regard time spent in the company of other people as 'torture'. He withdrew into a world of books. And in search of approval – or simply 'not to be despised' as he wrote in his autobiography – he turned to study. Aged sixteen, he moved to Moscow where he divided his time between the rented corner of a room and the public library. He read voraciously and, inspired in part by Jules Verne's classic *From the Earth to the Moon*, he became hooked on aviation and space. It was to find some expression in his own science fiction.

But it was in 1903, with the publication of *The Exploration of Space by Means of Reactive Devices*, that he produced his landmark work. Through a single equation that would one day bear his name, Tsiolkovsky described the relationship between the empty mass of the rocket, the mass of its fuel, the rocket's speed, and the specific impulse of the fuel, that is the amount of thrust produced by one pound of fuel in one second. Through calculation, rather than speculation and imagination, Tsiolkovsky was able to show that in order to achieve orbital speed a rocket would have to shed dead weight as it climbed and accelerated. To do this would require construction from multiple stages. At the point when the first stage expended its fuel it was jettisoned, giving way to a lighter, smaller second stage. This was powered by its own rocket engine that was, in effect, launched at altitude, with the benefit of a rolling start. The result was rockets assembled like layered wedding cakes of which only the smallest, uppermost section would ever reach orbit. Based on existing fuels and materials, Tsiolkovksy concluded that it wasn't possible to reach orbit without what became known as 'staging'.

As NASA decided on the configuration of the Space Shuttle, Tsiolkovsky's rocket equation remained the single greatest constraint on their ambitions. It was no more possible for them to build a spaceplane that, without staging, took off then flew itself into orbit before returning to Earth than it had been in 1903.

When President Nixon christened NASA's new spacecraft the Space Shuttle he may well have thought he was just talking about the spaceplane on top, but of course technically the term referred to the whole assembly: the spaceplane (the Orbiter), the big fuel tank and the two boosters.

Unable to afford the flyback booster John Young and others had favoured, NASA constructed the Space Shuttle stack around a huge disposable external fuel tank shaped like a cannon shell and filled with liquid hydrogen and liquid oxygen. In coming to a decision about the nature of the Space Shuttle's main engines already being developed by Rocketdyne, NASA opted for liquid hydrogen as the fuel that offered the greatest specific impulse. It was the most powerful fuel available. And it was light: a gallon of water weighs over fourteen times more than a gallon of liquid hydrogen. The trouble was that this came with a cost as liquid hydrogen is also the lowest *density* fuel. While it weighed relatively little, it took up a lot of space. By contrast, the liquid oxygen used as the oxidizer needed for the hydrogen fuel to burn in the vacuum of space weighs more but requires less volume, but room still needed to be found for both. Carrying the liquid oxygen and hydrogen inside the Orbiter itself, and finding room for a substantial payload bay, would have made it unacceptably large. The external tank that sat beneath the belly of the Orbiter was the result. Once empty, it was jettisoned and thrown away, consumed by fire on its re-entry into the Earth's atmosphere.

Riding the tank was the Orbiter itself, the winged space-plane that has since become synonymous with the term 'Space Shuttle'. The three rocket engines in its tail, gulping propellant and oxidizer from the external tank at a rate that would empty an average family swimming pool in twenty-five seconds, between them contributed over 1.2 million pounds of thrust at launch, but they were not, on their own, capable of lifting the Space Shuttle from the pad even a few inches. That required the help of the two large strap-on boosters that used solid fuel technology developed for America's nuclear missile arsenal.

Attached to either side of the external tank the two solid rocket boosters – the SRBs – were the largest ever built. They were designed to burn for a little over two minutes before being jettisoned, parachuting back into the sea, then being towed back for refurbishment and reuse. Together, the two 150-foot-high boosters delivered six million pounds of thrust – more than 80% of the total lift-off thrust.

The result of these various pieces of hardware, their conflicting demands and limitations was the now familiar configuration of Orbiter, external tank and solid rocket booster that at the time looked so strange alongside the tall, slim rockets that had preceded it. The whole effect was an assembly that's been memorably described as akin to 'bolting a butterfly on to a bullet'.

A programme as complex as the Space Shuttle was always going to require an unprecedented level of cooperation between NASA's different centres. As responsibility for the Apollo Command and Lunar Modules had belonged to Manned Spacecraft Center at Houston, so too did that for the Orbiter. Houston was also given overall responsibility for managing the integration of the entire Space Shuttle programme. That meant working hand in hand with Marshall Space Flight Center in Huntsville, Alabama, who under Wernher von Braun had developed the Apollo programme's giant Saturn V booster. Marshall was charged with the development of the main engines, external tank and solid rocket boosters. But there would be plenty of work for Hans Mark's Ames Research Center too.

From the outset it was clear that such a lumpy, awkward assembly of shapes, sizes, struts and protruberances was going to present a unique set of aerodynamic challenges as the Space Shuttle accelerated through the lower atmosphere. To test this, the facilities at Ames would be vital. With the programme approved, Max Faget's deputy at Houston, Milton Silvera, told his boss, 'We need to go to Ames and make him a friend, because we're going to need his wind tunnels.' Not only that, they'd need Ames's state-of-the-art simulators to develop the Orbiter's flight control software, and its high-temperature arc-jets to test the thermal protection system, for which Ames itself had primary responsibility for developing. Of all the areas in which Ames was involved, it was the latter, the heatshield, that had given Mark and his engineers most cause for concern.

After the programme's approval in January, each of the Center Directors was invited to give an assessment of both the contribution they might make to the building of the Shuttle, but also the risks and challenges as they saw them. In his letter, Mark cautioned that developing a reusable thermal protection system – the heatshield that would protect the Shuttle from the fierce heat of re-entry – was going to require 'a concentrated effort'. Furthermore, he warned, there were likely to be problems around the bonding of thermal protection tiles to the Orbiter's airframe, and 'particularly with the inspection of the bonds to determine that they are indeed secure after manufacture'.

His comments would prove to be prescient.

The challenge of building a heatshield for the Space Shuttle joined the development of the main engines as one of the long poles of the programme. No previous flying machine had ever had to contend with the extremes of performance and environment faced by the Orbiter.

As it was for Hans Mark, heat – and making sure the Shuttle was built to cope with it – was an overriding priority for engineer Tom Moser. Just weeks before *Apollo XI*'s flight to the moon, Moser had been asked to design a flag for the astronauts to plant on the moon. His ingenious, lightweight, space-saving design survived the journey intact, but as he watched Armstrong and Aldrin on the moon, trying to open up the telescopic rod from which the flag hung, he could see it was getting stuck and wasn't going to extend fully. It created the appearance that, in the vacuum of space, the flag was flapping in the breeze. While Moser cringed, others were so taken by it that they decided to replicate it on every subsequent mission and, in the process, inspire a thousand conspiracy theories.

Moser's embarrassment over the sticky *Apollo XI* flag had done nothing to impede his progress through the engineering hierarchy at Houston. After first producing sketches of possible Shuttle configurations in 1969 while still attached to the Apollo programme, he was now working out of Building 13 at Houston, responsible for the Orbiter's structural integrity.

He began by looking backwards.

NINETEEN

Houston, 1972

TOM MOSER TRAVELLED from Houston to Burbank, California, home of the Skunk Works, the top-secret division of the Lockheed Corporation that had designed and built the legendary high-flying U-2 spyplane. But it was the U-2's successor, a blazingly fast CIA reconnaissance jet called the A-12 OXCART, that was of interest to the bearded NASA engineer.

When it first flew in 1962, the A-12 was unique. Shaped like the flat blade of a broadsword extending forward from a pair of delta wings, each housing a long cylindrical engine that ran from the leading edge to trailing edge, the A-12 not only looked and flew like nothing else in the sky, it was built like nothing else. To endure the heat generated by flying at over three times the speed of sound, the OXCART and its Air Force counterpart, the SR-71 Blackbird, were built largely from titanium. Only titanium combined sufficient lightness and heat resistance to enable the spyplanes to operate at an altitude of 80,000 feet and stand up to temperatures that in places exceeded 1,000°F. The problem was that the only US source of the metal, the Titanium

Corporation of America, couldn't supply it with sufficient purity. US titanium was so brittle it could shatter if dropped. Lacking a homegrown source of high-quality titanium, the CIA looked for an alternative and found it in, of all places, the Soviet Union. Using third parties and front companies, the titanium needed to build a replacement for the U-2 spyplane was imported covertly from the Soviet Union by the CIA.

As a result of his success in building the A-12 and SR-71 Blackbird, Kelly Johnson, the jet's chief designer and boss of the Skunk Works, was top of Tom Moser's list of people to see.

The two things that most concerned the man from NASA were, as they had been for the OXCART design team, weight and heat. It's relatively easy to make a strong, heat-resistant aeroplane, but you risk it being too heavy to fly. It's also straightforward to make a very light aeroplane, but the challenge is then to ensure it has adequate structural strength. Compounding that, *all* aircraft designs tend to gain weight as they evolve. Moser decided he was going to allow for that from the outset. Commercial airliners are one and a half times stronger than they need to be, built to endure stress loads 50% greater than any they would ever expect to encounter in everyday use. Moser decided that for the Shuttle a 40% margin was enough. Uncertain of the wisdom of this, NASA had Moser's plan checked out by engineers from Boeing to 'make sure you really know what you're doing'. He was fine, concluded the experts.

The first hurdle cleared, Moser then set out to get to grips with the more complex issue of the thermal stress the Orbiter would face during a mission that took it from the warmth of sub-tropical Cape Canaveral to the freezing vacuum of space, before plunging back into the Earth's atmosphere in a manoeuvre generating enough heat to liquidize steel. With its metal airframe sheathed by the heat-shield being developed under Hans Mark at Ames, the Space Shuttle skin and structure wouldn't have to contend directly with those melting extremes, but it would need to cope with variations in temperature that were more than enough to ensure that the airframe would expand, contract, soften and harden throughout every mission.

At the Skunk Works, Moser was welcomed by Kelly Johnson before being briefed by Lockheed engineers. Moser crawled in and around SR-71 Blackbirds brought into the facility for deep maintenance. On the shop floor he saw engineers machining, forming and processing titanium. Moser kept asking questions. Titanium, he learned, had been nightmarishly difficult to work with. Bringing the metal to full strength took seventy hours. The hardness that was in so many ways advantageous also meant that new processes for machining and milling the metal had to be worked out; drill bits and tools designed to work with aluminium had simply disintegrated and snapped. At the same time, the new metal was extraordinarily sensitive, especially to chlorine. Ink from a Pentel pen could eat a hole through a sheet of titanium in twelve hours. Instead of the chlorinated municipal water that flowed from the Burbank taps, only distilled water could be used to wash the metal. At the end of his visit Moser asked them, 'Bottom line, if you were NASA and you guys were responsible for the Shuttle structure, would you build it out of aluminum or titanium?'

'Aluminum,' they answered without pause.

Next, Moser travelled to England, where he visited the

British Aircraft Corporation factory at Filton, near Bristol, to be briefed on Concorde, an aircraft he thought was perhaps the most beautiful piece of engineering he'd ever seen in his life. In contrast to Lockheed, the British and French engineers had built their supersonic airliner from aluminium-copper alloy. Because of a lower top speed, Concorde didn't have to contend with the same absolute temperatures as the American spyplanes Moser had studied. Instead, the challenge for designers was to accommodate extreme temperature differentials of the kind that cause a cold glass to shatter on first contact with boiling water. In order to maintain stability as fuel was burned, the cold fuel that remained in Concorde's tanks was pumped around the hot airframe to redistribute the weight. To deal with the thermal stress this caused, along with an airframe that at high speed and high temperatures was several inches longer than it was on the ground, Concorde's designers had incorporated complex expansion joints throughout the aircraft. They built in room for her to breathe, but the engineering necessary was expensive, elaborate and difficult to do.

To hell with it, Moser decided, *we're going to tackle thermal stress head on*. After talking to Lockheed and BAC, he decided on brute force. They were going to build a light aluminium structure and they were simply going to make it robust enough to accommodate the combination of flight loads and thermal stresses flying to space and back would generate. Back in Houston, he and his structures team modelled their plans mathematically and the numbers worked. 'OK,' Moser said, 'we can do that.'

Everything learned by engineers like Moser as they polished and whittled NASA's paper plane was fed back to the manufacturer given the task of making it real. And in July 1972, NASA announced that this would be North American Rockwell.

As plain old North American Aviation, the company's track record in aircraft design couldn't have been more illustrious. Rolling out of factories in Ohio, Kansas, Dallas and El Segundo, over 15,000 of the company's T-6 Texan training aircraft were built. Alongside them were the B-25 Mitchell with which Jimmy Doolittle had bombed Tokyo from the deck of the USS *Hornet* in 1942, and the legendary P-51 Mustang, the appearance of which over Berlin had convinced Hermann Goering that World War Two was lost.

After the war, from a new plant in Downey, California, their hot streak continued with America's first swept-wing jet, the record-breaking F-86 Sabre. There was the B-45 Tornado, America's first jet bomber, then the F-100 Super Sabre, the world's first supersonic fighter. In 1959, the unparalleled X-15 rocket plane first flew, and just five years later the extraordinary six-engined XB-70 Valkyrie supersonic jet bomber was towed out of the hangar at Downey. Designed to carry 25 tons of nuclear weapons at three times the speed of sound and an altitude of 19 miles, the Valkyrie was, at the time of its first flight, the heaviest, most powerful, most costly aircraft ever built. There was nothing, it seemed, that North American's design team, led by a brilliant, hard-charging maverick named Harrison 'Stormy' Storms, could not accomplish.

When, in 1961, NASA awarded them the contract to build the Apollo spacecraft they were on their way to the moon. Then, on 27 January 1967, during training, the crew of *Apollo I*, Gus Grissom, Ed White and Roger Chaffee, were killed by fire inside the capsule designed by North American. Storms was made the scapegoat for the tragedy and fired from the company he, perhaps more than anyone, had made pre-eminent. In the midst of the political fallout following the fire, North American, its reputation mangled, merged with Rockwell-Standard, an autoparts conglomerate based in Pittsburgh. But as North American was hauled over the coals, there were plenty in senior management at NASA who understood that the causes of the fire were not as simple as the sacking of a single aerospace executive suggested. It was clearer still that the Apollo design was, fundamentally, a good one. So when, five years later, North American *Rockwell* was one of four aerospace companies to submit a detailed bid to NASA to build the Space Shuttle – a machine that had the potential to be the crowning achievement of America's post-war aviation industry – they were at no disadvantage. Not only did Storms' old design team submit a bid that was $40 million cheaper than the next cheapest, it also beat Lockheed, Grumman and McDonnell-Douglas on the basis of NASA's confidence in its programme management.

In drawing up their proposal, the company had one other ace up its sleeve. Uncredited and out of sight, 'Stormy' Storms had worked on the North American Rockwell bid as a consultant.

In February 1973, armed with a stylish new corporate logo designed by Oscar-winning animator and graphic designer Saul Bass, the management of the new, merged company divested themselves of any mention of North American. The Space Shuttle would be built by a company simply calling itself Rockwell International.

But just two weeks before that, NASA had gifted Storms

and North American's earlier-generation spacecraft one last dance when they announced the crew for the final flight of their iconic Apollo capsule on a mission to meet their Soviet counterparts in space.

TWENTY

Houston, 1972–3

As DEKE SLAYTON heard it, *Marooned*, a 1969 movie starring Gregory Peck and Gene Hackman, helped inspire the mission that finally gave him a shot at spaceflight. The film, featuring cosmonauts coming to the rescue of a stranded American crew, had apparently impressed the Soviets. It certainly wouldn't have harmed NASA's cause when they reached out to their Russian counterparts in 1970 to suggest a joint mission. With political support from both sides of the Iron Curtain, an agreement was signed by Richard Nixon and the Soviet Premier Alexei Kosygin on 24 May 1972, and the Apollo-Soyuz Test Project was scheduled for 1975.

As one of NASA's original Mercury Seven, Slayton was disappointed not to be given command of the American crew. But after being cleared to fly again after more than a decade grounded by a heart arrhythmia, he took the view that *flying was flying* and Apollo-Soyuz, whether he was Commander or not, was his only chance.

After signing the Apollo-Soyuz agreement, Nixon addressed the Russian people over Soviet TV and radio. He pointed out that the USA and USSR had never been at war and that the two great nations 'shall sometimes be competitors, but ... need never be enemies'. Just days after NASA announced Slayton, Tom Stafford and Vance Brand as the American Apollo-Soyuz crew, however, the Soviet Union announced that an anti-satellite weapon it had been testing since 1963 was now operational. Seven test interceptions had seen five specially launched target satellites destroyed in a sleet of shrapnel. Some competition.

Inside the Air Force Satellite Control Facility – the Blue Cube – in Sunnyvale, the Soviets' success caught the attention of a group of young Air Force officers assigned to the Special Projects Office. If the new Soviet space weapon should ever close on a US spacecraft, they wanted, at the very least, to be able to take pictures of it. The Air Force had not previously tried using their reconnaissance birds to capture images of other spacecraft. Given the speeds, distances, angles and timings involved, the challenges of taking a photograph of one orbiting object from another in a different orbital plane were substantial. But presented with the threat of the new Soviet weapon the team of Air Force technicians thought that, with the enhanced capability of the new KH-8 GAMBIT-3 – or 'G³' as it was referred to inside the programme – just entering service, it was worth trying. They immediately got to work on developing new computer algorithms with which to control the G³ spy satellites.

In every area of space endeavour, computers were taking centre stage. And none more so than in the development of the Space Shuttle, where the computers, alongside the main engines and the heatshield, were deemed to be the third of the three long pole items that would determine the success or failure of the programme.

In 1960, his senior year at the University of Texas, Bob Crippen took the first class in computer programming ever taught there. During his time at college, IBM had announced it would no longer be using vacuum tubes in its computers. It was progress. The previous year the company had proudly unveiled the IBM 305 RAMAC, which needed a room of at least 30 feet by 50 feet to accommodate it, and weighed a ton. There were just 2,000 computers of any description in the whole of the country, but it was clear, Crip thought, that computers were the coming thing. Using mainframes and punch cards, he developed an interest in computing when it was still a field in its infancy. *Even better*, he thought, *it was fun*.

Naturally enough, while assigned to the MOL programme, Crip, alongside Dick Truly, had opted to work on the programming for the IBM 4Pi computer that lay at the heart of the Air Force space station's systems. In March 1969, just months before MOL was cancelled, NASA decided to follow suit, agreeing on the IBM 4Pi for their own *Skylab*.

In choosing what was essentially an off-the-shelf machine they were taking a wholly different approach to the path they'd followed for Gemini and Apollo. With weight and space so critical in those two smaller spacecraft, NASA had bespoke computers designed and built. Without room for duplicate back-up systems, the safety of the spacecraft depended on the reliability of the single computer. With so much at stake, a tailor-made computer – the Apollo Guidance Computer – was created by MIT in which the possibility of failure was simply designed out. Each component of the new processor was tested at every point in its manufacture. In the words of one NASA engineer, 'every piece of metal could be traced to the mine it came from'. It was an unprecedented undertaking, but resulted in a computer which simply refused to fail: 1,400 hours of operation produced not a single in-flight failure. Following the lunar programme's conclusion, a test conducted at MIT that saw one of the Apollo computers run continuously, cycling through endless commands in the hope of provoking a malfunction, was eventually abandoned. Everyone simply got bored waiting for something that was clearly never going to happen. But building, testing and qualifying the bespoke computer for manned spaceflight had been dauntingly expensive in terms of time, money and effort. Beyond the free-spending Apollo years, such a specific solution would be out of reach.

For *Skylab*, thankfully, NASA was able to take another approach. With space available in inverse proportion to money on the more austerely funded programme, an off-the-shelf computer was chosen. Already flight-rated, radiation-hardened and combat-proven in Vietnam aboard Air Force jets like the Republic F-105 Thunderchief and the Boeing B-52 Stratofortress, NASA went for the 4Pi. Two of them. NASA built in redundancy: if one of the computers failed, the other took over.

The first 4Pi was delivered to Manned Spacecraft Center in December 1969, four months after the MOL pilots first arrived at Houston. Assigned to work on the *Skylab* programme, Bob Crippen and Dick Truly felt immediately at home. Off the back of MOL, Crip looked at *Skylab* and saw similarities. *I can contribute here*, he thought. And he jumped in.

* * *

While Crip and Truly worked on the *Skylab* computers, the Space Shuttle Office faced a far more demanding computing challenge in which redundancy was just one part of the puzzle. As significant was the fact that the Shuttle was going to have a wholly digital flight control system. There would be nothing connecting the pilot's stick and rudder pedals to the aeroplane's flight control surfaces but binary code passed through the wires of electrical circuits.

It was untried technology until, on 25 May 1972, a white-painted Vought F-8C Crusader taxied out to Runway 04 at Edwards. At 10.14am the jet, decorated with a kinked, lightning-effect blue cheat line and bearing the legend 'DIGITAL FLY-BY-WIRE' along the fuselage, accelerated along the lake bed and into the air. Apart from the gleaming NASA paint scheme, the jet looked the same as the F-8s Dick Truly had so enjoyed flying during his time with a frontline naval fighter squadron. Inside Truly's jet, though, cables ran directly between the cockpit and the control surfaces. When he pulled the stick, his own effort was directly responsible for deflecting the ailerons. In NASA's jet, all of that had been removed. Instead, the pilot's control inputs fed straight into a computer which then told the ailerons what he intended them to do.

The flight marked the beginning of Phase I of the fly-by-wire programme. Using a single Apollo computer, NASA knew they could expect 99.9% reliability. But their ultimate goal was more ambitious than that: the aim was to achieve reliability of 99.9999999%. Even as the fly-by-wire F-8 flew for the first time, the team at the Dryden Flight Research Center, NASA's facility at Edwards AFB, were thinking ahead to a system that might get them to that figure. And that had them circling back towards the family of IBM computers so familiar to Crip and Dick Truly from the MOL and *Skylab* programmes.

By the time *Skylab* was ready for launch on 14 May 1973, Bob Crippen had spent so much time down at the Cape that he was catching flak in the Astronaut Office as the 'Mayor of Cocoa Beach'. After three and a half years commuting between Houston, Huntington Beach and Marshall Space Flight Center in Alabama, he, along with Hank Hartsfield, was part of the team taking the spacecraft through the final phases of integration and pre-launch checkout. And as members of the *Skylab* support crew, he, Hartsfield and Dick Truly were also working CapCom (Capsule Communicator) for the launch. It was their job to talk to the astronauts. NASA's first space station was ready to go and MOL pilots, using experience gained on the doomed Air Force programme, had been instrumental in helping get her there. Having worked so hard to prepare *Skylab*, Bob Crippen looked forward to finally seeing her go, then to manning CapCom once the crew joined the space station on orbit. By the end of the day, though, his keen anticipation of what was to come had gone.

TWENTY-ONE

Houston, 1973

ONE THOUSAND MILES to the east of Mission Control, in a storm of fire and thunder, Wernher von Braun's great Saturn V rocket cleared the tower of Pad 39A for the last time. At the top of the stack, the familiar shape of the Apollo spacecraft, capped with its probe-like escape tower, was absent. In its place was a thicker, blunter aerodynamic shroud covering the *Skylab* workshop beneath it. As the rocket slowly cleared the tower, control of the mission passed from Launch Control in Cape Canaveral back to Houston.

Dick Truly, sitting at the CapCom console, monitored the vast rocket's progress. Although plugged in at his station, today's mission, *Skylab 1*, offered little opportunity for him to contribute. The space station itself was launched unmanned. Truly would be fully involved the next day when *Skylab 2* launched on board a smaller Saturn 1B, carrying her crew, Pete Conrad, Joe Kerwin and Paul Weitz, to their orbital home. For the second flight, Truly was CapCom for ascent, rendezvous and docking.

As the Saturn V accelerated skywards, telemetry inside Building 3, home of Mission Control, suggested that the space station's micrometeoroid shield had deployed early. Initially it was dismissed as a false reading by flight controllers, then signals reached Houston that there was also a problem with *Skylab*'s large solar panels. Further telemetry from ground stations in Madrid and Carnarvon, Australia, confirmed the worst: *Skylab* was in serious trouble. Just how much remained unclear, but it looked as if both the solar wings needed to supply the bulk of the spacecraft's electrical power were gone. Alongside this, the micrometeoroid shield appeared also to have been lost. And while the chance of *Skylab* being struck by a meteoroid was remote, the shield, lined with gold foil, performed a more important function as a sunshade. Without it, the temperature inside the space station was expected to rise as high as 170°F. While NASA tried to interpret the data coming in, the launch of Conrad, Kerwin and Weitz was pushed to the right five days and the next launch window on 20 May, and then again to 25 May. From the ground, controllers manoeuvred *Skylab* to try to reduce its exposure to the sun. Sleep became scarce as engineers at Houston, Marshall Space Flight Center and McDonnell-Douglas worked round the clock to produce a solution to the heating problem.

At Vandenberg Air Force Base in California, engineers were dealing with technical problems of their own. Operation number 4338, the latest G^3 spy satellite mission, had been scheduled for launch on 14 May, but because of a fault with the power supply to the bird's camera-focusing mechanism, on 15 May, when America woke up to newspaper headlines flagging NASA's woes, the Titan 3-B launch vehicle was still on the pad. Inside the Air Force's Blue Cube Satellite Control

Facility, the launch of the NASA space station had stoked a few regretful memories of the MOL programme. But when Captain John Earl Essing, working for a Programme Control Division within the GAMBIT programme office, read the reports about *Skylab*, the thought struck him that, once they got 4338 flying, they might be able to help. The work to develop a satellite-to-satellite – 'sat-squared' – imaging capability prompted by the Soviet announcement of its anti-satellite weapon was ready.

Initially, Lee Roberts, the Colonel in charge of the GAMBIT programme, said no, he wasn't prepared to risk compromising the 4338's primary intelligence-gathering mission. The sat-squared team persisted, though, and in the end were able to persuade the Colonel to share the proposal with Major-General David Bradburn, Director of the Air Force's Special Projects Division at the Blue Cube. If Bradburn said yes, the Colonel told Essing and his comrades, he would eat his hat.

At first, Max Faget's design team at Houston planned to fly the Apollo spacecraft in close formation with *Skylab* while one of the astronauts, standing in the hatch, rigged a sail-like sunshade, made from orange parachute nylon, to the space station. Under intense media scrutiny, the team worked on their design in the old centrifuge building, but in the end their efforts were usurped by rival effort. Also conceived inside Houston, the successful design was for a device that worked like a parasol. Four spring-mounted arms supported a silver shade made from a mylar, nylon and aluminium material used to make spacesuits. With the arms folded around a central pole, the parasol could be pushed through one of *Skylab*'s small-diameter scientific airlocks before, arms extended, the square sunshield was lowered back down into position.

As intense as the effort to improvise the sunshield had been, it still left unresolved the issue of *Skylab*'s electrical supply. *Skylab 2*, the first twenty-eight-day mission, would be able just about to get by on power from the distinctive windmill-like solar array that was used to power the solar telescope and the fuel cells in the Apollo CSM. But without power from the main workshop's solar wings, the longer missions planned for *Skylab 3* and *4* just weren't possible.

Engineers in Houston and Huntsville explored ideas for portable solar arrays and even the launching of a specially designed solar module that might be docked with the space station. But there was also a glimmer of a chance that the first *Skylab* crew might be able to fix her up themselves. Data suggested that part of the micrometeoroid shield was preventing one of the solar arrays from extending. If that were the case, the astronauts, using tools, could maybe free it. Rick Nygren, an engineer who'd worked alongside Dick Truly and Bob Crippen at McDonnell-Douglas during the space station's development, was in Building 5 at Houston, trying to assess the merits of various tools, including large bolt cutters, that might be launched with Conrad and his crew. It was a frustrating process, hampered because he and his colleagues didn't fully understand what had actually caused the malfunction. In wrestling with it, the difficulty faced by engineers was that it was unclear, based solely on the telemetry, just what sort of a tangle of wires and broken machinery would face the astronauts arriving at *Skylab* for the first time.

* * *

In his office at Sunnyvale Air Force Base, Major-General Bradburn listened to the pitch made by his junior officers. If they took the decision to lose nine days from the first part of the planned GAMBIT-3 mission and dropped one of the two film capsules early, then assuming that Blue Cube mission controllers were successful in bringing the bird's Eastman-Kodak camera to bear on *Skylab*, it would be possible to get the images to NASA before the first crew launched on 25 May. Bradburn liked it. *Skylab*, he thought, was a national asset, its success inextricably linked to America's position in the world. To get the pictures, though, would mean flying the satellite outside its design limits. To forestall any possibility of protest from the bird's manufacturers, angry at seeing an Air Force decision jeopardize the financial incentives they earned from successful missions, Bradburn brought them inside the tent. All said they'd waive their right to protest. And Bradburn made the case to John McLucas, the Director of the National Reconnaissance Office.

At 16.40 GMT on 16 May 1973, Mission 4338 began with a successful launch from Vandenberg that placed the satellite into a 110° inclination orbit. With a perigee – the point closest to Earth – of 86 miles and an apogee – the furthest point – of 247 miles, there was no stage at which the reconnaissance satellite's orbit would pass outside that of *Skylab*. It meant that to capture an image of the space station, mission controllers would have to point their bird's cameras away from Earth and towards outer space. For one day only, a KEYHOLE satellite was going to play the part of space telescope. It seemed a fitting challenge for the Air Force Satellite Control Facility, whose unit crest, beneath an orbiting eagle, carried the Latin motto 'Inveniemus

Viam Vel Faciemus' – 'We will find a way or we will make one.'

By 21 May, Marshall's Space Simulation Branch had done their best to vandalize an underwater model of *Skylab* with flailing cables and twisted metal. With four days to go until the launch, Paul Weitz, one of the space station crew, donned a spacesuit before being lowered into the neutral-buoyancy tank at Marshall to test the tools being evaluated for the imminent repair mission. Of the hopes that the astronauts might be able to effect a repair, *Skylab*'s Programme Manager said 'we're not too optimistic'. But as Weitz practised freeing the model's trapped solar array in the water tank, the Air Force was ramping up its effort to help.

At 9.20 in the morning on the 21st, John McLucas telephoned the Director of the CIA and left a message saying that unless he heard otherwise he would authorize the recovery of Mission 4338's first film bucket that afternoon. Its early return would cost the NRO about 15% of the expected imagery of what he described as low-priority targets. But with the closest passes between *Skylab* and the GAMBIT-3 having taken place on the 18th and 19th, David Bradburn needed to see what, if anything, his men had managed to capture.

Before lunchtime on the 21st, a Lockheed JC-130B of the 6594th Test Group took off from its home base, Hickam AFB in Honolulu. It flew east before circling over the Pacific. High on the big turboprop's back, behind the cockpit, was a distinctive blister containing the direction-finding equipment it needed to perform its mission, which was one of the most classified and specialized roles in the Air Force. Since 1959, the 6594th had been retrieving film recovery capsules ejected from orbiting US reconnaissance satellites as they descended towards the Pacific by parachute. The unit's crest bore the motto 'Catch a Falling Star'.

At 16.01 EST, trailing a recovery loop from the ramp at the rear of the aircraft, the JC-130B crew lassoed what they referred to only as 'the system' and reeled it into the aeroplane's cargo hold. En route back to Oahu, they locked their prize into a specially built shipping container.

After returning to Hickam, with 'the system' already flying east on board an Air Force C-141 jet, the JC-130B's captain bought two rounds of drinks for his crew, one at the officers' club and another in the enlisted men's club. Twelve hours later the undeveloped film arrived for de-spooling, developing and inspecting in Eastman-Kodak's lab in Rochester on the shores of Lake Ontario. Seven hours after the spooled film had arrived at the lab, hopes that Mission 4338 had captured *Skylab* were beginning to fade. Then, as two photo interpreters hunched over light tables scanning a 2-foot-long frame from the satellite's take using small magnifying loupes, the mood changed.

In an effort to try to cover all eventualities, the collection of tools Rick Nygren and the team in Building 5 were assembling was starting to look like a weight problem. Then two men in dark suits and ties came in, escorted by a NASA engineer. One of them was carrying a Manila envelope. Nygren and his colleagues were ushered over and invited to look through the sheaf of 8-by-10-inch photographs it contained. All NRO satellite imagery carried the security classification 'RUFF'. These pictures bore a more exceptional label, 'RUFF

SENSITIVE', meaning they were spy satellite pictures of another spacecraft. All the prints showed the damage to *Skylab* in detail. For the struggling engineers it was as if someone had turned on a light.

As they pored over the images, the nature of what was causing the problem became immediately clear. It looked like good news. A single piece of aluminium strapping had snared one of the solar arrays. Some kind of telescoping tool, *like those used to work on power lines*, Nygren thought, with a jaw cutter on the end was all they needed.

The men in suits returned the photographs to the envelope and, before they left the room, elaborated on the nature of their contribution to treating *Skylab*'s sickness. 'You never saw those photos,' they said, 'and we've never been here.'

Back in California, at the Air Force space HQ where the MOL astronauts had enjoyed such high hopes of establishing a manned Air Force presence in space, *it was as if*, thought David Bradburn, *the unmanned world sent a salute to the manned world*. And, as NASA looked ahead, it wouldn't be forgotten.

Three days after the NRO shared the images of *Skylab* with NASA, the *Skylab 2* mission launched from Cape Canaveral carrying the space station's first crew. By the time the men from the Department of Defense arrived in Houston, the plan to repair *Skylab* was already well advanced, and while the photographs had only confirmed visually what NASA's telemetry was already telling them, they provided useful detail. Enough, at least, to give Nygren pause for thought. *It would be interesting to know*, he wondered, *what would have happened if we'd never seen those pictures*.

The Air Force effort was significant in other ways. First

of all, as NASA had noted, it demonstrated that Blue Cube controllers had developed the ability to take sat-squared images; but at the same time it also underscored the continuing lack of a real-time photo-reconnaissance capability from space. Even with the decision to sacrifice part of the planned intelligence mission, it had taken days for the pictures to get to NASA. If DoD needed to know what was happening in the Soviet Union *now*, it still couldn't be done. Ironically, that kind of urgent, real-time intervention was exactly the kind of capability that might have been offered by a crew of military astronauts flying aboard the Air Force's cancelled Manned Orbiting Laboratory. Had Mission 4338 not already been waiting on the pad, had NASA required information more immediately, and had that information been more critical, the NRO would have been powerless to help. It was clear that if RUFF SENSITIVE pictures were going to be available whenever they might be needed, a satellite capable of delivering real-time imagery was still a necessity.

The first *Skylab* crew blasted off from Pad 39B at Kennedy on board their Saturn 1C rocket in the early hours of 25 May. Seven and a half hours after launching they at last had the space station in their sights. 'Tally-ho the *Skylab*,' reported Conrad, 'we got her in daylight at 1.5 miles, twenty-nine feet per second.'

'Roger, Pete. Copy,' acknowledged Dick Truly at CapCom. But the day's fun had barely started.

As the Apollo spacecraft held formation with the space station, Conrad described the damage. 'As you suspected,' he told Truly, 'solar wing one is completely gone off the bird.' The second solar array appeared to be partially deployed but trapped.

Trying to capture decent television footage, though, was proving to be a headache. 'A worse possible frigging place to point this thing . . .' Paul Weitz complained as he tried to manipulate the camera from within the confines of the Command Module.

Yet despite the difficulty, the astronauts' inspection and the glimpses they'd conveyed to Mission Control suggested it might be possible to release the remaining solar wing.

The crew pulled on their helmets and gloves, and from Mission Control Dick Truly cleared them for a 'local flight' tucked in alongside *Skylab*. Inside their own cramped spacecraft, all three astronauts were now enclosed in full spacesuits.

Three hours after they'd first joined on the space station, Conrad carefully manoeuvred the CSM close enough for Weitz, with his legs held by Joe Kerwin, to reach out of an open hatch to attempt a repair using a cable cutter and universal tool. As they passed across the northern Pacific, out of reach of NASA's relay stations, the Public Affairs Officer in Houston explained 'they should be, according to the timeline, just in the process of closing the hatch'. But when the crew were reacquired by the California tracking station it was immediately obvious that things were not going according to plan.

'Is there a black piece that's curled up due to the lip next to the green?' asked Conrad from inside the Command Module.

'I don't understand what you're talking about,' replied Weitz.

Conrad was stumped: 'Hell, I can't see it,' he told his crewmate, 'the fucking hatch is in the way.'

Listening to Conrad and Weitz from the CapCom

console, Dick Truly chuckled as he reminded them, pre-occupied as they were with the task, that the world was once again listening in. They might want to tone it down a bit.

After struggling on for another ten minutes without success, Weitz conceded defeat: 'I hate to say it, but we ain't going to do it with the tools we got . . .'

Forced to abandon the repair, Kerwin pulled his crew-mate back in through the hatch before they repressurized the capsule in preparation for an attempt to dock with the space station.

Two hours later they were still trying. Four attempts to hard-dock with *Skylab* had been unsuccessful, the mechanism failing to engage and lock. Running out of ideas, Mission Control hastily worked out plans to have the dog-tired crew back away from *Skylab*, park the CSM in an identical orbit 20 miles behind the space station, and get some sleep. But there was one final option, and Pete Conrad thought it was worth trying, 'because,' he told Dick Truly, 'if we ain't docked after that, you guys are out of ideas'.

If Conrad's crew once again donned their helmets and gloves it was possible to crawl up into the tunnel hatch of the CSM and start snipping wires – a procedure to bypass the automated latching sequence that they'd had explained to them on the ground just once. The job done, Conrad guided the CSM in to mate with *Skylab* a fifth time. 'Give it ten seconds,' Truly had told them. The crew counted to an agonizing seven before, with the rat-a-tat of a machine gun, the twelve latches finally engaged, connecting their space-craft to the space station. They'd rolled a six with their last throw of the dice.

From the moment the Saturn 1C cleared the tower, every

Above: Space Cowboy. A young Bob Crippen on his first horse, Sugar, beside the house in Porter, Texas, where he grew up.

Right: In the early 1950s *Collier's* magazine laid out German rocket pioneer Wernher von Braun's vision for space, serving as an inspiration for many who later became part of the US space programme.

Left: Last of the Gunfighters. During the Cuban Missile Crisis Dick Truly flew Vought F-8 Crusaders from the USS *Enterprise*, the US Navy's first nuclear-powered aircraft carrier.

Below: With dreams of becoming a test pilot, Bob Crippen earned his wings of gold as a naval aviator before flying A-4 Skyhawks off the deck of the USS *Independence* during the Cuban Missile Crisis. Crip is kneeling third from the right.

Above: Bob Crippen trained as a test pilot at Chuck Yeager's Aerospace Research Pilots School at Edwards Air Force Base in 1965. One of his instructors was Dick Truly. Crip is back row, far right.

Right: John Young trained at the US Navy's test pilot school at Patuxent River in 1959. Three years later he was a project pilot evaluating the fleet's new interceptor, the F-4H-1 Phantom II.

Above: In 1962 Young took part in Project HIGH JUMP, establishing a number of time-to-climb world records in a stripped-down, lightweight Phantom. He was sure his involvement secured his selection as an astronaut.

Right: John Young (closest to camera) and Gus Grissom in the *Gemini III* capsule before making the spacecraft's first manned flight in 1965.

Left: US Air Force test pilot Joe Engle in the cockpit of the North American X-15, an experimental rocket plane which he was to fly into sub-orbital space in 1965.

Below: Launched from beneath the wing of a B-52 mothership, the X-15 reached speeds of Mach 6.7 and heights of over sixty-seven miles.

Left: In the late fifties and early sixties the USAF pursued plans for a small, winged spaceplane called the X-20 Dyna-Soar. A mock-up was built.

Right: Project LUNEX. Nor was Dyna-Soar the summit of the Air Force's ambitions in space. A 1961 USAF plan to put men on the moon included this diagram of their proposed spacecraft and lander.

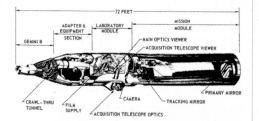

72 FEET

MISSION MODULE

ADAPTER & EQUIPMENT SECTION

LABORATORY MODULE

GEMINI B

MAIN OPTICS VIEWER

ACQUISITION TELESCOPE VIEWER

PRIMARY MIRROR

CRAWL-THRU TUNNEL

FILM SUPPLY

CAMERA

TRACKING MIRROR

ACQUISITION TELESCOPE OPTICS

MOL BASELINE SYSTEM
FIGURE 1

Left: MOL. Unconvinced by USAF arguments for Dyna-Soar, Secretary of Defense Robert McNamara cancelled it in 1963. In its place he approved the development of a military space station labelled the Manned Orbiting Laboratory.

Below: Manned Orbiting Laboratory mission patch.

Left: Dick Truly was assigned to the MOL programme in 1965. He's pictured here with a scale model of the space station, capped with the Gemini capsule the crews were to use to travel to and from orbit.

Below: The MOL pilots in 1966. *Back row*, left to right: James Taylor, Francis Neubeck, Dick Lawyer, Al Crews, 'Mac' Macleay, Jack Finley, Dick Truly. *Front row*, left to right: Bob Crippen, Bob Overmyer, 'Bo' Bobko, Gordon Fullerton, Hank Hartsfield.

Left: An unseen picture from 1966 of Dick Truly, Mike Adams and Jack Finley at Wright-Patterson Air Force Base. The three had been training to use the tunnel linking their Gemini-B capsule to the MOL space station in simulated zero-g conditions aboard an Air Force KC-135 christened 'Weightless Wonder'.

A USAF Titan IIIC rocket launching from Kennedy Space Center on 3 November 1966. Carrying a facsimile of the MOL space station below a Gemini-B capsule used to test the spacecraft's modified heatshield, it was the only occasion on which MOL hardware was ever flown.

Above: A KH-9 HEXAGON reconnaissance satellite nears completion in a clean room at Lockheed's facility in Sunnyvale, California. The four domes are re-entry buckets used for film recovery.

Right: After being dropped from the KH-9, the re-entry buckets floated towards the Pacific beneath parachutes before being snatched from mid-air by the specially modified C-130s of the USAF's 6593d Test Squadron flying out of Hawaii.

Above: The MOL Guys. The seven refugees of the cancelled Air Force space station programme joined NASA in 1969. *From left to right*: 'Bo' Bobko, Gordon Fullerton, Hank Hartsfield, Bob Crippen, Don Peterson, Dick Truly and Bob Overmyer.

Left: *Apollo XVI* Commander John Young gives a 'big ol' Navy salute' while leaping from the surface of the moon in April 1972. Young was on the moon when Mission Control told him the Space Shuttle programme had been approved.

Above: NASA engineer and space-craft designer Max Faget demonstrates the principle of the Space Shuttle using a balsa-wood model that he'd built in his garage.

Above: NASA's original plan was to launch a straight-winged Orbiter from the back of a fully reusable winged booster. In this picture the design – christened DC-3 by Max Faget – is tested in the wind tunnel.

Left: As the design evolved – influenced by the needs of the Air Force – a delta-winged Orbiter emerged. As seen here, plans to use either a reusable winged booster or the first stage of a Saturn V moon rocket were both considered.

Below: 5 January 1972. Richard Nixon and NASA Administrator James Fletcher discuss a more familiar-looking Shuttle configuration on the day the President announced the decision to proceed with the programme.

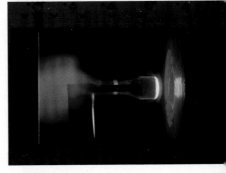

Above: Dr Hans Mark during his time as Director of NASA's Ames Flight Research Center. Once sceptical about the value of manned spaceflight, he became one of the Shuttle programme's most committed supporters.

Right: Ceramic tiles designed for the Shuttle's heatshield were tested inside the arc jet facilities at Ames used to simulate the hot-gas flow of re-entry.

Above: One of the Shuttle's giant solid rocket boosters blasts out nearly 3,000,000 pounds of thrust on the test stand.

Right: A masterpiece of engineering. An RS-25 Space Shuttle main engine undergoing testing. Much of the engine's early development was plagued by explosions, fires and violent disintegrations.

Below left: A bearded Bob Crippen emerges from the Skylab Medical Experiment Altitude Test in 1972 – fifty-six days sealed inside a pressure chamber designed to simulate life on board *Skylab*, America's first space station launched the following year.

Below right: John Young in his office at Johnson Space Center, Houston. In 1974, following the retirement of Alan Shepard, the first American in space, Young became head of the Astronaut Office.

word of Houston's communication with the crew had been spoken by Dick Truly. Over twenty-two hours of it. Given what the crew themselves had endured, he knew he was hardly in a position to gripe, but he was exhausted. With the completion of a successful rendezvous, though, he was off duty, finally relieved after a marathon stint on CapCom.

With Bob Crippen, he grabbed a six-pack of beer and headed outside. Getting away from the space centre's buildings, the two friends made their way to a bench overlooking Clear Lake, adjacent to the centre, and talked.

After the time he'd spent on the workshop up at Huntington Beach, Truly had felt at one point that he was in with a chance of winning a place on one of the *Skylab* crews. *They'll come to their senses and let me fly*, he thought, but it had turned out to be a pipe dream. Both he and Crip had now been astronauts for over seven years without getting anywhere near orbit. Inevitably, there were occasional low points, but they enjoyed the work and, more importantly, the maths was clear enough. Crip may still have felt like they were second string next to the Apollo astronauts, but with the end of the lunar programme, that group's number was becoming increasingly depleted as they retired from NASA in search of new challenges. The Astronaut Office was small and shrinking, and as a result the MOL guys were moving up the ladder.

Of their fellow MOL veterans, Hank Hartsfield, Don Peterson and Bo Bobko had switched their attention to the Space Shuttle as soon as their support of the Apollo programme came to a close. All three were assigned to work on developing the Orbiter's ground-breaking flight control systems. At the end of 1972, Gordo Fullerton was down at the Cape to shut the hatch on *Apollo XVII*, the last moon mission, before being given responsibility for the layout of the Orbiter's cockpit. *The Apollo capsule*, he thought, *the worst cockpit ever designed by man*, and he relished the prospect of doing better. When the last crew returned home from the *Skylab* programme, Crip and Truly would be working alongside him at the heart of the Shuttle programme. *In on the ground floor*, thought Crip.

But that was for another day. A shattered Dick Truly managed to finish only a single beer before hitting a wall. He just about had the energy to get home and crash out.

The next day, after entering the space station, the *Skylab 2* crew successfully erected the ad hoc parasol, which slowly began to bring the temperature inside the workshop down to manageable levels. For nearly two weeks they struggled with the lack of power caused by the absence of the two solar wings, switching off lights, eating food at room temperature and restricting their scientific work until, on 7 June, Conrad and Kerwin climbed into their spacesuits and left the space station. Armed with tethers, hooks, the cable cutter and even a dental saw taped to the front of one of the spacesuits, they worked for three hours to free the trapped solar panel. It was hard physical work. At one point, Conrad's pulse raced to 150 beats per minute with the exertion. But after Kerwin managed to gain real leverage by standing on the hull of the space station secured by a rope, they overcame the big wing's resistance and it suddenly came free. Thrown from *Skylab* by the force of his own effort, Kerwin floated out into space attached to an umbilical line. He looked back and smiled with relief. They'd done it. Seeing that solar panel fully deployed and beginning to warm up in the sun was, he thought, *the prettiest sight I've ever seen in my life.*

In California, at NASA's Ames Research Center, Hans Mark had followed the *Skylab* crew's labours with a keen interest. If his time spent on the *Apollo XIII* Accident Investigation Board had been a turning point for Mark in his appreciation of the value of manned spaceflight, the job done by Conrad, Weitz and Kerwin had reinforced his change of heart. Once sceptical, the Ames Director was fast becoming a champion. Looking forward to the Space Shuttle programme, Mark would be more than just a convert to the cause; he would become a powerful advocate for what humans might add to the nation's future space endeavours.

In nurturing and manhandling *Skylab* to life, NASA had pulled triumph from near disaster. It was a worthy companion to *Apollo XIII* in any list of the agency's greatest hits. But less than a month after the second crew were launched, the Astronaut Office was forcibly reminded of the inherent risks of leaving *terra firma* when news spread that one of their best and brightest was fighting for his life in intensive care at the University of Texas Hospital. This time one of the crew of *Apollo XIII* had *really* hurt himself.

TWENTY-TWO

Houston, 1973–4

FRED HAISE KNEW he'd made his break on the lead aeroplane early. Now, as the two aircraft approached Galveston field in formation, he was too close. As he lost sight of the leader below the nose of his own machine he decided that, rather than risk overrunning the runway, he'd best take the old Vultee BT-13 Valiant round the circuit for another approach.

From inside the cockpit, the 'Vibrator', as she was known to a generation of World War Two American airmen trained to fly in them, felt familiar enough. From the ground, though, it was a different story. The little piston-engined trainer was one of nine BT-13s bought by Twentieth Century Fox to impersonate Imperial Japanese Navy 'Val' dive bombers in the movie *Tora! Tora! Tora!*. Sporting Japan's distinctive red rising sun insignia and wearing spats around the fixed undercarriage, the old warbird looked the part. Now she belonged to the Confederate Air Force, and Haise, along with his friend Joe Engle, was one of a handful of volunteer pilots who flew re-enactments of the Pearl Harbor attack at air shows around the country. Flying mock dog-fights was what astronauts did for fun.

Haise advanced the throttle and felt the familiar buzz of the twin-bladed propeller as his machine accelerated along the track of the runway. Then, just as he approached the end of the strip, the 450-horsepower Pratt and Whitney engine quit on him. In an instant, he was processing his options. He was too late to put her down on the runway. Ahead there was shallow water which he knew would catch the aeroplane's fixed undercarriage and flip her straight on to her back, submerging the cockpit. Drowning didn't appeal. Trading speed for height, he banked to the right towards what looked like a cattle pasture surrounded by trees. To avoid them he decided to eke out his glide to reach a dirt field beyond. From here on it was straight-forward. Haise had flown unpowered landings in everything from gliders to F-104 Starfighters. The old trainer settled into her approach with Haise, relaxed and on top of the situation, gently guiding her in. No drama. But what had looked from the air like an empty field was actually a con-struction site. Because of the angle of his approach, Haise was unable to see the deep foundations already dug into the ground. As he gently dead-sticked the BT-13 on to the deck, one of the main wheels dropped into a trench and threw the little aeroplane into a cartwheel before it landed upside down, skidding backwards along the surface. Then it caught fire. And the cockpit was jammed shut.

He was lucky not to break any bones, because now, with flames licking around him, Haise had to kick his way through the Plexiglas canopy to get out of the burning air-craft. By the time he was free he'd suffered second- and third-degree burns across 65% of his body. Ahead lay three months of agonizing treatment in the University of Texas Hospital as Dr Larsen, an army-trained specialist from the next-door Shriners Burns Hospital for Children, kept him swathed in bandages twenty-four hours a day, removing them only to scrape off the scabs to prevent immobilizing scarring forming. During the critical first three weeks, friends, family and the Astronaut Office worked a three-shift system to make sure that Freddo was never on his own. For Haise himself, throughout it all, the worst of it was the thought that he might not fly again.

Without an engine, as Freddo's crash-landing illustrated, an aeroplane cannot stay in the air. It's coming down, it's just a question of when. Every aircraft design has its own glide ratio, which represents the relationship between the height it will lose over a given distance of forward flight. A modern, lightweight, long-winged competition glider might achieve a glide ratio of as much as 50:1. If it were launched from a mile-high mountain it would land 50 miles away. An airliner like the 747 would glide to a point 15 miles away, while a puddle-jumper like the Cessna 172 could cover 9 miles before running out of sky. In the lower atmosphere, the Space Shuttle, a marriage of a big slab-sided fuselage with a short wingspan, would manage a glide ratio of just 4½:1. Those large-area delta wings the Air Force wanted, while they might provide extra lift at hypersonic speed and on the fringes of space, meant that as she approached an airfield the Orbiter lost height like a winged brick, and without engines there was no possibility of a second chance if a first attempt to land went wrong.

That seemed like a big risk to take with a returning Space Shuttle, and there were early plans to equip the Orbiter with air-breathing jet engines that could swing down after re-entry and allow her to fly like a normal aeroplane. It felt as if that added an obvious margin of safety. But it also added weight. Too much. Chris Kraft, the hard-nosed, tough-talking Director of the Houston Space Center, described the problem succinctly. 'With jet engines,' he said, 'we couldn't lift a damn pound into orbit.' The Orbiter would be unpowered.

In the opening credits of the hit TV show *The Six Million Dollar Man*, pilot Steve Austin, played by Lee Majors, sustains massive injuries in a plane crash. 'We can rebuild him,' the voiceover intones, and Austin emerges better, faster and stronger from his experience. But the footage used at the beginning of each show of an aircraft tumbling across the ground was real, and the test pilot in the cockpit, Bruce Peterson, didn't have access to the technology enjoyed by his fictional counterpart. He suffered a fractured skull, serious facial injuries, and would ultimately lose the vision in his right eye as a result of his 1967 crash at Edwards in a 'lifting body' – an aircraft without wings that used lift generated by the aerodynamic properties of the fuselage alone to glide, unpowered, safely back to the ground after being dropped at height from beneath the wing of another aircraft. Peterson's accident was the exception in what had otherwise been a long and successful programme. Inside NASA, and especially within the Astronaut Office, there was confidence that jet engines were a belt-and-braces luxury the Space Shuttle could do without.

A decision on how the Orbiter was to land, though, addressed only half the problem. Without engines of her own, there was still the question of how she was supposed to get from Rockwell's California factory to Florida, where she was to be launched. Transportation by sea was a possibility, but each fourteen-day journey was costed by Rockwell at $20 million. And nor did it offer any solution to getting an Orbiter to and from the high desert where Edwards was located. NASA's Langley Research Center supported a proposal for a vast, specially designed transport aircraft called VIRTUS that would have been the largest heavier-than-air flying machine ever built, with a wingspan nearly two and a half times greater than that of a 747, but this was dismissed as too big, too ambitious and too costly, even putting aside the time it would take to develop. It was unnecessary in any case. Anticipating the requirement to air-launch a prototype Shuttle in a series of unpowered flight tests in 1976, John Kiker, a NASA engineer and self-confessed aviation enthusiast, had already been hard at work on an alternative. And his inspiration came from an unlikely source.

In the late 1930s, the range of Britain's new Imperial Airways Short Brothers C-Class Flying Boats was insufficient to provide a transatlantic airmail service. The airline's technical manager, Major R. H. Mayo, realized that if he could mount another, smaller aircraft on top of the big C-Class, then launch it off the back at the limit of the bigger aeroplane's range, he might have a combination that was capable of bridging the gap. The resulting Short-Mayo Composite, as peculiar a concept as it seemed, worked as advertised.

Air-launched American aircraft, like Yeager's X-1 or the X-15, had always been dropped from beneath the carrier aircraft, but, inspired by Mayo, Kiker started to explore whether the Major's concept could be successfully adopted by the Shuttle programme. He corresponded with Major Mayo's widow, who was only too happy to share her late husband's reports. There were just two contenders for the job of giving the Orbiter a piggyback: the USAF's new C-5A Galaxy transporter, then the biggest aeroplane in the world, and the Boeing 747 Jumbo Jet, the world's biggest airliner. Kiker was given data from both companies and realized that in all three key areas, propulsion, aerodynamics and structural strength, both of the heavies were up to the job. Encouraged by Houston's engineering boss Max Faget, Kiker took his proposal to the Orbiter Programme Manager, Aaron Cohen, whose reaction was straightforward. 'This,' he said, 'is absolutely the dumbest idea I've ever seen in my life.'

Kiker was undeterred. And, although his hands were tied, Faget remained encouraging. It was exactly the kind of uncomplicated engineering solution he liked. 'I can't give you any money, but,' he suggested, 'if you can do it with the models . . .' – and he authorized Kiker to take home whatever materials and instrumentation he needed. An internationally respected radio-controlled model-maker and pilot, the same thought had also occurred to Kiker himself. He was already hard at work on a radio-controlled version of the Shuttle using a one-fortieth-scale plastic foam model which he planned to launch from the back of another radio-controlled plane.

Fortunately, Kiker didn't just have Faget on his side; there was also Hewitt Phillips, a model-making friend, who just happened to be Head of Flight Dynamics at NASA's Langley Research Center. In frustration at Cohen's resistance, Kiker called Phillips and urged him to intervene. Although reluctant to get involved, Phillips knew Kiker was right, and shortly after he sent a message to Houston a smiling Chris Kraft told Kiker, 'Well, I think you're crazy, but we've been directed to do it. Apparently we'll do it.' It worked like a dream. And NASA got to work on a full-size version.

Boeing and Lockheed were commissioned to produce studies into using the 747 airliner and C-5 military transport respectively. While in the summer of 1974 NASA requested a pair of C-5s from the Air Force, the decision was soon reversed. There were worries that bolting a Space Shuttle to the back of a C-5 would blank its high tailplane and make it uncontrollable. At the same time, the Air Force would always have had first call on the C-5s and, in the event of war, could have pressed their jets back into military service. But in the end, money made the decision easy. The mid-seventies oil crisis had forced a downturn in commercial aviation which put NASA in a position to buy not rent. Lack of demand meant nearly new Jumbos were being mothballed in lines at aircraft boneyards like Roswell. As a result, NASA picked up a nearly new 747-100 from American Airlines at the knockdown price of just $16 million. One careful owner, not many miles on the clock.

In Seattle, Boeing brought a member of the big jet's original engineering team out of retirement to oversee the work of converting her from an airliner into what NASA had imaginatively christened a Shuttle Carrier Aircraft – or rather, of course, an acronym of that: SCA.

Meanwhile, Astronaut Joe Engle – still Houston's go-to guy for a stick-and-rudder job – travelled to Seattle to fly the Orbiter off the back of a 747 in Boeing's simulator. It provided further proof of concept. Yet despite all the evidence, there were still people inside both NASA *and* Boeing who just couldn't quite bring themselves to believe that trying to fly and launch an Orbiter off the back of a 747 wasn't going to result in a fireball.

TWENTY-THREE

Houston, 1974

IT WASN'T THE kind of request you often heard in the Astronaut Office. So when, after moving to the Shuttle programme, Crip said, 'Hey, I'd like to work on the computers,' his boss, T. K. Mattingly, wasn't going to say no. Mattingly, bumped off the crew of *Apollo XIII* after being exposed to German measles, had subsequently flown as John Young's Command Module Pilot on *Apollo XVI*. Now he was putting his workaholic tendencies to good use running the engineering and development side of the Astronaut Office.

'Go ahead, Crip,' Mattingly told him.

His eagerness to have Bob Crippen devote himself to interface between man and machine reflected a degree of unease among the pilots that the Shuttle's flight control would be handled entirely using fly-by-wire technology, which was still very much in its infancy. But in Crippen he had the right guy on the job. The IBM AP-101 computer chosen for the Shuttle may have been, Crip thought, *essentially obsolete*, but it was also familiar, being the latest iteration of the 4Pi computer that had been at the heart of both the Air Force MOL space station and *Skylab*. What the old processor lacked in sophistication it made up for in robustness and dependability.

The key to achieving the required levels of reliability using an affordable, off-the-shelf computer like the AP-101, though, was redundancy. NASA calculated that if they used three synched computers, they could expect computer failure to cause the loss of a Shuttle three times every one million flights. If they added another, however, the odds against a catastrophic failure looked dramatically more favourable. Harnessing four computers together pushed that probability back to *one in two hundred and fifty million* flights. Still not entirely happy to trust in the machines, NASA also included a fifth, back-up computer, programmed not by IBM but separately by Rockwell. If a fatal software bug, endemic to the work done by IBM, somehow slipped through the net, the uncontaminated Rockwell machine would bring the Shuttle home. It all served to underline just how dependent she was on those computers. But at the same time, she had to be. The Orbiter was unique in being at once a launch vehicle, a spacecraft *and* an aeroplane. There had never been such complex control requirements combined in a single vehicle, and it put her beyond the capabilities of unaided human pilots. Without those computers, the Orbiter was unable to function, fly or survive – a dodo.

With the selection of the AP-101, it was unlikely that the reliability of the hardware was going to be a major issue. Instead, software engineering was thrust into the spotlight.

The complexity of the programming task that lay ahead was compounded by the limited memory they had to play with. It was a story that was familiar enough to Crip and Truly from their work on *Skylab* when the 16K memory available across both computers continually hampered the programmers' efforts. Forty years later, an average smartphone would offer around 65,000 times the memory available to *Skylab*'s software engineers. Recognizing that it had underestimated in the past, NASA tried to be smart when it came to anticipating the requirements of the Shuttle. When the Orbiter's manufacturer, Rockwell, said in its winning bid that its spaceplane would need 32K of memory, NASA doubled it: across the four primary computers, NASA expected the Orbiter to have 256K of core memory. It was soon apparent that even this wasn't close to being enough. As the software requirements grew, so too did the capacity of the computers until each of the AP-101s offered 104K of memory. It still wouldn't have got you far trying to play *Grand Theft Auto*. Nor, as it turned out, was it enough to get the Orbiter into space and back. The 300-strong team of IBM programmers working on the Shuttle's flight control software would eventually need over 700K of memory, stored on magnetic tape mass-memory units and loaded on to and dumped by the AP-101s as required for different phases of flight.

The coding language they used to sap that precious storage was called HAL/S. Given the challenges involved, it was little wonder that some drily speculated that it had been named after HAL, the evil, sentient computer star of Stanley Kubrick's 1968 movie *2001: A Space Odyssey*.

While work progressed on the computers in Houston, in California NASA's Ames Research Center was making a crucial contribution to another of the Space Shuttle programme's long pole items: the heatshield. At the programme's inception, Hans Mark had drawn attention to the challenges posed by the development of a thermal protection system for the Shuttle. Now much of the responsibility for tackling them lay with the engineering teams at Ames, and that was just the way their ambitious Director wanted it.

Prior to his arrival there had been rumours that the centre had been earmarked for closure. As the new boss, he had no compunction about using concerns over that to force through his own agenda. His aim was to ensure Ames became indispensable. He led by example, creating an air of urgency. Displaying a prodigious capacity for hard work, Mark was at his desk by six every morning, firing out memos known as 'Hans-o-grams' that left no one in any doubt about his attention to the detail of their work. He set the pace with twelve-hour days, and seven-day weeks, determined to foster a more robust, forceful approach to the way Ames went about its business. Plugging gaps in his own knowledge, he learned to fly to better understand the work being done by his people, and read the classics on the red-eye special to and from Washington in order to have at his fingertips the killer literary reference or aphorism that might win a debate.

Mark instigated a plan to make Ames's unique collection of wind tunnels, simulators and mainframe computing power into, in all three cases, the best in the world. All were vital to the development of the Space Shuttle. Since 1973, the centre's 3½-foot hypersonic wind tunnel, capable of generating airflows over scale models of the Space Shuttle of up to ten times the speed of sound, had been operating two shifts per day, every day, in pursuit of the tens of thousands of hours of tests that were necessary. The simulators allowed different flight controls and configurations to be explored and the new computers meant a leap forward in the country's ability to predict complex fluid dynamics – a capability that was, again, critical to understanding the Shuttle's aerodynamics beyond where they could be physically tested.

It was the centre's thermal protection branch, though, that had the primary responsibility – as they had for the Apollo capsule – for establishing the best way to protect the Shuttle from the heat of re-entry. It would make the job of safeguarding the returning lunar spacecraft look like child's play.

The bottom of the Apollo capsule was the size of a large dining table. The area needing protection on the Space Shuttle was more akin to a tennis court. And when it came to re-entering the Earth's atmosphere, that made a vast difference to the numbers NASA had to contend with.

The kinetic energy carried by the 85-ton Orbiter orbiting at 17,500 mph was roughly the same as that of a Nimitz Class aircraft carrier displacing 100,000 tons. *If it were travelling at over 500 mph.* When, on re-entry, the spacecraft made contact with the atmosphere, most of that staggering amount of energy would be converted to heat, generating temperatures that, in a compression shockwave pushed out ahead of the Orbiter's passage, would be similar to those found on the surface of the sun.

Just to compound the problem, the Shuttle's heatshield had to be reusable, and that precluded the use of the kind of one-shot ablative shield used for Apollo. The process of ablation harnessed the furnace of re-entry itself to burn the outer surface of the heatshield into a carbon-rich, low-conducting layer that then protected the material below. Hot gases generated in the process provided an additional layer of protection, as well as carrying away heat. In successfully doing its job, though, it was destroyed. That wouldn't work for a reusable spacecraft.

In Houston, the possibility of using some of the exotic metal alloys was quickly abandoned when tests revealed the extent to which the metal buckled and flexed when subjected to extremes of temperature. Not only would that have led to gaps in the integrity of the heatshield, equally importantly it would have disturbed the smooth flow of air over the surface of the Shuttle at hypersonic speeds.

That left just one game in play: ceramics. Rockwell proposed to cover the Shuttle in tiles made from an aluminium silicate called Mullite, found almost exclusively on the Isle of Mull off the west coast of Scotland. Rival manufacturer Lockheed had a similarly unlikely-sounding suggestion: they wanted to use sand.

In the early sixties, the company had begun research into heat-resistant materials made of compacted silica fibres. In contrast to metal, silica offered lightness, low heat conductivity, and was as indifferent to extremes of heat and cold as the surface of the Sahara Desert. To find silica sand of the necessary quality, Lockheed turned to Johns-Manville, the largest fibreglass manufacturer in the world. Using sand from Missouri, Johns-Manville produced 99.9% pure silica fibres which Lockheed then mixed with water to produce a slurry. Contained in moulds, it was then fired in kilns to produce ceramically bonded blocks of a material that weighed just 9lb per cubic foot. Ninety per cent of each block was porous. Just one-seventh the density of water, its extreme lightness was, like balsawood or pumice stone, unexpected, but this wasn't its only remarkable characteristic. Within seconds of being removed from a kiln, and while the centre was still glowing red hot, the tiles could be picked up by their edges with unprotected fingers. Similarly, they could be plunged straight from a furnace into a bucket of cold water without damage.

Five months of competitive evaluation produced a clear winner in Lockheed's silica tile, surviving all that testers at Ames could throw at it and emerging intact. And yet even the silica tile still didn't offer sufficient protection for the parts of the Shuttle exposed to the greatest temperatures: the nosecone and the leading edges of the wings. For these, something of Dyna-Soar, the old Air Force spaceplane cancelled a decade earlier, survived.

Faced with the same need to protect Dyna-Soar, a material called reinforced carbon-carbon was developed. A composite made of graphite and carbon, it was hard and lightweight but brittle, although it had the unusual – and useful – quality of increasing in strength as it got hotter. It was happy to endure temperatures above 3,000°F.

With silica tiles on the bottom and reinforced carbon-carbon protecting the nose and wing leading edges, the only remaining area for concern was the top of the Orbiter – the leeside in terms of exposure to air friction. And here, Chief Engineer Max Faget didn't think it needed anything at all.

Faget had said the same thing about the Apollo capsule. The designer kept hold of a small piece of blue wrapping plastic to remind him. Gut instinct had told Faget that the leeside of the capsule wouldn't need heat protection during re-entry, but he was persuaded to play it safe by adding a thin layer of thermal protection. When the capsule returned from space, however, the fine mylar dust sheet covering that had protected it prior to flight was still happily in place.

Faget was wrong, though, about the Shuttle. In March

1975, the Air Force would share the classified results of test programmes that showed heating on the top of the vehicle would be greater than expected. While the temperatures on the leeward side would be far less severe than those suffered by the bottom, they would still be more than the 350°F that could conceivably be handled by unprotected aluminium. As a result, the least vulnerable areas of the Shuttle's airframe, like the tops of the wings or the payload bay doors, were blanketed in a heat-resistant Nomex felt.

When construction began of the first, prototype Orbiter in Rockwell's Downey plant it seemed that, with the materials for the Shuttle's heatshield settled, at least one of the programme's long poles was well on track. What no one was paying attention to, though, was another issue highlighted by Hans Mark in his 1972 letter: that of bonding the heatshield to the Shuttle's airframe. Nor would they until there was an airframe they could use to actually put it to the test.

Construction of that first, prototype Space Shuttle Orbiter started on 4 June 1974. It would be another year before final assembly began.

TWENTY-FOUR

Houston, 1974

To PRACTISE LANDING on the moon *on Earth*, John Young and the other Apollo Commanders had trained using a vertical take-off machine called the Lunar Landing Training Vehicle. To simulate lunar gravity, the thrust of a single jet engine created the impression that the vehicle weighed only a sixth of its Earth weight and allowed the pilot to use smaller peroxide thrusters to control its descent from 300 feet. It was a triumph of function over form, its unshrouded pipes and plumbing suggesting someone had found a way to allow the engine room of a ship to take to the air. Nicknamed the 'Flying Bedstead', John Young just called it 'The Thing'. But, in tandem, the simulations provided by the LLTV and the ground-based Lunar Module simulator ensured that by the time Young touched down in the Descartes Highlands in *Orion*, the *Apollo XVI* Lunar Module, his technique was so polished and familiar that his heart rate never exceeded ninety beats per minute.

Other than that, though, and the sub-orbital space-flights of the X-15, and the moon landings themselves, there had been no occasion on which any astronaut had needed to practise landing. They fell to Earth backwards in capsules, hoping for the best, before splashing down into the sea beneath a trio of colourful parachutes. With the introduction of the Shuttle, the crews were going to have to land every time. And perfectly. NASA needed to find an aeroplane that could *pretend* to be a Space Shuttle in which the astronauts could become as familiar with landing it as Young had been with *Orion* on the moon.

Training to fly one aircraft by using another to mimic its flying characteristics wasn't an entirely new idea. While early efforts to manipulate the feel of the way an aircraft flew had begun during World War Two, it was 1957 before Cornell University's Aeronautical Laboratory first flew an aircraft that would go on to become the longest-lived test aircraft in history. The variable-stability Lockheed NT-33 was a simple straight-winged jet trainer fitted with an analogue computer and additional flight controls that allowed it to emulate the flying characteristics of nearly anything. By the mid-seventies the little jet had provided convincing impersonations of aircraft as wildly diverse as the rocket-powered X-15, the British BAC TSR-2 strike bomber, the Fairchild A-10 tank-buster, the McDonnell-Douglas F-15 and F/A-18, and the General Dynamics F-16. Ahead lay impressions of the F-117 stealth fighter and even the F-22 Raptor. Not bad for an aircraft designed in the 1940s. That all this was possible using a single super-annuated jet trainer didn't help NASA narrow down their choice for a flying Space Shuttle simulator.

While a tandem-seat jet like the T-33 or T-38 could be made to perform like the Orbiter, its slim cockpit could never provide any sort of facsimile of the spacecraft's flight deck. The Shuttle was the size of an airliner. It made sense, maybe, to try to employ one. Both John Young and Joe Engle flew up to Seattle to fly Boeing's 737, the company's now practically ubiquitous short-haul workhorse. The Shuttle, big, draggy and low on lift, plunged to Earth in a way that was entirely at odds with the requirements of a passenger jet, but the astronauts found that, if they flew the 737 in as dirty a configuration as possible with the wheels and drag flaps down, then applied reverse thrust to the two Pratt and Whitney JT8D-15 turbofans in flight, they could reproduce the steep descent of the Orbiter. The Boeing jet might have been ideal, but, unlike the big 747 bought to carry the Shuttle on its back, it wasn't available at a knock-down price. That put it out of reach.

An entirely different proposition was the Grumman A-6 Intruder, a two-engined all-weather US Navy attack jet. Unlike most of its peers, the A-6 was unusual in seating its crew side by side like the Shuttle. Built like a tank, there was no doubt that the big carrier-based jet was robust enough to stand up to the stress of the repeated 30° dives ending in a 2g pullout that the Shuttle landing training required. Some inside NASA had their doubts about whether anything but a military fast jet would be up to it. But despite the pilots sitting next to each other, the relatively cramped cockpit of an A-6, with all-round visibility from its big glasshouse canopy, again didn't feel much like the flight deck of the Orbiter.

The choice, ultimately, came down to one of two different executive jets: the twin-engined Gulfstream G-II, or the four-engined Jetstar proposed by Lockheed. In some respects, the latter held the advantage. Lockheed had already converted a Jetstar into a variable-stability flying simulator, but where it fell down was in its inability to deploy reverse thrust in flight. Short of trailing a parachute behind it, there was no other way of simulating the high drag of the Orbiter. And so NASA selected the G-II and, wearing nothing but green primer, a pair of the jets flew from the Gulfstream plant in Savannah, Georgia, to Grumman's factory on Long Island for conversion.

Soon, John Young had plenty more to worry about than choosing a Shuttle Training Aircraft.

With the retirement of America's first man in space, Alan Shepard, in 1974, the mantle of Chief Astronaut passed to Young. He wasn't sure he was NASA's first choice as Shepard's replacement, but then self-aggrandizement had never been John Young's style. Even his high-school year-book had suggested 'with your quiet poise, you seem a bit modest'. Young himself just put it down to age. 'If you're the oldest guy there, they let you run it.'

Certainly the astronaut corps was becoming increas-ingly depleted as Apollo and Gemini veterans moved on, but for all his deflections, Young was the perfect candidate. No sooner had he returned to Houston after his first space-flight aboard *Gemini III* than he was asking Deke Slayton to put him on the next available flight. Now a veteran of four space missions, his desire to fly in space remained undimmed. So too did a deep conviction that the first job of any astronaut, whether on the ground or on orbit, was to pre-serve the life of other astronauts. Young brought a characteristically forensic eye to the detail of the Shuttle programme. In his new role he wanted the Astronaut Office to have a say on every aspect of the Shuttle's design and construction, and he would lead by example, picking and probing with sharp, droll memos at anything he thought was a cause for concern.

Before he could devote his complete attention to the Shuttle, however, Young first had to oversee the success of the Apollo-Soyuz Test Project. From behind the desk of his new corner office in Building 4 he couldn't help a wry smile as he considered the $100 million price tag for American and Russian engineers developing a new androgynous docking system. It had been deemed important that the new device avoided either side having to, as he put it, 'get the business' from the other when the two spacecraft mated – in the toe-to-toe stand-off of the Cold War this *really* mattered. While that problem was worked at – and considerations of who was getting screwed by whom aside, the docking module *did* need to be able to harmonize the different atmospheres used by US and Soviet spacecraft – Young's astronauts embarked on a series of exchange visits.

For the MOL guys it represented another step closer to the heart of things. Bob Crippen, Dick Truly, Bo Bobko and Bob Overmyer were named as support crew for the Apollo-Soyuz mission. And for a group whose experience was so firmly rooted in the Cold War, it was a tantalizing prospect.

TWENTY-FIVE

Moscow, 1975

EACH OF THE astronauts had made his own preparations for the visit to the Soviet Union. Without much hope of success, Dick Truly and Gene Cernan asked whether their wives, Cody and Barbara, could join the delegation on what felt like a once-in-a-lifetime chance to travel behind the Iron Curtain. After weeks of silence following the request, Truly had more or less given up when, shortly before the trip, it was suddenly approved. Meanwhile, Bob Overmyer had more material concerns. Alarmed at the prospect of being fed caviar for breakfast he packed a suitcase full of all-American snacks and peanut butter to take with him to the USSR.

In fact, the Soviets had done all they could to make the Americans feel at home, building a sort of replica Holiday Inn for them at the Star City training base. It was a welcome gesture, although there were occasional gaps in the standard of service. And evidence of much closer attention in other areas. While unpacking after arriving at the hotel, Cody Truly opened the room's wardrobe to find it bare. 'There are no coat hangers,' she told Dick. Five minutes later there was a knock on the door and a maid handed over a stack of hangers. From then on, before bed, Cody circled round the room talking to the furniture, unsure of where the bug was hidden. 'Dasvidanya, lamp! Dasvidanya, bed!'

After Cody returned home, the four ex-military astronauts saw with their own eyes locations like Star City and the Baikonur launch site in Kazakhstan that, within the confines of the NASA Astronaut Office, they alone had studied in detail from classified National Reconnaissance Office satellite imagery while assigned to the Air Force MOL programme. So familiar were they that Dick Truly was convinced he could have sketched the layouts of the Soviet's launch complex from memory.

At the conclusion of their official duties they were invited to stay on for the May Day celebrations. The MOL crew jumped at the opportunity. As they walked across Red Square before the parades, Overmyer – still, while detailed to NASA, on active duty with the Marine Corps – turned to Air Force pilot Bobko and said, 'I never doubted that I'd be here. I just always thought it would be at two hundred feet and in full afterburner . . .'

During a return visit to Houston by the cosmonauts, Apollo-Soyuz crew member Deke Slayton invited his new friend Alexei Leonov – who, after Gagarin, was the USSR's most celebrated spaceman – to look at the work being done on the Space Shuttle ground-based simulator. It was a chance for the Russian to learn more about what was the most advanced, exciting manned spaceflight programme in the world. Much to Slayton's surprise, though, Leonov displayed no interest whatsoever. His reaction struck Slayton as strange and off key. He could only speculate that the cosmonaut was under orders not to get involved in any conversation about the Shuttle so as to avoid awkward questions about Soviet plans.

At dinner, John Young asked the delegation's chief engineer more directly. 'Aren't your engineers working on building a Shuttle?'

'Of course,' he replied.

Young noticed a man sitting to the other side of the engineer – *KGB*, guessed the American astronaut – dig his elbow into the man's ribs.

'Well, *maybe* we are,' the engineer backpedalled.

In Moscow, the reaction to NASA's pursuit of the Shuttle had, at first, been curious and sceptical, but it soon became paranoid. Nor were Soviet concerns entirely without foundation. The idea of an orbital spaceplane had originally gained prominence as a bomber.

Austrian engineer Eugen Sänger first pitched the idea of a rocket-powered intercontinental bomber to the Austrian military in 1933. Just three years later, the Luftwaffe invited Sänger to set up a rocket research laboratory to rival that of Wernher von Braun's equivalent Wehrmacht facility. While von Braun poured his efforts into developing the V-2 missile, Sänger aspired to build a machine he called *Silbervogel* – 'Silverbird' – which, with the United States' entry into the war after the bombing of Pearl Harbor, acquired the name 'Amerika Bomber'. Launched by a rocket sled along a 2-mile monorail, *Silbervogel* was to be hurled to a height of 12,000 feet before lighting its own main engines. After accelerating to a speed of 14,000 mph and a cruising height of over 500,000 feet, it was then supposed to skip off the edge of the atmosphere to extend its range across continental USA to a landing site in Japanese territory in the Pacific. Sänger even highlighted potential targets and annotated maps with the projected destruction from a bomb dropped at hypersonic speed by the *Silbervogel*. In truth, the Amerika Bomber was never more than a pipe dream, and had it been properly developed it would have faced – among other things on a long list – insurmountable thermal problems. But Sänger's rakish-looking Silverbird became the starting point for any post-war discussion of orbital spaceplanes.

After the war, Walter Dornberger, the ex-head of the rival Peenemunde rocket research facility, tried to tempt Sänger to the United States. Employed by the Bell Aircraft Company, Dornberger was working on a top-secret project known as BoMi, a contraction of 'Bomber Missile', for the USAF. All Dornberger's early designs drew on Sänger's *Silbervogel* research. Sänger declined the invitation to move to the US, deciding that he'd rather stay in Paris where he was feted by the French and appointed the first president of the International Astronautical Federation. The Americans, though, weren't the only ones who'd got their hands on Sänger's work.

Enthralled by the idea of the *Silbervogel*, Joseph Stalin directed the NII-I design bureau to explore the possibility of developing the idea as a means of delivering Soviet nuclear weapons. Such was Stalin's impatience, however, that while that study was ongoing, he dispatched his son Vasily, along with an Air Force officer, to kidnap Sänger and spirit him back to Moscow. The operation fell apart because of Vasily's distraction by Parisian high society and the defection of his Air Force comrade to the UK. Blissfully unaware that his abduction had even been a possibility, Sänger stayed put.

The failure to capture Sänger didn't end Soviet interest

in spaceplanes. In the late sixties, the Mikoyan-Gurevich design bureau, better known as MiG, conducted work on a small spaceplane – or *kosmolyot* – similar in concept to the USAF's cancelled Dyna-Soar design. Known as 'Spiral', four different versions of the vehicle were envisaged including one capable of carrying a 3,500lb space-to-surface nuclear missile. While some work on the programme continued into the seventies it was effectively sidelined in favour of establishing a permanent lunar base, considered by its supporters to be a proper riposte to the brief visits to the moon achieved by the Apollo programme. As the Communist Party's Secretary for Defence Matters, Dmitry Ustinov, pointed out in 1974, 'the task of conquering the Moon remains especially important to us . . . our main general task'.

But the Shuttle kept nagging away at the Soviet leadership. First of all, a June 1974 study calculated that there would be no cost savings from choosing a reusable Shuttle over expendable rockets. Compounding this, Mstislav Keldysh, now president of the Academy of Sciences, then concluded: 'we do not see any sensible scenario that would support the shuttle for scientific uses'. The goals of the Shuttle programme announced by NASA seemed to make no sense, so they drew their own conclusions.

It clearly, decided a senior figure in the Ministry of General Machine Building that had responsibility for any Soviet Shuttle, *had a focused military goal*. The Generals convinced themselves – largely on the basis of the USAF requirement that their Shuttle be capable of launching from Vandenberg and returning after a single orbit – that this goal was to deliver a first-strike nuclear weapon. The idea of a Communist moon-base was soon rapidly losing ground to a Soviet Shuttle that would be capable of doing whatever the US Shuttle could do, whatever that might be.

In the summer of 1974, Valentin Glushko, the new head of the USSR's leading rocket design bureau, NPO Energiya, reluctantly set up a department dedicated to studying Shuttle-type reusable spaceplanes, knowing full well that to build one would likely end his own hopes of a permanent Soviet presence on the moon. Such was the way Cold War decisions got made.

Back in the US, at a party at Bob Overmyer's house, Bob Crippen was talking with Vladimir Dzhanibekov, one of the cosmonauts assigned an Apollo-Soyuz support role similar to his own. After becoming a cosmonaut in 1970, the Russian had enjoyed some involvement with a training group set up at Star City in 1966 in anticipation of the development of the Spiral spaceplane. Dzhanibekov kept that piece of information to himself. Instead, he raised his glass and proposed a toast: 'Crip, may our guns never cross!'

Just in case they did, though, on 25 June 1974, the Soviets had launched a space station into orbit, *Salyut 3*, armed with a 23mm-calibre rapid-fire anti-aircraft cannon capable of firing 850 rounds per minute.

The Cold War, as much as it had given birth to Apollo in the first place, drove the decisions made by East and West in the seventies. The threat of the now cancelled Mach 3 North American B-70 Valkyrie bomber had prompted the design of a new Soviet interceptor, the MiG-25 'Foxbat', capable of similarly high speed and altitude. This brutally powerful aeroplane quickly claimed one of the time-to-climb world records set by John Young back in 1962 during Project HIGH

JUMP. In response to the big MiG, the USAF in turn demanded even greater performance from their new fighter, the McDonnell-Douglas F-15 Eagle. And, in early 1975, the F-15 seized back for America honours won by the Foxbat, reaching a height of over 98,000 feet from a standing start in less than three and a half minutes. But the 'Streak Eagle', as the modified, bare-metal F-15 was dubbed, was already something of an anachronism. The Eagle was the last new American frontline military aircraft design to set outright world records with its flight performance alone. For a generation of pilots and engineers who'd been cast in the heat of the fifties and sixties, when every new design that tumbled out of an American factory – Skystreaks, Sabres, Skyrays, Super Sabres, Crusaders, Voodoos, Delta Darts, Starfighters, Phantoms – seemed to fly higher and faster than what had come before, only the Shuttle, yet to fly, offered to raise the bar again. Almost a distillation of everything that had been learned over the course of an exhilarating quarter of a century, it was still as big a challenge as the American aerospace industry had ever faced.

At seventy-five major sub-contractors spread throughout every state bar Alaska, tens of thousands of people were busy working to meet it. Alongside them, NASA facilities across the country saw the Shuttle occupy ever greater resources in personnel, time and effort, and Ames Research Center in California was no exception.

In the early days of the space programme, the resistance of material to extreme heat was tested by placing it in the path of a rocket engine exhaust. By the time of NASA's decision to use silica tiles for the Space Shuttle's heatshield, Ames Research Center was using arc-jets: small, hot-gas wind tunnels that used an arc of electricity to superheat the air to above 10,000°F. In effect, they were generating their own lightning. And to do it, their 20-megawatt arc-jet used enough electricity to supply a town with a population of 7,000 people. It had been possible to test the Apollo heat-shield using samples little bigger than a postage stamp. The Shuttle needed more. In 1971, Hans Mark secured funding for the construction of a brand-new 60-megawatt arc-jet capable of testing full-scale tile panels at Mach 6 for half an hour at a time – twice as long as they'd be subjected to heating by a returning Space Shuttle. Once it came online four years later, it, like the wind tunnels, was soon running double shifts.

At Marshall Space Flight Center in Huntsville, Alabama – made famous under Wernher von Braun's leadership – work continued on the development of the Space Shuttle main engines. Just building test facilities for the new engine had cost Paul Castenholz, the Rocketdyne engineer who'd led the successful bid for the contract, his job when major cost overruns and delays pushed back the start date for testing. Although 1975 saw the first ignition of the advanced new engine, the test-firing lasted just half a second. As for delivering an operational, man-rated rocket engine that in a split-second would transform liquid hydrogen from a temperature near absolute zero into a flame hot enough to boil iron? They were still at the foot of the mountain. More obvious progress was being made elsewhere on the two other elements that completed the Shuttle stack: the external tank, or ET, and the solid rocket boosters, always known as SRBs.

In Michoud, Louisiana, 9 miles east of New Orleans, at the Martin-Marietta plant building the Shuttle's huge external fuel tank, the first money spent was on a mural.

James Odom, the Programme Manager, commissioned a $7,000 full-sized painting of the tank on the wall of the plant. He regarded it as money well spent.

Of the three elements that made up the Shuttle stack, the ET was the easiest to take for granted. It neither spewed fire during the ascent, nor did it return to Earth. Instead, tracked by the Department of Defense, the disposable tank burned and broke up on re-entry, depositing what remained from its blazing descent in the Indian Ocean. It was also going to be the biggest single structure ever launched into space. Over 150 feet long and, at 27½ feet, 6 feet wider than the fuselage of a Boeing 747, it could more or less have engulfed the Saturn 1B rocket being prepared to carry the Apollo-Soyuz crew into orbit. In commissioning the painting, Odom wanted to remind the Michoud workforce of the size and complexity of the job they had to do. Not only did the tank have to carry the liquid hydrogen and oxygen that slaked the main engines, it also acted as the backbone of the whole stack. As well as being lightweight, it had to be strong enough to cope with the complex, asymmetric loads generated by the two SRBs attached to its sides with explosive bolts, and the Orbiter itself. And unlike the similarly large Saturn V first stage which, with its five engines mounted directly underneath, had a clean, simple, vertical thrust vector that ran straight along the length of the whole rocket, the ET had to accommodate the thrust of the Orbiter's three main engines, coming in from the side, offset diagonally from the centreline. As an additional challenge, Odom's team also had to engineer a structure that could cope with extremes of temperature. Beneath the skin, the liquid hydrogen and oxygen were stored at cryogenic temperatures below their boiling points of –297°F and –423°F, and yet the surface of the tank had to endure the temperatures generated by its acceleration to hypersonic speeds. It would require a third of an acre of thermal protection to insulate it.

Odom soon followed his $7,000 investment in art with rather more substantial spending. As the only element of the Shuttle configuration that would be used once then thrown away, he needed a production line. Around $1 billion was invested in tooling early in the programme to generate the capability to build ten to fifteen tanks a year. If flight rates required more, he had the space to ramp it up. At least he didn't have to worry about trying to load it on to the back of the Shuttle Carrier Aircraft and flying it around the country. While a study by Boeing had concluded that it was feasible to strap an ET to the back of a Jumbo Jet, the decision, in the end, was taken to carry the big tank from Michoud to Cape Canaveral by barge.

With the Orbiter being ferried by air, that just left the job of getting the two solid rocket boosters to the Cape. And the relationship between this last element of the Shuttle stack, and *their* chosen method of transportation, was a subject often explained with what was purported to be a history lesson.

Even members of NASA's senior management liked to tell people that the 12ft 2in diameter of the boosters – the biggest ever built – was dictated by the width of a horse's backside. As they told it, America's railroads used the same gauge as those in Great Britain, where the distance between the tracks was based on the width of a cart track, which in turn was based on the width of two horses. It was a good story, and it didn't really matter whether or not it stood up. What was certain was that the big solid rockets were going to travel to Florida by rail from their manufacturer, Thiokol, in Utah.

The SRBs were made up of four separate segments, each given its own railroad truck, that were then assembled one on top of the other like cylindrical building bricks, into a finished near 150-foot-tall booster rocket. Long lead items for the SRBs were ordered alongside the design and engineering of the tooling needed to build them. The previous year, drop tests had been conducted from B-52 bombers to test the parachutes that would carry the spent boosters down into the Atlantic where they were to be retrieved by boat and towed back for refurbishment and reuse.

Houston, meanwhile, remained lead centre for the Orbiter itself. And they were happy to welcome back one of their own, now also repaired and returned to work.

When Fred Haise rejoined the Shuttle programme office after his accident, there was little outward sign of his ordeal, but beneath his trousers he now looked as if he permanently wore socks. The protection from the fire afforded him by his thick leather flying boots had created a clear line between where he had been burned and where he had not. Freddo's return was a welcome one. Without him around, Houston was just a touch less energetic. People would have to remember once more to be mindful of the possibility that, as they reached for a door handle, Freddo might burst through in the other direction, but his humour and committed approach to everything he did was good to have around again. So too was his capacity to absorb, learn and process information fast. In the programme office it was coming at him from every direction.

Each morning at 7.30, Haise met with Orbiter Programme Manager Aaron Cohen and the two of them would cycle through the progress of every one of the sub-system divisions contributing to the development of the Orbiter. Everything funnelled through Cohen's office and nothing could be changed without his say-so. Decisions made had to be binding or would lead to increased cost or delay. What Haise and Cohen filtered through were called RIDs – review item discrepancies – passing only those they deemed likely to impact schedule, budget or safety up the line to Bob Thompson, the Space Shuttle Programme Manager, and John Yardley, the Associate Administrator for Manned Spaceflight at NASA HQ in Washington. The final Critical Design Review undertaken prior to Rockwell cutting metal on the first Orbiter prototype produced around 15,000 RIDs. Not for nothing did Cohen sit below a sign on the wall of his office reminding people that 'better is the enemy of good'. It was certainly proving to be the case with the software.

A fellow astronaut once claimed Bob Crippen had 'the best memory of any man I've ever known'. Then one day, as the two of them made their way to the simulator, Crip, poised to enter a PIN number on to a keypad to get in the building, just stood there at a loss. Eventually he admitted, 'It's gone.'

'What are you talking about?'

'That number's gone,' Crip said. 'I'll never remember it again.'

His colleague figured *he'd just pushed too much stuff into his head*. Crippen was enduring a similar problem with the computers. Along with Gordo Fullerton, focusing on the Orbiter's cockpit, he worked with the pilots to establish what they wanted to see on their displays, then he and the software team did their best to accommodate them; but it often proved too much for the limited memory available from the computers. The way T. K. Mattingly boiled it down, the crews wanted video games, the engineers needed the memory that required to actually make the vehicle fly. There was only one decision to be made.

As much as anyone else in the Astronaut Office, Crip wanted the displays on which he was working to show comprehensive, easy-to-read schematics of the Orbiter's systems. You wanted to see at a glance how current was flowing from the three fuel cells, or where valves in the hydraulic system were opening and closing. But when the effort to illustrate it caused the software design to outgrow the computer's memory, it was Fred Haise who managed the scrub to get the software back within the capabilities of the computers, while Crip had to work out what *was* possible. The two of them worked through the problem together. Too often the solution came down to having to display the raw data numerically, pushing the burden of interpretation back on the astronauts. Every time, Crip and Haise then had to sell that to the crews. 'This,' Haise told them, 'is what we're going to have to live with.' Unless they did, they weren't going to fly. And that meant that once the imminent Apollo-Soyuz mission had been and gone, American astronauts would be grounded.

On 15 July 1975, a flawless launch from Pad 39B at Kennedy seemed to be a fitting way to end the fourteen-year career of Wernher von Braun's Saturn rockets. And on CapCom duty in Houston, it was Dick Truly who talked to Tom Stafford, the Commander of the NASA crew, as the American and Soviet spacecraft circled the globe in formation.

'I've got two messages for you,' Truly began. 'Moscow is "go" for docking, Houston is "go" for docking. It's up to you guys. Have fun.'

With that, he'd initiated the last major act of the Apollo programme. The immediate future of American manned spaceflight now lay with, and only with, the Space Shuttle. And more than any other group within the Astronaut Office, that programme belonged to the MOL guys.

PART TWO

Mojave

'*The hangar doors opened, this wonderful ship came rolling out, and the Air Force band played the theme from* Star Trek *. . . I want to thank NASA for giving us the* Enterprise, *Gerald Ford for naming it the* Enterprise, *the* Star Trek *fans who insisted that it should be named the* Enterprise. *To the scientists and engineers who make these wonderful things happen, I say to all of them and the* Enterprise: *live long and prosper.*'

Leonard Nimoy

TWENTY-SIX

Houston, 1976

CHRIS KRAFT HAD never been shy of asserting his authority over the Astronaut Office. At the heart of mission control since the earliest days of the US space programme, he'd made sure that men like Scott Carpenter or Wally Schirra, once they'd incurred his ire, never flew in space again. Since taking over as Director of the Manned Spacecraft Center at Houston in 1969, Kraft had remained as irritated as he'd ever been by what he saw as the astronauts' inclination to play their cards too close to their chests. He understood why, on occasions, test pilots might be reluctant to ask for help or open up – to relinquish control to non-flyers outside the fraternity – but spaceflight demanded total transparency. *If you don't tell me everything*, Kraft thought, *there's not a thing I can do to help*. Kraft had always admired the way Deke Slayton performed his role as Director of Flight Crew Operations, but he was also a test pilot, and with the conclusion of the Apollo-Soyuz mission and Slayton's long-awaited and well-earned spaceflight, Kraft saw an opportunity to move his own man into position.

First, he gave Slayton the job of managing the upcoming Approach and Landing tests at Edwards, the flight test programme that would see the prototype Orbiter launched off the back of the 747 currently undergoing conversion at Boeing. With Slayton moved on, in January 1976 Kraft told George Abbey, his technical assistant, 'You're the new Director of Flight Operations.'

For an ex-Air Force pilot inspired to get into the space programme because he liked Buck Rogers, Abbey's roots were unusual. Although born in Seattle, George Abbey considered himself Welsh-American. His mother was from Laugharne in South Wales, home of Dylan Thomas. Her cousin, the local milkman, used to drink with the poet every afternoon at Brown's Hotel. Dick the Milk was a pallbearer at Thomas's funeral. Abbey's mother and father met in London before emigrating to Seattle where Abbey watched planes fly in and out of the Boeing plant. After training as an Air Force pilot he studied for a Masters in Electrical Engineering at the Air Force Institute of Technology, then, on graduating, returned to Seattle, assigned to the cancelled Air Force Dyna-Soar spaceplane programme before a detail to NASA followed. He never left, impressing successive bosses with the depth and breadth of his insight into the work being done at Houston.

As technical assistant, he'd been Kraft's eyes and ears; before that, he had performed the same job for Kraft's predecessor as Center Director, Robert Gilruth. Abbey, Kraft reckoned, knew 'where all the bodies were buried'.

In his new role, it was now Abbey's job to make the final decision, alongside Chief Astronaut John Young, on flight assignments. It's unlikely that meetings between the two men were characterized by too many wasted words. Neither had a reputation for being garrulous. At the top of their list as they began work together was choosing the crews for the upcoming Approach and Landing test being run by Deke Slayton.

They discussed the possibilities at length, considering the merits of every member of the Astronaut Office, but in the end only two criteria drove their decision: who was most qualified to command and who would best complement them.

Young and Abbey were quick to agree that the two obvious Commanders, because of their track records as test pilots at Edwards prior to becoming astronauts, were Fred Haise and Joe Engle. No one else could offer their depth of experience in aerodynamic flight-testing, and with Freddo now returned to Class 1 Flight Status after his accident, it was time for him to get back on the horse. Sitting in the right-hand seat as Pilot – NASA had chosen to label the Shuttle crews Commander and Pilot, rather than bruise egos by calling anyone a co-pilot – they chose Dick Truly and Gordo Fullerton, because of their familiarity with the Orbiter's systems. Two other MOL veterans, Bo Bobko and Bob Overmyer, were picked as support crew. Over the year ahead they'd work hand in hand with the two primary crews and, during the test flights themselves, man CapCom or fly chase planes alongside the Orbiter as she swooped down from 25,000 feet towards Rogers Dry Lake Bed.

After the desperate disappointment of losing his seat on *Apollo XVII*, Joe Engle was thrilled. Anxious to see what he could apply to the Shuttle programme from his experience flying the X-15, he couldn't quite suppress a feeling that this – developing a first winged orbital spacecraft – was what he had always been at NASA to do. Inevitably, there were some who felt shut out. After working so closely on the Orbiter's flight control system, Hank Hartsfield's unhappiness at not winning a place on one of the crews was extreme. He couldn't understand why he hadn't been selected. His frustration didn't come as a surprise to Dick Truly, as delighted at his own selection as Hartsfield was downcast. All of them had been at NASA for the same length of time. *I lucked out and got it*, thought Truly. But it wasn't luck.

In Engle and Haise, Young and Abbey had a pair of test pilots whose track records made them demonstrably among the best in the world. But they could only fly the test programme if the Shuttle's computers, electronics and hydraulics were working. Those systems, as Joe Engle put it, 'needed to be nursed and cared for during the flight'. Truly and Fullerton could do that. With that requirement in mind, perhaps only Bob Crippen had real cause for feeling he'd missed out, but he didn't see it like that at all. He was surprised enough to see his friends Truly and Fullerton get the nod, and happy for them. Not least because it seemed to bode well for when the time came to pick crews for the Shuttle's first flight into space.

Under Deke Slayton's management, Haise, Engle, Truly and Fullerton parcelled out the work that needed to be done in preparation for the first glide back to Edwards in the summer of 1977. Joe Engle took the lead on the flying aspects of their preparation, working on the hardware and flight profiles they'd need to train for the flights. He was also helping prepare Houston's new Shuttle Main Simulator. Essentially a cockpit mounted on hydraulic pistons, it reproduced the movements of the Space Shuttle in flight, lending an extra layer of realism to the simulation. Truly, with Bob

Crippen, followed the software development. Haise and Fullerton, meanwhile, ferried to and from the Rockwell plant in Palmdale, California, in T-38s, working with the company's engineers to prepare the Orbiter herself. And she was something. Her vertical tail towered near 60 feet above the ground, requiring modifications to the Palmdale facility just to get her in and out. Unlike previous spacecraft, it seemed clear that she'd been built to do a job – to *work* – not just provide the bare minimum required to get astronauts from one place to another.

Three weeks after NASA announced the crews, Rockwell finished final assembly of the prototype that Haise, Engle, Fullerton and Truly would launch off the back of the big 747 carrier aircraft. Just four years on from Richard Nixon's announcement, NASA had something to show for its efforts. And two of the MOL guys had their first flight assignments.

In the same week as NASA announced the names of the crews for the Approach and Landing tests, the Kremlin gave in to their nervousness about America's motives for building the Shuttle. In a decision that amounted to little more than a desperate state-sponsored effort to keep up with the Joneses, they finally committed to the development of a Soviet 'Reusable Space System'. The decree issued by the Central Committee of the Communist Party and Council of Ministers stressed that special importance was attached 'to increasing the defence capabilities of the country'. The first three reasons for developing a Soviet Shuttle were given as:

- counteracting the measures taken by the likely adversary to expand the use of space for military purposes;

- solving purposeful tasks in the interests of defence, the national economy, and science;
- carrying out military and applications research and experiments in space to support the development of space battle systems using weapons based on known and new physical principles.

'Purposeful tasks' may have been vague in the extreme, but there was no doubting that the Soviets anticipated the main role of their Shuttle to be military, a reflection of their belief that the US Shuttle, launched from Vandenberg, would, through performing a fleeting 'dive' into the atmosphere before recovering to orbit, deliver a first strike against Moscow or Leningrad more quickly than Soviet forces could respond.

Just over a week before the first American Orbiter emerged from the hangar at Rockwell's Palmdale plant, the President changed her name. In recognition of the country's 1976 bicentennial celebrations, the new prototype was to be called *Constitution*. Then an orchestrated campaign by *Star Trek* fans deluged the White House with over 100,000 letters arguing that *Constitution* should be named *Enterprise* in recognition of Captain James T. Kirk's iconic starship. Faced with what appeared to be the will of the people, Gerald Ford went with it. And on 16 September, in front of a banner announcing that she was ushering in a 'New Era in Transportation', the finished prototype, christened *Enterprise*, emerged into the California sun, towed behind a squat tractor that seemed overdressed in a special Stars and Stripes bicentennial paint job. Nearby, NASA displayed an Apollo capsule, highlighting just what a dramatic advance the new machine represented.

Fred Haise, Joe Engle, Gordo Fullerton and Dick Truly, all wearing crisp blue NASA flightsuits, posed for photographs. Also there for the occasion were Mr Spock, Mr Sulu, Bones, Lt Uhura, Mr Chekov and Scotty – or at least the actors who played them – sporting the finest in safari suits and seventies fashion that dollars could buy. The occasion may have helped capture the imagination of the public – and with *Enterprise*'s first flight still nearly a year away, NASA needed that – but within the Astronaut Office, the decision to name her after a spacecraft in a TV show was, for those who cared, the source of a good deal of eye-rolling and scepticism. Fred Haise wondered whether the Trekkies had realized that this *Enterprise* wasn't ever going to go into space. Dick Truly, who'd flown F-8s with VF-33 Naval Air Squadron off the deck of the *real* USS *Enterprise*, couldn't help but feel slightly annoyed by it. And so he just decided not to accept it, instead sticking to his own story that she was simply the latest vessel in a long, illustrious line. The third USS *Enterprise* had, after all, fought with distinction against Barbary pirates and, later, the Royal Navy. Truly's denial at least gave him a welcome opportunity to tease his Air Force crewmate Joe Engle with impromptu quizzes on naval history during dead time in the simulator. There was certainly plenty of that.

Unlike the Apollo simulators, the Space Shuttle sim used the same flight computers and software as the real thing. If a problem revealed itself in the sim, it also existed in the vehicle itself and needed to be isolated and fixed. There was no shortage of them. The crews could usually bank on getting no more than about two hours' flying into any four-hour sim session before they heard 'Sorry, guys, we gotta stop; one of the computers has died and it's going to be fifteen minutes . . .' At least the simulator cockpit was warm. Sitting 25 feet away within the three-storey-high bay, the simulator supervisors – known as SimSups – who ran the training were wrapped up in winter coats at their consoles even as the Houston summer was delivering 100°F and 100% humidity outside. To keep the banks of computers cool, it was always 55°F inside Building 5. Such was the frequency with which the simulator crashed that Engle and Truly christened her 'Lucille' after a recent Kenny Rogers hit. With each new glitch or freeze, the sound of the two idle astronauts massacring 'You picked a fine time to leave me, Lucille . . .' crackled through the SimSups' headsets.

It was frustrating for all concerned, but the simulator also offered Engle the chance to fly the Orbiter in ways it would never be possible to do while sitting in the Commander's seat of the real thing. If the SimSups tinkered with the simulator's settings, Engle could fly circuits, landing, taking off and landing again, something the unpowered Orbiter was incapable of doing. And when asked if Jovial Joe ever repeated the trick he pulled in the X-15 by rolling the Shuttle through 360°, other astronauts would just smile admiringly. It wasn't for them, perhaps, but they liked the idea that someone had the audacity to do it.

Nor was this the only way the simulator gave Engle the opportunity to demonstrate just how exceptionally gifted a stick-and-rudder pilot he really was. It was possible to detach the rotational hand controller – the joystick – from its mount in front of the pilot. Engle could pick it up, turn it backwards and upside down, then continue to fly and land the Orbiter so smoothly and precisely that the SimSups could barely tell the difference. It had them shaking their heads in awe, unable to understand how he managed to process something so counter-intuitive. One of them, Jerry Mill, reckoned Engle could have brought a clawfoot bathtub back to a safe landing. Gene Cernan, who should have been Engle's Commander on *Apollo XVII*, claimed his former Lunar Module Pilot 'could probably have flown a lawn-mower'. There's no doubt that Engle would have relished the chance to give either a go. They just wouldn't, along with his stints flying round the hangar at the Bell Aircraft Corporation with his personal jetpack strapped on, have been recorded in a logbook. Engle stopped keeping one when the volume of flying he was doing just made it impractical to try to keep up. As far as he was concerned, flying the sim was a breeze.

The trouble was that his ability was actually masking a potentially serious problem with the Orbiter's flight controls, which only the development of a new Shuttle Training Aircraft was beginning to reveal.

TWENTY-SEVEN

Houston, 1976

THOSE AT THE meeting remembered it well. The first day of April 1976. The Shuttle Training Aircraft (STA) programme was running near six months late and 50% over budget. Houston Center Director Chris Kraft was furious, angry enough to cuss like a docker as he bawled out the Grumman vice-presidents responsible before throwing them out of the room, promising them that if he had anything to do with it Grumman would *never* get another contract out of NASA.

The main challenge had been the software needed to get a digital autopilot borrowed from a big McDonnell-Douglas DC-10 jet airliner to make the two ten-seater business jets fly like Space Shuttles. The Gulfstream G-IIs that had been chosen for the job were heavily modified. The most obvious external evidence of this was a pair of large rectangular vertical fins bolted underneath the aeroplane to help reproduce the sideways forces the Shuttle could generate. Inside, the left-hand half of the cockpit was rebuilt to look like the Shuttle flight deck from the Commander's seat, while behind the cockpit in the main cabin a console was installed for the computer tech running the simulation. The G-II's wing flaps, which normally extended backwards and down to provide drag and lift on approach, were engineered to hinge up instead, providing the same drag, but dumping lift. Operating the Rolls-Royce engines' thrust reversers in flight at 90% power was causing worrying levels of vibration. It had some wondering whether the Gulfstream jet was going to be up to the job until further modifications managed to bring it under control.

There was something else though. Despite Joe Engle's lack of concern, the modified Gulfstream was displaying a worrying tendency to get trapped by a control problem called pilot-induced oscillation, or PIO. And there was a growing feeling that that was potentially serious.

Up at Edwards, someone once labelled PIO 'the JC manoeuvre', because 'Jesus Christ!' was the involuntary exclamation from pilots when it happened to them. It was a characteristic aggravated by the emerging fly-by-wire technology and amplified by pilot pressure.

If a pilot concentrating intently on, for instance, performing a perfect landing makes a control input and doesn't feel its immediate effect, he's likely to repeat the action. If there's any delay between his inputs on the stick and the action of the aeroplane's flight surfaces then in repeating himself he'll have doubled up before the first demand's been carried out. Then, having commanded the plane to change its attitude twice as much as he had intended, he's got to respond just to bring it back to where he wanted it. If he feels no immediate response to this correction, he's going to double up on that too. Soon, like a driver trying to wrestle a fishtailing car on a skidpan, he's losing track of which control input is having what effect. He's behind the plane.

In the Shuttle, and the Gulfstream training aircraft that was designed to mimic it, the tendency to suffer from PIO was exacerbated by the characteristics of the delta wing and the huge barndoor-sized elevons required by the Orbiter to grip the thin air at very high altitudes. At approach speed, when you pulled back on the stick, the effect of the elevons deflecting up to raise the nose was also to deprive the wing of a substantial chunk of its overall area and lift. And this meant that, before the Orbiter climbed, as commanded by the pilot, it fell.

It was a combination that might have been designed to provoke PIO, and yet, while Joe Engle was aware that the STA might have a tendency towards it, it never caused him any trouble. In fact he was convinced it was an issue with the simulation, not with the Shuttle's own flight control software. In the ground-based sims in Building 5, which after all were using exactly the same software as the Orbiter, he'd simply never provoked it.

Engle was a low-gain pilot. Like Charles Lindbergh or Chuck Yeager he barely moved the stick, anticipating the need to do so and making small, necessary corrections in plenty of time. His inputs were smooth and progressive, never snatching at the controls. And in the relaxed environment of the simulator it was even less likely he would do that. In order to try to prove that the STA – and any PIO it suffered – was an accurate simulation, the Programme Manager analysed and compared control stick movements on three different simulators: a fixed-base cockpit that was bolted to the floor, the motion-based sim on hydraulic rams at Houston, and a vertical motion simulator that moved through hundreds of feet in a hangar at Ames. With every extra dose of realism – and with it stress – so the level of gain from the pilot – the firmness of his flying – increased. The simulation provided by the Gulfstream was valid. But if that was the case, then flaws in the way the Gulfstream flew were also going to be shared by the Space Shuttle itself. Left unchecked, this had the potential to lead to the loss of a Shuttle. For now, though, a work-around was concocted that kept the STA programme on track. And that meant the astronauts could get a taste of what the new machine could offer them.

Although sitting in the left-hand captain's seat, John Young was just a passenger as he accelerated down the runway at El Paso International Airport in the Shuttle Training Aircraft for the first time. Next to him, Dave Griggs, the instructor pilot, was flying the Gulfstream from the right-hand seat, facing the half of the instrument panel carrying the jet's standard dials and displays. Griggs took off and turned north towards White Sands Space Harbor, New Mexico. Ten minutes later, at 35,000 feet above the lake bed landing point, Griggs lowered the STA's main landing gear, sending a low rumble through the airframe. Next he engaged reverse thrust and set the rpm at 92%, provoking a more forceful shudder as the two Rolls-Royce Speys fought the direction of travel. Griggs activated the hand controller on Young's side of the cockpit and told NASA's senior astronaut that the simulation was running.

In a steep glide, holding his airspeed at 280 knots, Young steered the jet down to a point 12,000 feet up and 7 miles out from the strip and rolled out on to long finals on a flightpath towards the centreline of the runway. A little forward pressure on the stick and he bunted over into a 20° dive, using the attitude of the G-II to settle at 300 knots.

The picture ahead was not one he was used to. Although Young had shot these steep approaches into White Sands in the T-38, its gear down and airbrakes flared to slow her progress, in the cockpit of the little jet trainer it was easier to get your bearings. Now, through the letterbox of glass provided by the Gulfstream's flight deck, he could see no sky, just the desert floor as they speared towards it. Then Young's approach came unstuck, forcing Griggs to take back control and climb out.

Chris Kraft credited his Chief Astronaut with 'incredible flying skills', but in his first fifty flights in the new training aircraft Young aborted nine times. It was a measure of how singular and unfamiliar the Shuttle approach was even to seasoned test pilots. When Haise, Engle, Truly and Fullerton separated from the 747 aboard *Enterprise* there would be no option to abort. Young's initial batting average of four-out-of-five wasn't good enough in an aeroplane with no engines. By the time the Approach and Landing test programme took flight at Edwards, landing the Orbiter had to be second nature. But with the modified Gulfstream finally operational in spring 1977, the two crews had the tool they needed to ensure it was.

Joining NASA's squadron of T-38s and the two Gulfstream STAs was the Shuttle Carrier Aircraft, the SCA, which, after its conversion from a standard 747, first took to the air after emerging from Boeing's Seattle plant on 2 December 1976. NASA, though, wasn't the only organization introducing new machinery to its fleet. Seven hundred miles to the south, at Vandenberg Air Force Base, December also saw the first flight of another brand-new vehicle.

There was no press release sent out to mark the

occasion, but the launch of the new vehicle was no less significant for that. For the National Reconnaissance Office, it represented a revolution. Their new spacecraft was lofted south into a polar orbit of 96.93° aboard a Titan IIID rocket from Vandenberg's SLC-4E launch pad on the 19th of that month. At first, those who took note of such things assumed it was the launch of another KH-9 HEXAGON. Then veteran space observers pointed out that the new spacecraft's orbit was a good deal higher than 'Big Bird' had so far flown. In fact it was flying higher than *any* previous American spy satellite.

Five years after President Nixon had authorized development of the KH-11 KENNEN digital spy satellite, the first of the new birds was on sentry.

A month later, at 3.15pm on the day after his inauguration, the new President, Jimmy Carter, was being briefed in the White House Map Room by the CIA's Acting Director, Hank Knoche. Knoche had brought with him a sheaf of 6-by-6-inch black and white photographs captured by the KH-11. He spread the pictures out on the big map table in the middle of the room for the President to examine. Although Carter had been briefed on the capabilities of the new satellite after winning the election the previous November, this was the first time he'd seen evidence of what it could do for himself. In fact, what he was looking at were the first images taken by the KH-11 KENNEN since, after a month of testing and checkout, it had become operational. The President leaned in and looked at the pictures in silence before a wide smile played across his face. The significance of the imagery was immediately clear. Taken just the previous morning, it had already been in the CIA's possession for over twenty-four hours. The NRO was now capable of showing the President what he needed to see, when he needed to see it. The American intelligence community finally had the access to real-time intelligence from space it had so craved. The danger that a Six-Day War might be fought and won before the NRO could respond was over. America's new commander-in-chief laughed approvingly and requested more imagery to share at the first meeting of his National Security Council the following day.

Since coming to Ames, Hans Mark had hardly become removed from the black world of classified military pro-grammes into which he'd first been inducted while at Livermore Laboratory. The Ames wind tunnels had been used to test early stealth fighter designs for Lockheed's Skunk Works. There was a highly classified programme to mount a high-intensity laser on a Boeing KC-135 for which Ames worked on fire control, systems engineering and, again, wind tunnel testing. And since 1971, when the first KH-9 HEXAGON spy satellite was launched, they had all been flown in to Moffett Field for integration at Lockheed's Sunnyvale facility before being carried by road to Vandenberg AFB.

At the time of the MOL space station's cancellation in favour of the unmanned Big Birds, Mark had been of the firm opinion that human involvement in space science and exploration was an unnecessary complication. He now advocated manned spaceflight with the zealousness of the convert he happily admitted to being. He'd even managed to add a now iconic human dimension to two unmanned spacecraft built during his time at Ames.

Deep space probes *Pioneer 10* and *Pioneer 11* were

launched in December 1972 and 1973 respectively; the former, after a journey that would take it past Jupiter and Pluto, was on course to reach Aldebaran, an orange giant star, in approximately eight million years' time. The programme had been led by a team at Ames, and in the eyes of one of Mark's former students at Berkeley, Carl Sagan, the two spacecraft presented a unique opportunity.

Mark had followed Sagan's career with great interest, admiring both his ascent to a position as perhaps the world's best-known communicator of space science and his abilities as a scientist. With *Pioneer 10* already at Kennedy being prepared for launch, Sagan telephoned him at Ames to pose a question: 'What if somebody finds it?'

'What are you talking about?'

'Well, what if somebody finds it?' Sagan repeated.

Soon, Mark was smiling, happy to go along with the suggestion that, in case it was discovered by intelligent life elsewhere in the universe, the probe should include a message telling them who we are and where we come from. Sagan produced a design that included a line drawing of an unclothed man and woman alongside celestial maps and representations of the spacecraft itself, which Mark then had made up on lightweight metal plates at a local engraving shop in LA. Knowing that any formal request to NASA HQ would see the idea smothered by paperwork and analysis, Mark simply flew down to the launch site and attached it to *Pioneer 10* without asking the question.

Mark had come to believe that mankind's destiny was to colonize space. His conviction, first sown in conversation with Wernher von Braun, was germinated by a study, funded by NASA and conducted at Ames by Princeton University physicist Gerald O'Neill, into the feasibility of vast, wheel-shaped space stations similar in appearance to those seen in *2001: A Space Odyssey*. In Kubrick's movie, delta-winged Orion III spaceplanes wearing Pan-Am logos shuttle people between Earth and space like airliners criss-crossing the Atlantic. Spaceflight is no longer the exclusive preserve of test pilots.

The Shuttle offered a step in that tantalizing direction and it was an aspiration which Mark himself had been directly involved in advancing. As a member of a committee identifying flight crew requirements, he'd successfully argued for a new category of astronaut, the Payload Specialist, for which years of dedicated training as part of the NASA Astronaut Office would not be required. When a report by two of his researchers at Ames later concluded that, unlike Mercury, Gemini and Apollo, medical requirements for a Shuttle flight were only that someone should be in 'reasonably good physical shape', it confirmed Mark's belief that the Shuttle would dramatically broaden access to space. His wife, Marion, he imagined, might quibble at the idea of him being described as a romantic, but when it came to space, he was prepared to concede that he was. 'Even poets,' he wrote, could travel into space, 'then share the experience with everyone.' He certainly hoped that he would one day fly aboard the Shuttle himself.

At Dryden Flight Research Center, NASA's facility at Edwards Air Force Base, the *Enterprise* Approach and Landing test programme remained firmly the preserve of men with the Right Stuff. Fred Haise, Joe Engle, Gordo Fullerton and Dick Truly were all graduates of Chuck Yeager's Aerospace Research Pilots School, while leading the programme was Deke Slayton, the last of the original Mercury Seven still working at NASA. There was no danger of any of the men who flew the Shuttle having to endure the taunts once levelled at Slayton and his comrades that they were just 'spam-in-a-can'. The winged Orbiter they were testing was a machine of which Joe Engle knew even the famously dismissive Yeager approved; not one from which, as the veteran test pilot put it when talking about NASA's capsules, you had to sweep out monkey shit before climbing in. There were no plans to fly *Enterprise* using chimpanzees.

TWENTY-EIGHT

Edwards Air Force Base, 1977

MOJAVE'S DRY LAKE beds could be deceptive. Twelve years earlier, in the days before each X-15 flight, Fred Haise had raced up and down them on a motorcycle, looking for anything that might prohibit their use as alternate landing sites. After flying in on board an old Air Force Douglas C-47 Gooney Bird, he and his fellow members of the support team – chase pilots and engineers – hammered spikes down through the crust of the lake bed to make sure it wasn't soft and muddy beneath the surface. They had a heavy steel ball that they'd drop at points along the length of the strip, then check the depth of the indentation. If the surface cracked or deformed it might also tear the gear off a landing rocket plane. All being good, though, they'd finish by laying asphalt stripes to mark out the boundaries of the runway. These days it was done by radar. While Haise and the other Approach and Landing crews had trained to fly *Enterprise* using the Shuttle Training Aircraft and sim, the support crews, led by Bo Bobko and Bob Overmyer, scanned the Edwards lake bed looking for water running below the surface, checking the geology beneath the friable skin of the ancient salt pan.

Now, on the morning of 12 August 1977, Bobko was 1,400 miles away from the high desert, manning CapCom in Mission Control back in Houston. Alongside him, the rest of the Flight Control team sat at consoles ready to monitor the telemetry relayed from *Enterprise* as she took wing for the first time. Like a space mission, the Shuttle's Approach and Landing tests would be directed from within Building 30 at Houston.

It was dawn up at Edwards. Inside NASA's Dryden Flight Research Center, three of Bobko's MOL contemporaries were in their flightsuits, preparing to get airborne in support of the Shuttle's first flight. Using the callsign Chase One, Bob Overmyer, flying with Dick Truly in the back seat of his jet, would lead a small squadron of T-38s. Hank Hartsfield was on the team too. Keeping a watchful eye on it all was Chief Astronaut John Young, flying another one of NASA's six-strong chase fleet.

In the pale, brightening light of the early morning, *Enterprise* and the Shuttle Carrier Aircraft beneath her cast long shadows across the Dryden ramp, the sun catching and reflecting off the 747's bare aluminium fuselage. Still visible, like old foundations from the air, was a ghost of the old American Airlines logo that had once been painted on her side.

The main briefing had taken place the night before. Now, with take-off scheduled for 0800 local time, Fred Haise and Gordo Fullerton were going to fly the Orbiter for the first time. After a short weather briefing alongside the chase pilots, Haise and Fullerton walked outside on to the Dryden ramp. For this, her maiden flight – and for the two that would follow – *Enterprise* wore a streamlined white shroud, a bulbous white fibreglass duck's arse, to reduce the turbulence and drag caused by the rocket nozzles at her rear.

A little over an hour before the flight, the two astronauts stepped off the high platform of a hydraulic cherry picker and into the open crew hatch on *Enterprise*'s port side. Wearing blue NASA flightsuits, Haise and Fullerton climbed up from the mid-deck into the cockpit and strapped in with help from ground crew. From the Orbiter's flight deck, perched nearly 60 feet from the ground on top of the 747, Haise and Fullerton had a commanding view of their surroundings.

Any lingering doubts about this outrageous-looking combination – described by some as the world's biggest biplane – had been assuaged earlier in the year by a series of captive flights in which *Enterprise* had remained firmly bolted to the back of the SCA. Crowds had gathered to watch those earlier sorties, but now they were like nothing Edwards had ever seen before. One thousand reporters and as many as 70,000 spectators, who had clogged the roads since well before dawn, claimed their positions on Rogers Dry Lake Bed and waited in anticipation.

Haise and Fullerton removed the pins from their ejection seats, stowed them, and confirmed to the ground crew that the chairs were now armed. The technicians wished them luck and stepped back across the high gap into the cradle of the cherry picker, then closed and locked the crew hatch before being lowered to the ground.

From the cockpit of the SCA, unable to see the Shuttle mounted above and behind them, pilots Fitz Fulton and Tom McMurtry started the 747's engines, gently dipped the brakes, then taxied out to the threshold of Runway 22. Riding the brakes, Fulton advanced the throttles. As the engines spooled up towards full power, the stack began to tremble. The roar from the four 50,000lb thrust Pratt and Whitney turbofans rolled across the dry lake.

In Houston, Bo Bobko counted down to the beginning of the take-off roll: 'Five, four, three, two, one . . . release.'

Fulton came off the brakes. Cleared for take-off, he asked for full power and the SCA and its 150,000lb piggyback passenger began to accelerate slowly down the runway.

On the flight deck of the Orbiter, Haise and Fullerton were still getting used to the unfamiliar feeling of moving seemingly unaided and looking down from so lofty an eyrie, along the path of the runway. From their seats inside *Enterprise* it was impossible to see any part of the aeroplane carrying them. *Like a magic carpet ride*, thought Haise. The first time he'd experienced it, during one of the series of captive test flights, he'd thought, as Fulton had pulled the nose of the 747 into the air to rotate, *we're not going fast enough to make it off the ground* . . .

At 8.48am and a little over 30,000 feet, Fitz Fulton bunted the SCA over into a shallow dive, accelerating to 240 knots, then, chopping the power to idle and extending the airbrakes, configured his jet to release the Shuttle. Mounted at such a high angle of attack on the big airliner's back, the Orbiter's thick wings bit greedily at the air. With the 747 sloughing off as much lift as possible, now *Enterprise* was actually carrying it beneath her rather than being held aloft. The 32-feet-per-second separation produced as a result would ensure the clean getaway that, at first glance, looked so unlikely.

'Carrier ready,' declared Fulton.

Above him, in *Enterprise*'s flight deck, Haise reached forward and mashed the SEP button. With a report that shuddered through the Orbiter's airframe, seven explosive bolts blew to set her free for the first time.

'Chase One, sep,' reported Bob Overmyer from the T-38 flanking the SCA's starboard wing as *Enterprise*, relieved of the burden of the 747 underneath, seemed to leap into the air, white trails of condensation streaming off her wingtips drawing lines against the clear blue sky.

In *Enterprise*'s cockpit, before the blast of the separation had even finished resonating through the fuselage, an ugly green 'X' flashed up on the CRT 2 display between the astronauts, signalling the failure of one of the Shuttle's flight control computers. As Haise flew her for the first time, the harsh tone of the master alarm provided a soundtrack. In the centre of the Orbiter's instrument panel, a red light flicked on in the middle of the matrix of warning lights.

Oh my God, thought Orbiter Programme Manager Aaron Cohen, following the flight test from a control room in nearby Palmdale, as the telemetry recorded the computer crash. And he bit his pipe in half.

In the cockpit, Fred Haise concentrated on flying the vehicle. Next to him, Gordo Fullerton got his head down, flipping quickly through his cue cards to find the correct procedure to follow. He twisted round to pull circuit breakers on rate gyros and turned off accelerometers feeding the toppled computer, manually reconfiguring the flight control system in case of further failures.

'Chase Two, clear.'

'Roger, pushing over.'

Having been told by the chase plane to his left that he was no longer in close proximity to the carrier aircraft, Haise nudged forward on the hand controller, lowering the nose and using gravity to accelerate to 300 knots. As he rolled gently away to the right, the silver 747 banked and dived away to port, putting further distance between them. Haise, using the three remaining computers to fly the aeroplane, was struck immediately by the feeling that, in flight, *Enterprise* felt crisper than the simulator.

In the five minutes it took for the Orbiter to glide to Rogers Lake Bed Runway 17, he didn't have long to enjoy it. Gordo had missed most of the first minute of the flight, but with the computer malfunction contained and *Enterprise* performing well, Haise handed him control as they flew the downwind leg of the circuit into Edwards.

This thing is flying good, thought Fullerton as he descended towards the turn on to base leg. Moving the stick to the side to pull *Enterprise* into the 90° turn, the aeroplane's nose suddenly lurched to the side, nearly banging Fullerton's head on the wall of the flight deck. *Son of a gun*, he thought, *they were right after all*. Because the cockpit was so far ahead and above the Orbiter's centre of rotation, a steep turn had the effect of flicking the nose round at a greater speed than that of the turn itself. It had been predicted in the simulator, but still caught Gordo by surprise. Thankfully, as Freddo took back control to bank on to finals to land, it was the last of them.

Flanked by a T-38 jet on her port wing, *Enterprise* touched down gently on her main gear after a free flight of just five minutes and twenty-two seconds. As they rolled to a halt along the lake bed under light braking, the tip of the long white air data probe attached to the Orbiter's nose resonated through two or three feet like an angry giant white tuning fork.

During this first free flight of the *Enterprise*, Bob Crippen was on the ground upholding a NASA tradition by watching the flight with the crew's wives, answering their questions, explaining and reassuring. When, following the return to Houston, the failure of the computer was investigated, he was relieved to discover it had been a broken solder inside the processor, not a software failure, that had led to it falling over. At the same time it had also shown that the redundant computer sets worked as advertised. Losing one was entirely manageable and was hardly remarked on in the press coverage that followed. The headline news was that the Orbiter flew. And that she flew well.

TWENTY-NINE

Edwards Air Force Base, 1977

DELAYED FOR TWO weeks because rain had turned the lake bed to mud, Joe Engle and Dick Truly flew the second free flight a month after the first; then, just ten days later, after engineers had worked nights and weekends to repair a hydrazine fuel leak, Haise and Fullerton flew a third flight.

With everything going according to plan, and under pressure from headquarters to bring the Approach and Landing test programme to a successful conclusion, Deke Slayton took the decision to fly the next flight without the low-drag tailcone. As reassuring as *Enterprise*'s performance had been with the shroud on, it wasn't representative of how the Orbiter would handle when returning from space, with the collection of rocket bells that protruded from her tail exposed. Free flight four fell to Engle and Truly. This was the one the two of them had been looking forward to. Their job was to try to confirm a mass of aerodata predictions in a minute and a half of what amounted to a steep dive. It would be test-flying at its most intense.

Engle had prepared by asking his friend, fellow Apollo astronaut Tom Stafford, now Commander of the Air Force

Flight Test Center at Edwards, if he could beg a flight in the new General Dynamics YF-16 fighter prototype being tested at Edwards. Stafford was happy to help. Flying the light-weight jet, still sporting a distinctive red, white and blue paint scheme, Engle realized that its digital fly-by-wire flight control system allowed for a technique that could make all the difference to the task he and Truly had to perform. Not only that, it was something the two of them had already explored in the sim in Houston.

Along with singing to them when the simulation crashed, Engle and Truly had found another way to wind up the SimSups. If, while Engle was flying from the left-hand seat, Truly pushed the 'Flight Control Power' button to activate the controls on his side of the cockpit, both then had equal authority over controlling the Shuttle. With prac-tice, they were able to act as one. With Engle taking responsibility for rolling left and pitching up and Truly for rolling right and pitching down, they could land the Shuttle together without the SimSups even noticing. And if, with sufficient feel and care, Engle moved his stick to the left while Truly pulled his in the opposite direction, the net effect was that, cancelling each other out, the Shuttle con-tinued to fly straight and true until the moment Engle piped up to say to SimSup Jerry Mill, 'Jerry, look at my stick', by now fully deflected left without any apparent effect on the aeroplane's attitude, 'something's not right . . .'

If they used the same technique in *Enterprise*, by flying her together and at the same time they could get far more done in the short time available in a single flight. Rather than having to let the Shuttle settle between each manoeuvre, Truly, from the right-hand seat, could set the required angle of attack and hold it, while Engle pulsed the hand controller from the left-hand seat, overlaying his own control inputs on Truly's. In both the simulator and Gulfstream training aircraft, they repeated their brief flight over and over again, leaning over each other's shoulders to try out different ideas. By the time they were scheduled to fly they had it worked second by second. Truly likened it to choreographing a ballet.

At dinner in the 'O' Club at Edwards the night before the flight, Engle and Truly had been urged by Flight Surgeon Lon Bergman to quit drinking and go to bed. When the doc himself finally headed off back to his room, the astronauts stayed behind, assuring him that they'd pick him up at six the next morning.

The two astronauts called it a night and went to bed a few minutes later. After a good night's sleep, Engle and Truly met before dawn. They needed to be up early to set the trap. Before driving off, they opened cans of beer and scattered empties around the car before deliberately approaching to pick up Bergman from a direction that suggested they were pulling up following an all-nighter in the nearby town of Lancaster. They lurched to an untidy stop alongside the medic, who was waiting for them outside the Edwards Bachelor Officer Quarters.

'Get in, Lon!' they slurred loudly from the car's open window. 'We're late. Come on!'

Bergman climbed into the front seat, immediately inhaling the strong musty tang of beer. As they sped off towards Dryden Flight Research Center to prepare for the flight, the doc was speechless.

In reality, Engle and Truly couldn't have been better prepared for the flight. The major concern was that, by removing the tailcone from the back of the Orbiter, air churned up by those rocket bells would cause such turbulence over the tail of the big Boeing that the buffeting and vibration could be potentially destructive. The configuration had been tested in the wind tunnel, but fluid dynamics remained hard to predict. As a precaution, the comms loop between the 747 flight deck, *Enterprise* and Mission Control was kept open throughout the take-off roll. At any sign of trouble, Fitz Fulton, the Boeing's captain, could hit the brakes.

Just having Fulton at the controls of the carrier aircraft inspired confidence. By 1977, he was perhaps the most experienced large aeroplane test pilot in the world. He'd flown the B-29 mothership for Chuck Yeager's Bell X-1; ushered both the B-52 and supersonic, delta-winged Convair B-58 Hustler into Air Force service; dropped the X-15 from beneath the wing of NASA's B-52; and flight-tested the six-engined Mach 3 North American XB-70 Valkyrie bomber prototype. On exchange in the UK he'd added two British bombers, the Avro Vulcan and Vickers Valiant, to his logbook. And Concorde.

When Fulton and his co-pilot, Tom McMurtry, received the 'go for take-off' call from Houston, they were looking down the length of Runway 04. Not only did it offer 15,000 feet of hardtop, but a further 9,000-foot extension on to the lake bed. Four and a half miles. If, as the SCA took flight, Fulton judged the vibration to be unsustainable, he'd have the option of a straight-ahead landing on what remained of the long runway in front of him.

The curtain went up on Engle and Truly's performance at 15.49 on 12 October, when Fitz Fulton put the 747 into a shallow dive 25,000 feet above Mojave. Oxygen masks clamped over their faces, Engle and Truly prepared themselves.

'On speed . . .' Fulton called.

'Three, two, one . . .' As he spoke over the radio, Joe Engle reached forward and, at 20,300 feet above Highway 58, he punched the separation button. 'Sep.'

This time, as *Enterprise* and the SCA parted company, Truly resisted the temptation to call 'Fitz away!' as he had during his first free flight with Engle. It may simply have been the view from the flight deck that stopped him. He and Engle had practised hot and cold separations from the SCA in the simulator, so they were prepared for leaving too steep and fast or too shallow and slow. But this one was something else. At the end of Engle's pushover, they were pitched down at 28°. *Holy shit*, thought Truly as he looked down through the cockpit glass at the lake bed. But as Joe Henry Engle always said: 'It doesn't get to be fun until we drop the 747 . . .'

As the speed increased to 210 knots, Truly applied gentle back pressure to the stick, pulling *Enterprise* out of the steep dive and setting her up at a medium 7° angle of attack. As soon as they were stable, Engle pulsed his own controller, layering his own commands over those of his pilot. With precision and an almost rhythmic beat, he followed one movement with another, ticking off points on the aerodata graphs with each stick input. Once complete, Truly relaxed the pressure on his stick, unloading the aeroplane, checking her progress then holding her at the planned 4° angle of attack. Again, Engle worked fast through a carefully sequenced progression of stick inputs, each one confirming the Orbiter's real-world performance against predictions. In the seconds before Engle had to set up *Enterprise* to land, the crew ran through a final set of stick inputs with the speed brakes extended.

A minute and a half after leaving the back of the 747, Truly lowered the landing gear as Engle held her steady before flaring to land. *Enterprise* touched down at a speed of 189 knots. After gently lowering the nose, Engle pushed his feet hard against the rudder pedals to test the Orbiter's response to heavy braking. An alarming chatter that had prompted Gordo Fullerton to tell Haise 'Get off the brakes – we're going to break something!' when they first tested maximum braking during their second flight had been corrected. *Enterprise* slowed smoothly, allowing Engle and Truly to test her nosewheel steering as they rolled towards wheels stop.

An unqualified success, free flight four had at every test point confirmed the predictions of the aerodata book. As they emerged, laughing, from the Shuttle's crew hatch on to the ramp stairs, Engle and Truly were wearing Snoopy-style leather flying helmets and goggles. The only thing missing from the ensemble was an aviator's white scarf.

All that remained was for the programme to show that the Orbiter could be brought in to land at a pre-determined point on one of Edwards' concrete runways. Six months earlier, Deke Slayton, thinking about the history of landing X-planes on the lake bed, pointed out to the press that 'twenty-four research aircraft made thirty-eight landings before we had enough nerve to come down on concrete. I'm not sure we have enough guts to land the Orbiter on concrete on flight five.' But that's exactly what they were going to do, with responsibility for the last flight swinging back to Fred Haise and Gordo Fullerton. Dick Truly wasn't sure that the objective this time was a particularly smart one. The secret to landing a glider, he knew, was energy management, not spot landing.

Before Haise and Fullerton got the chance to test the theory at Edwards, however, the Astronaut Office played host to a royal visitor to Houston in the shape of an immaculately dressed Prince Charles. After a tour of Mission Control, the Prince, a pilot himself, was given the chance to fly in the Shuttle motion-base simulator. At the instructor's console in Building 5, he took off his suit jacket, handed it to one of his entourage, and climbed into the flight deck. Sitting in the Commander's seat alongside him, Fred Haise told the SimSups: 'Just a regular session. You don't have to *not* have any failures.'

It was an opportunity, then, thought the instructors, to show off the Shuttle's ability to cope with things going wrong and keep going. Carrying out a simulated launch and ascent, the SimSups began with a flight computer crash then followed it with the loss of an engine. As they heaped failure upon failure, Haise was faced with such an overwhelming task that even his formidable flying skills couldn't save the Shuttle from destruction. The instructors apologized, but the damage was done.

'I guess a kill's a kill,' replied Haise with a wry smile.

Given the opportunity to try his own hand at landing the Shuttle, the twenty-eight-year-old Prince twice let the Orbiter get away from him, porpoising wildly as he struggled in vain to regain control. Crashing into the ground, the heir to the British throne was embarrassed by his failure.

'Naw, it's cos you're not used to the stick and everything,' Gordo Fullerton reassured him.

'Everyone does it,' Joe Engle chipped in.

And it was true: at some point the Shuttle's distinctive flying characteristics had bitten them all.

Over the lunch that followed it was realized that the Prince's schedule took him to California at the same time as the final Approach and Landing test flight. Someone suggested that he might like to come to Edwards and see how it was done. He jumped at the chance. Joe Engle was asked to escort him on the day.

Two days later, Engle showed the Prince around the Dryden hangar. Relaxed and engaged, the royal visitor was clearly fascinated by the exotic hardware on display. And perhaps a little envious, having himself only enjoyed the briefest career as a frontline helicopter pilot with the Royal Navy. You're lucky, he told the astronaut, 'because you don't have to do any of this royal stuff'. But that, of course, was the price he paid to get a ringside seat next to the runway. It only remained for Fred Haise and Gordon Fullerton to kiss the runway in *Enterprise* and bring the ALT programme to a successful conclusion.

Diving in formation with the Orbiter, Bob Overmyer, John Young and Dick Truly flew chase aboard a trio of T-38s.

From the Commander's seat on the Shuttle's flight deck, Haise could see he was hot and high. *Enterprise* was proving to be a little more slippery through the air than predicted. Determined to hit the bullseye marking the touchdown point, he opened the airbrakes to 80% to scrub off speed and pulsed the elevons to try to pitch the nose down.

On board Chase One, Overmyer thumbed the transmit button to call the altitude as *Enterprise* passed through 2,500 feet.

In the cockpit of another T-38, Bob Crippen was at a holding point on one of Edwards' web of taxiways, his engines idling, waiting for *Enterprise* to land before he headed home to Houston. He watched as the Orbiter flared at 2,000 feet, her flightpath flattening, trading speed for height, in anticipation of a smooth 185-knot touchdown.

Haise was still too fast and high. In an effort to make sure he didn't overshoot the mark, he tried the speed brake again and made small inputs on the controller to correct his approach. Then he felt the starboard wing dip a touch. Instinctively, he nudged the controller to the left to level the wings. There was no reaction. He nudged the stick again. *Nothing.*

Haise faced a convergence of problems. There was an inherent lag in the fly-by-wire control system between the commands of the pilot and the reaction of the control surfaces. A quarter of a second at most, but enough for a pilot under pressure to notice. The position of the Shuttle's flight deck relative to the centre of rotation meant it was harder to feel the aircraft's movements through the seat of the pants. Most importantly, the Orbiter's computers didn't have the capacity to cope with every input made by the pilot. The software prioritized pitch – up or down – over roll; if it was busy with the former, the latter had to wait its turn. *Enterprise's* flight control system was saturated.

Crippen watched as Freddo nearly got her down, 10 knots fast, just beyond the aiming point, then saw *Enterprise* balloon 10 feet into the air again. It looked like classic pilot-induced oscillation, the flight control problem that had plagued the development of the Gulfstream Shuttle Training Aircraft. Haise was making further corrections before the plane had responded to his last ones. And as he got behind the Shuttle's movements, Haise's inputs were becoming increasingly firm-handed, pushing the hand controller hard through 50% of its arc of travel.

Three hundred yards further down the strip, *Enterprise* came down hard on her starboard main gear, the heavy hydraulic oleo compressing under the load, then she bounced back 10 feet into the air, before pitching down. Haise was getting behind her.

Sitting alongside him, Fullerton, not, unlike Haise, completely absorbed in wrestling the controls, saw the problem. 'If you get off the stick, maybe it'll damp,' he advised his Commander. Just letting go would de-saturate the computers and allow the Orbiter to settle. Haise relaxed his grip on the stick. Another 300 yards down Runway 22, *Enterprise* touched down safely and rolled to a standstill.

After the flight, a smiling Prince Charles pointed out that much the same thing had happened to him in the simulator. 'You made me feel good,' he told Haise and Gordo. 'You did as badly as I did.'

It didn't help Fred Haise's mood much. Deke Slayton ribbed him, suggesting that he could log an extra landing for the flight. That evening, at the post-flight drinks in Lancaster, a representative from Rockwell presented the astronaut with a model of a broken-winged Orbiter. It was better than sympathy, at least.

Pilot lore has it that a good landing is any landing you can walk away from, but Haise was disappointed not to have put her down on the spot as planned. Gordo Fullerton was quick to point out that it wasn't anything he hadn't seen a hundred times during commercial flights. There was an upside, though. Haise's messy effort to touch down on the mark had given grateful engineers working on the landing gear valuable data that none of the other landings had provided. More importantly, it had exposed a potentially catastrophic problem with the flight control software. The last seconds of the fifth and final flight of the programme had served a more critical function than simply proving the Orbiter's ability to land on a dime.

Between them, Haise, Fullerton, Engle and Truly had amassed just twenty-one minutes and twelve seconds of flight time in *Enterprise*. It was enough to win them the 1977 Iven Kincheloe award, the highest prize for flight-testing given by the Society of Experimental Test Pilots. For their unpowered flights above the Mojave, they were also made honorary members of the World War Two Glider Pilots Association. And although she would tour the US and even, in years to come, visit Europe on the back of the big Boeing carrier aircraft, those short minutes represented the sum total of *Enterprise*'s career as an independent flying machine. Although never to fly free again, in proving the predictions of the aerodata book she had given NASA huge reassurance that the work they were doing was on track. The 150,000lb Orbiter had muscled her way to the ground untroubled by the ripples, eddies and shears that could unsettle the long-winged Shuttle Training Aircraft. All that was encouraging. But during the Approach and Landing test programme, *Enterprise* descended from a maximum altitude of 26,000 feet and reached a top speed of 294 knots. Her successors would return to Earth from an altitude of over 140 miles at more than twenty-four times the speed of sound. When it came to pushing the envelope, *Enterprise* hadn't even come close.

For public consumption, despite the wobble at the end, NASA announced that the final flight had gone 'exactly as planned'. At the same time, however, they instigated a major flight test programme to tackle the issue of the

Orbiter's susceptibility to PIO. Using Dryden's digital fly-by-wire F-8 Crusader, test pilots replicated the effects of saturating the Shuttle's flight control system as Haise had, eventually solving the problem by filtering out the extreme control inputs from the pilot. The software, however, was hardly the only serious challenge faced by NASA engineers in responding to an exhortation from John Yardley, the manager in Washington with overall reponsibility for the Shuttle's development, to 'Get that son of a bitch in space. Get that thing in space or we're going to lose the programme.'

PART THREE

In the Balance

'We risk great peril if we kill off this spirit of adventure, for we cannot predict how and in what seemingly unrelated fields it will manifest itself. A nation that loses its forward thrust is in danger, and one of the most effective ways to retain that thrust is to keep exploring possibilities. The sense of exploration is intimately bound up with human resolve, and for a nation to believe that it is still committed to a forward motion is to ensure its continuance.'

James A. Michener

THIRTY

Houston, 1978

Bob Crippen was in his office in Building 4 when the phone rang. He picked up to hear George Abbey's laconic voice on the other end of the line. 'Hey, Crip,' the Flight Ops boss began, 'the *Enterprise* is coming through; why don't we go out there and take a look.'

With the conclusion of the Approach and Landing programme, *Enterprise* was on her way to Marshall for a series of vibration tests, but with an Orbiter on her back, a direct flight from Edwards to Huntsville was beyond the range of the 747 carrying her. Instead, the 700,000lb combination staged its way across America, guzzling a gallon of jet fuel every time she flew the length of her own fuselage, and took pit-stops along the way. On 10 March 1978, she landed at Houston's Ellington Field.

Crippen and Abbey jumped into his old pick-up and drove to the nearby Air Force base. As the two of them walked around beneath the big carrier aircraft they looked up at the Orbiter towering above them. It was a more familiar sight now, but when he'd first seen it, bolted up there, Crip's immediate thought was *we've screwed up; this is the most awkward-looking aviation configuration I've ever seen.*

Then Abbey got to the point. 'Crip,' he asked flatly, 'how would you like to fly the first one?'

The Flight Operations boss, ever inscrutable, didn't smile. He hadn't even assumed Crippen would say yes. But, starting with Crip, he'd decided he would always assign flights by posing a question, just to make sure that an astronaut was comfortable taking it on. No one once said no.

Crip's broad grin was enough for both of them. The astronaut's only discomfort was in trying to suppress an urge to perform cartwheels and handsprings on the Tarmac. He'd not had any inkling that this was the purpose of Abbey's invitation to see *Enterprise*. Now the feeling was almost overwhelming. *Mindblowing*. He'd be flying with John Young, Abbey told him. That made sense. He'd always imagined that Young would command STS-1 – Space Transportation System One – as the first flight was labelled by NASA, but he'd expected there would be an experienced astronaut like Fred Haise sitting alongside him. Crip never asked why they'd picked him as pilot, but Bob Thompson, the Shuttle Programme Manager, did.

Abbey and Young had had Crippen in mind for over a year. Young, with a record of spaceflight firsts behind him, was, even seniority aside, the obvious candidate. Alongside him Abbey wanted someone who knew the Orbiter's systems, and no one else in the Astronaut Office was more deeply immersed in them than Bob Crippen. The 'Tiger Team' leading the software effort had been headed up by a much-respected Apollo Flight Director by the name of Phil Schaffer. But when stress-related illness forced Schaffer to leave NASA, it was Crippen who took up the reins.

When Thompson asked 'Why him?', the answer was clear.

'Crip,' Abbey told him, 'knows the avionics better than any other person.'

'They said I could go fly the first one!' Crippen was still bubbling when he got back to the Johnson Space Center, his joy infectious. He discovered Joe Engle and Dick Truly had been chosen as the reserve crew for STS-1. The two friends congratulated each other, Crip detecting no hint of Truly's initial disappointment at not winning a seat on the first flight for himself. It hadn't lasted long and, in any case, once Truly had processed it, he knew Crip's selection made sense. There were no hard feelings when an elated Crippen invited everybody to the bar after work to celebrate. One of the MOL guys was going to fly at last. He wasn't going to be the only one.

On 22 March, in front of royal-blue curtains and the Stars and Stripes, NASA introduced the crews of the first four orbital test flights – OTFs – to the press. The eight men sat at microphones behind a long wooden desk emblazoned with the triangular Space Shuttle programme patch. At the right of the line, in a fashionably cut light grey suit and waist-coat, Bob Crippen sat next to John Young. The crewcut Crip had worn in his first few years at NASA had now grown out into a smooth side parting. Tanned and relaxed, he looked like the kind of spaceplane pilot Hollywood might cast. To the right of the STS-1 crew were Joe Engle and Dick Truly, then Fred Haise, chosen as Commander of the third crew, alongside his Pilot, *Skylab* veteran Jack Lousma. Vance Brand, who'd flown with Tom Stafford and Deke Slayton on Apollo-Soyuz, America's last manned mission, had command of the fourth crew. His Pilot was Gordo Fullerton. Three of the first four Space Shuttle flights would have MOL veterans on board. More specifically, it would be those who had worked most intimately with the Orbiter's systems, displays and computers. The seating plan at the press conference seemed to suggest a pecking order, but in fact only Young and Crippen were officially attached to a flight. There was the possibility that there might still be a need for some chopping and changing. Especially as the mission NASA had earmarked for Fred Haise and Jack Lousma came with a deadline that couldn't be extended.

Two weeks before the Shuttle astronauts were announced, NASA had revisited a programme many in and out of Houston had thought was finished. Since February, an eight-man team from Johnson and Marshall had been based at Kindley Field, a US Naval Air Station in Bermuda, preparing to re-establish contact with America's first space station, *Skylab*. Prior to returning to Earth after eighty-four days aloft, the spacecraft's last crew had boosted her into a higher 237-nautical-mile orbit. Since then she'd been hibernating, her systems shut down, circling the Earth in an endless 50° inclination orbit. Except it wasn't endless. Far from it. Since 1974, *Skylab* had descended to an altitude of 215 nautical miles. Now, because of exceptional levels of sunspot activity, that fall was accelerating. The solar turbulence heated the Earth's outer atmosphere, causing warmed gases to rise to an altitude where they clung at *Skylab*, dragging her towards an uncontrolled re-entry that had the potential to rain chunks of what remained of the 85-ton spacecraft on to towns and cities below. In September 1977, NASA had taken the decision to try to save her. And on 6 March the following year, the team in Bermuda, using NORAD air defence radars to find her, pointed a UHF radio signal in her direction that brought her back on line. Back in Houston, Fred Haise and Jack Lousma began training for the rescue mission.

A handful of different options were examined, including pushing or towing *Skylab* into a higher orbit from where she could either be safely de-orbited over the Pacific or reactivated for use with the Shuttle. The latter was gaining momentum among the *Skylab* team, who drew up a plan that might ultimately see the space station recharged, refurbished and expanded to accommodate a crew of as many as eight astronauts. Either way, though, they first needed to arrest and reverse her orbital decay. The method chosen was an inelegant-looking dump-truck-sized assembly of fuel tanks, thrusters and engines, controlled from the flight deck of the Orbiter using closed-circuit TV for guidance. A machine designed with just one function: to provide thrust.

The mission required Haise to manoeuvre the Space Shuttle to within a few hundred feet of *Skylab* before the radio-controlled booster, ejected from the payload bay by springs, was then remotely piloted by Lousma to dock with the slowly turning space station, its docking port describing circles in the sky. Tricky, for sure, *but it was doable*, reckoned Lousma. Furthermore, neither he nor Haise could resist the flight's inherent appeal. *Classic space rendezvous*, enthused Freddo, who likened it to the lunar orbit rendezvous he'd trained for on *Apollo XIII*. Haise said no to a chance of a flight on Apollo-Soyuz because, for all its historical significance, it was not, as far as he was concerned, *a real challenging or exciting, testing mission*. Not like this one. It was his Pilot, though, who best summed it up. The *Skylab* rescue, Lousma had to admit, 'was kind of Buck Rogersish'. Nothing this complicated had been expected to be part of the Shuttle's repertoire at this early stage in the programme. And the two astronauts were in a race against the sun.

Every morning Haise and Lousma arrived at work to be greeted with a picture of the sun, and one of *Skylab*. With every solar event, they could see a corresponding decline in the space station's altitude. Commander and Pilot divided responsibilities.

At Houston, Haise focused on rendezvous procedures and proximity operations needed to orbit so close to the space station. Without a rendezvous radar on board the Orbiter, Haise had to plan the rendezvous with *Skylab* through old-fashioned calculation – a combination of orbital mechanics and dead-reckoning navigation in which a single degree of inaccuracy would lead to a separation between two orbiting objects of a mile in just twelve seconds. If that wasn't complicated enough, the only way to catch an object orbiting ahead is to slow down; speeding up will only have the effect of moving you into a higher orbit, and further away from the object you're trying to run down. Haise anticipated it would take days of phasing to synchronize orbits with the space station.

Lousma, meanwhile, had lead on the remote-controlled booster, travelling between Martin-Marietta, who had the contract to build the booster, in Denver and Marshall Space Flight Center, where engineers had built a rudimentary simulator out of a TV monitor and hand controller.

If the astronauts pulled it off, the Shuttle would, at a stroke, prove its versatility and usefulness. It was exactly the kind of mission it was always envisioned she might perform. But NASA was in danger of getting ahead of itself. As Haise and Lousma trained for a mission they relished, there were sharp differences of opinion about the levels of solar activity that were expected – and with that, when *Skylab* would fall to Earth. NASA was criticized for drawing on data stretching back to the seventeenth century to make their solar predictions, and the first flight of the Shuttle would come no sooner than the summer of 1979. In an effort to try to square the circle, Haise and Lousma were told their rescue mission was going to be bumped up the flight roster to become STS-2 – the second Shuttle flight.

The success of the Approach and Landing tests at Edwards had been headline news around the world. They looked great on television. The message inferred by many was that the Shuttle's first spaceflight was just around the corner. That was misguided. *Enterprise* might have looked like the real thing, but in reality she was a long way from being a spacecraft, let alone a starship. The tiles of the heatshield were entirely absent; the parts of Lockheed's silica tiles were played by polystyrene facsimiles, and the reinforced carbon-carbon nosecap and leading edges by fibreglass. The flight control software loaded on to *Enterprise*'s four computers covered just a fraction of what the orbiting Shuttle was going to have to do. Most conspicuously, the big cluster of rocket bells at the back were fakes, and there was ballast at the back of the aircraft to compensate for the absence of the three heavy main engine powerheads. It was the areas in which she was not representative of the orbital Shuttle – heatshield, flight control software and main engines – that had been identified, from the outset, as being the biggest challenges faced by the programme, and so it was proving. None more so than the engines.

Just successfully completing the first five-second ignition sequence for one of the ground-breaking new Space Shuttle main engines had taken five years since Rocketdyne had won the contract in the spring of 1971. Twelve weeks after that, the engine test rig at Marshall Space Flight Center was able to run an engine at 50% power – the engine's minimum power setting – for a shade over three seconds. It was another year before an engine reached full power. By the time of a review of the engine's progress in the autumn of 1976 it had been hoped that they would have amassed a total of 8,000 seconds of running time. They'd managed a little over 2,000. And by the end of 1977, when a schedule laid out in 1973 had expected the engines to have accrued a total of 38,000 seconds, Marshall and Rocketdyne had managed to claw their way to just over a third of that. The struggle so far, though, paled in comparison to what lay ahead.

The sight of *Enterprise* swooping down from cloudless skies over Edwards had provided a welcome and necessary boost for the programme. It was just as well, because, as alluded to by John Yardley after the completion of the Approach and Landing tests, the Shuttle's challenges weren't just technical. In Washington, the programme really did have a fight on its hands.

THIRTY-ONE

Washington, 1977

AFTER THE ELECTION of President Jimmy Carter in 1976, Hans Mark, whose résumé combined long experience of classified defence research and cutting-edge aerospace technology with being a lifelong Democrat, was offered the post of Under-Secretary for the Air Force. When he accepted in spring 1977, Mark assumed the most senior civilian position in the flying branch. As had his predecessors in the role, he also became Director of the National Reconnaissance Office.

When he arrived at the Pentagon in 1977, Mark found the Air Force and CIA engaged in their traditional turf war, slugging it out between the East and West Coasts like rival rappers. In California, based at Los Angeles Air Force Station, was Major-General John Kulpa, Director of Special Projects for the Secretary of the Air Force and Deputy Commander for Space Operations, Space Division. In Langley, fighting for the CIA, was Leslie Dirks. Mark had high regard for both men, but each wanted to retain control over his own slice of the NRO's pie. And both shared a reluctance, despite it being national policy to do so, to use the Space Shuttle as their launch vehicle. In 1973, a second study into using the Shuttle as the launch vehicle for the KH-9 HEXAGON 'Big Bird' had concluded that it was too costly to make the necessary modifications to the satellite and plans to use the Space Shuttle were abandoned. With the Shuttle yet to fly and a new Titan-34D under development offering even greater lifting capacity than existing Air Force rockets, it was proving too easy for both spooks and bluesuiters to drag their heels.

As a result, the new KH-11 KENNEN digital spy satellite had not even been designed for launch on board the Shuttle. And yet the Shuttle offered a more substantial payload capability than the expendable rockets the Air Force were already using. So resistant were they to the Shuttle that, rather than build a next-generation spy satellite, known as Advanced KENNEN, that made full use of the Orbiter's capacious payload bay, they'd chosen instead to compromise its size and capability so that they could use existing expendable rockets. Using the Shuttle, the new satellite had the potential to carry another 4 tons of the fuel it used for manoeuvring in orbit, which would dramatically extend its time on station and operational flexibility. With the extra fuel on board the satellite became too heavy for a Titan IIID rocket to carry into polar orbit, but it was *exactly* what in 1971 the Air Force had told NASA they needed out of the Shuttle.

'Sorry, fellas,' Mark told the two antagonists, 'that's crazy. You deliberately compromised capability that you could have, because for reasons I don't understand you don't want to use this launch vehicle.'

Ensuring that the hugely complex, sensitive and vastly expensive new reconnaissance satellites were compatible with the Shuttle was far more complicated than just being a question of moving it from on top of a rocket to the payload bay of the Shuttle. It was like redesigning a wedding cake to be supported from the side, rather than sitting on its base. It would come at a cost, they told him, but Mark forced Dirks to have the next-generation Advanced KENNEN spy satellite redesigned, amused by their futile resistance to his decision. *It was their own fault*, he thought; *had they designed the goddamn things for the Shuttle in the first place* ... He wasn't going to allow them to approach the development of a new radar reconnaissance bird the same way.

First, though, he was forced to play Solomon. For all the advanced capability of the KH-11, its cameras were no more able to see through cloud cover than any previous generation of optical spy satellite. To do that, you needed radar. Leslie Dirks proposed adding radar to the existing KENNEN spacecraft. Mark liked the elegance of the CIA's approach. It's possible that John Kulpa did too, but he didn't let that stop him arguing that the Air Force needed its own, stand-alone radar reconnaissance satellite.

That the KH-11 KENNEN was a CIA creation also had a bearing on the decision. Some had been critical of Mark's leadership at Ames for its lack of diplomacy, but he was more than capable of employing it when he needed to. There was certainly merit in not placing all the NRO's eggs in one basket, but keeping the Air Force on side also had value. The problem was the extra cost that giving the Air Force a standalone system would incur. And here, Mark drew on his experience at NASA.

While he was Director of Ames, NASA had worked on the design of a vehicle called the Multimission Modular Spacecraft. The idea was to produce a flexible satellite core, compatible with the Space Shuttle, that would mean each new satellite didn't have to be designed from scratch. If, he thought, the bluesuiters were able to use NASA's common bus as the skeleton of their new radar reconnaissance satellite it might bring the cost down to a level that was competitive with the CIA's proposal. From the outset, the new spy satellite, intended to complement the optical capabilities of KENNEN, was designed to fly on board the Shuttle. Although the programme, codenamed LACROSSE, didn't ultimately follow the route suggested by Mark, it was enough to get it greenlighted. And, initially at least, that meant a further, valuable tie between the Air Force and the Shuttle.

NASA had been sufficiently concerned about the USAF's commitment to building their own Shuttle launch facility at Vandenberg to raise it directly with the Secretary of Defense in 1974. Three years later, with barely lukewarm support from an Air Force that had never been happy in its arranged marriage with NASA, the Shuttle was vulnerable to proposals from the Office of Management and Budget to trim the programme. Just a month after Fred Haise and Gordo Fullerton brought *Enterprise* to a standstill for the last time, the OMB suggested building just three Orbiters instead of the five proposed, and ending the development of the West Coast launch site at Vandenberg. Taken with the latter idea, the General Accounting Office drew up a plan to launch the Shuttle into polar orbits from Kennedy Space Center by employing what it labelled a 'dog-leg' ascent. If the Shuttle launched to the east then rolled on to a northerly trajectory as it climbed, it would be capable of reaching the high orbital inclinations required by the Air Force. The GAO consultants argued that low numbers of Shuttle flights arcing over the mainland United States was a risk worth taking. In February 1978, Mark made it crystal clear that jettisoning two 90-ton solid rocket boosters near Cleveland, Ohio, was not in any way sensible or acceptable, and the plan was abandoned.

At the same time he made the pursuit of the Shuttle launch site at Vandenberg a priority. In support of that, he argued that rather than rely on Houston, the Air Force should develop their own mission control facilities inside Blue Cube at Sunnyvale Air Force Base. He believed that only if he could provide the Air Force with a greater sense of ownership, independence and control over the DoD missions might they learn to love the Shuttle.

On his desk in Room 4C-1000 Mark reputedly displayed a model of a 'blue' Shuttle – an Air Force Shuttle. There was no doubt at all about where he had staked out his own position. With NASA facing technical challenges, delays and cost overruns, they were fortunate to have such a staunch champion for the programme inside the Pentagon. The ex-Ames Director's advocacy hadn't been enough to retain a five-strong Shuttle fleet though. While Vandenberg survived the budgetary mauling, only four Orbiters – the minimum number required to keep the Shuttle alive as a viable military launch system – made the cut. It was better than the three the Washington bean-counters had wanted.

THIRTY-TWO

IN THE LAST few days before the launch of *Apollo XVI*, Command Module Pilot T. K. Mattingly liked to drive down to the pad to take a look at the towering Saturn V booster on which he'd begin his journey to the moon. Stuck in quarantine with little to do but watch TV and videotapes with his crewmates John Young and Charlie Duke, each night, after dinner, Mattingly would slip off to climb a little further up the gantry to get to know their rocket. One night, after reaching the instrument unit near the top of the stack, he spotted a light glowing inside. Mattingly peered in, startling a technician who was working late. The astronaut introduced himself to the worker who, recovering from his surprise, enthusiastically explained the detail of the job he was doing. The two men talked about the mission itself. Responsible for just a tiny part of it, the tech was awestruck by the magnitude of the overall undertaking. 'I really don't know how you're going to get back,' he admitted to Mattingly, 'it seems a long ways off. You know, I don't understand that, but you can bet it won't fail because of me.'

The STS-1 crew, John Young and Bob Crippen, appreciated the importance of this. It didn't matter how big or small the job was, from the testing of the beyond-the-state-of-the-art main engine turbopump to the hand-sewing of the quilts used to provide insulation beneath the Orbiter's thin aluminium skin, if every single person working on the Space Shuttle could say the same thing – *it won't fail because of me* – then the first flight would not fail. Built into the astronauts' schedules were gruelling numbers of visits to contractors, where they met the men and women building the Shuttle and its systems. They called them widows and orphans tours, but a visit from an astronaut was a powerful incentive to everyone working on the vehicle. It personalized the importance of each and every contribution. When Crip talked, encouraged and thanked them, reminding them, with a laugh, that 'it's our butts attached to that vehicle', it made a difference. It cut both ways, though: the astronauts always came away feeling as buoyed by the experience as the people they met. Crippen left each plant knowing *they're going to do their best to make it work right.*

John Young and Bob Crippen flew to Hill Air Force Base, Utah, in a T-38 in October 1978. From there, an Air Force chopper flew them out to the test range outside Brigham City. There had already been two successful development firings of the big boosters, but Young and Crip wanted to see it for themselves. Manufacturer Thiokol invited them to come out to their Wasatch Division facility to see their third live firing: DM-3.

The two astronauts, still wearing their royal-blue NASA flightsuits, were bunkered at a safe distance along with members of Thiokol's management team. Then, in the lee of the mountain range behind, they lit the candle. Crip was wide-eyed at the power on display as over one million pounds of burning ammonium perchlorate, aluminium and iron oxide tore up the mountainside behind the test mount beneath an angry, toxic thunderhead of acidic hydrogen chloride exhaust that rapidly billowed hundreds of feet into the sky. The noise ripped across the scrubland, echoing off the slopes behind, heard more by their whole bodies than their ears. One of the contractors, proud of the awe-inspiring sound and fury just produced by his company, walked over to John Young, looked in the direction of the burning SRB and asked, 'How'd you like to fly one of those?'

'I'd rather fly two,' Young told him.

Each generating nearly three million pounds of thrust, the two solid rocket boosters that flanked the external tank delivered over 80% of the Space Shuttle's launch thrust. And the truth was, Young didn't entirely trust them. Nor was he alone in that.

Unless you count real or imagined pioneers like Chinese nobleman Wang Hu who, legend has it, launched himself into the air on a chair powered by forty-seven black powder rockets in AD 1500, solid rockets had never been used for manned flight. While the intention had been to use them to augment the thrust of the Titan IIIM rocket used to launch the MOL crews into orbit, that prospect evaporated with the programme's cancellation. Partly, this lack of enthusiasm was because, compared to liquid-fuelled rockets, they're inefficient, producing less thrust for an equivalent weight of propellant. But it was also because there's very little to distinguish them from fireworks except for their size. Once they're lit, there's no turning them off or throttling them back. Whatever happens, you're going *somewhere*. But

Right: To Boldly Go. After a campaign by *Star Trek* fans to have the prototype Shuttle named *Enterprise*, Spock, Sulu, Bones and Scotty all attended the roll-out ceremony, but chose to wear casual clothes rather than Starfleet uniforms.

Left: Without jet engines of its own the Shuttle had to be ferried from landing site to launch site by other means. One ambitious, stillborn proposal was for the enormous VIRTUS carrier aircraft built from two B-52 fuselages.

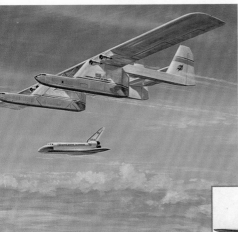

Below: In the end inspiration came from the Short-Mayo Composite, a British idea from the 1930s that launched one aircraft in mid-air from the back of another.

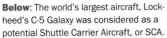

Below: The world's largest aircraft, Lockheed's C-5 Galaxy was considered as a potential Shuttle Carrier Aircraft, or SCA.

Above: A downturn in commercial aviation allowed NASA to acquire a Boeing 747 cheaply. Their ex-American Airlines jumbo is seen here prior to its conversion into its new role as a Shuttle Carrier Aircraft.

Right: To test the principle, a remote-controlled scale model of the 747/ Space Shuttle was built and flown.

Left: To train astronauts to land the Shuttle, two Gulfstream business jets were modified – a process that included the two rectangular fins below the fuselage – to replicate the distinctive flying characteristics of the Shuttle.

Right: The 1977 Approach and Landing test programme would see *Enterprise* fly, launched from the back of the SCA, for the first time. Astronauts Fred Haise, Gordon Fullerton, Joe Engle and Dick Truly were chosen as crew.

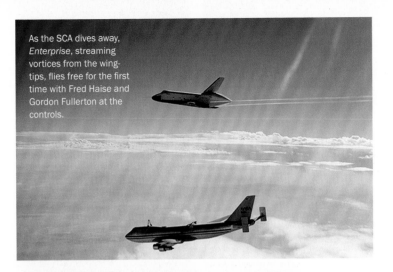

As the SCA dives away, *Enterprise*, streaming vortices from the wingtips, flies free for the first time with Fred Haise and Gordon Fullerton at the controls.

Below: The Approach and Landing test mission patch.

Above: *Enterprise* banks over the Mojave Desert. The fibreglass shroud over the rocket engines at the back was to make her more aerodynamic.

Right: Fred Haise guides her in for her first landing on Rogers Dry Lake Bed at Edwards Air Force Base.

Above: Job done. Haise, Fullerton and *Enterprise* after their successful first flight.

Right: *Enterprise* enjoys an airborne salute from the 747 carrier aircraft and one of the fleet of NASA T-38 jets that had accompanied the first flight as chase planes.

ALT-4. Joe Engle and Dick Truly were in the cockpit for the fourth flight, and the first without the shroud. Truly couldn't help but be surprised at the steepness of the Shuttle's descent.

Right: The fifth and final ALT test flight exposed dangerous problems with the Shuttle's fly-by-wire control system when *Enterprise* began to roll and porpoise wildly as Fred Haise tried to bring her in to land.

Above: NASA introduces the crews for the first four Space Shuttle orbital test flights. John Young and Bob Crippen, far right, were named as the primary crew for the STS-1, with Joe Engle and Dick Truly as reserve.

Above: The astronauts spent endless hours in the simulators, training to be ready for any contingency. Here Fred Haise, *Apollo XIII* mission patch to the fore, sits in the Commander's seat.

Haise and his Pilot, Jack Lousma, trained for a 'Buck Rogers' mission planned to return NASA's vacant *Skylab* space station to a safe orbit, but delays to the Shuttle programme saw it scrubbed. *Skylab* was lost when she burned up on re-entry in 1979.

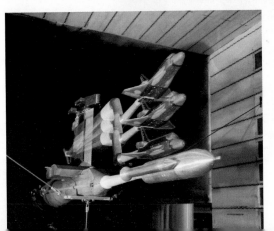

Left: No stone unturned. While the astronauts practised abort procedures in the sims, NASA used wind tunnels to test whether in an emergency it might be possible to separate the Shuttle from the stack at hypersonic speeds while the SRBs were still burning. They thought not.

Above: Showstopper. *Columbia*, the first Shuttle built for spaceflight, arrives at Kennedy Space Center on the back of the SCA in 1979. As can be seen in the picture, the tiles making up the heatshield were falling off, a problem which, if not fixed, would ground the Shuttle.

Right: Once it was believed a solution to the tile-bonding problem had been found it had to be tested. To do this NASA used the new process to fix tiles to F-104 and F-15 jet fighters and then flew them at high speed.

Left: With the fix tried, tested and proven, most of the tiles making up *Columbia*'s heatshield had to be removed then re-attached.

Below: NASA's nervousness about the robustness of the Shuttle's heatshield led to the development of an on-orbit tile repair kit, shown here being tested by Dick Truly.

Above left: The eighty-five-ton *Columbia* hangs from the ceiling of the Kennedy Space Center's vast Vehicle Assembly Building prior to being mated to the external tank and SRBs.

Above right: Max speed: one mile per hour. The Space Shuttle stack begins its 3½-mile journey from the VAB to Pad-39A on the back of an eight-track, 3,000-ton crawler-transporter.

Below: Ready to go. *Columbia* sits waiting on Pad-39A.

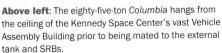

STS-1 Mission patch.

Above: Mission Commander John Young in the T-38 jet he flew from Houston to KSC for the launch.

Left: Mission planning. After a five-mile run Pilot Bob Crippen sits down with Young to study the STS-1 flight plan.

Right: 'Any landing you can walk away from …' Bob Crippen steps out at KSC after another long session shooting practice approaches in Shuttle Training Aircraft in the days prior to launch.

Below: Looking like test subjects in the lab, Bob Crippen and John Young suit up for the mission.

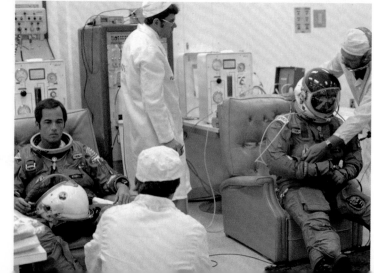

if one of the Shuttle's big boosters were to fail to ignite, 2.8 million pounds of thrust from the other would send the Shuttle stack cartwheeling off the pad.

John Young knew better than most that solids had not always enjoyed the best reputation for reliability. His wife, Susy, had worked on the USAF's Minuteman ballistic missile programme in the early sixties and remembered only too well their tendency to self-destruct. That the Shuttle would be using two only further complicated the situation. One of the biggest challenges in developing solids into dependable, predictable rocket motors was in getting them to ignite and fire evenly. Each SRB had to be capable of going from nothing to 1.9 million pounds of thrust in less than half a second. Beyond that, tolerances for any ebb and flow in thrust were measured in tens of milliseconds. Facing the same sort of issue, Wang Hu reportedly addressed the problem by employing forty-seven people to light his rockets in unison. Thiokol simply used another solid rocket as an igniter. The size of a dustbin, the igniter sprayed fire down the hollow centre of the SRBs' tubes of fuel, torching the surface of all the propellant at once to ensure it burned evenly.

Young would be relieved once the SRBs had successfully served their purpose, had been jettisoned and were on their way back to the Atlantic beneath parachutes. They were an all-or-nothing solution that had been largely driven by cost. He'd signed up to them, but it didn't mean he had to like them. He was happier wrestling with systems to which he could bring more influence to bear. 'If you don't like grungey engineering jobs,' he maintained, 'I don't think most people would care much to be an astronaut. You gotta like messing with systems.' He did. As well as ferrying around the country visiting contractors, Young and Crip were putting in serious classroom time together for lectures and briefings by subsystems experts. While they and the other three named crews trained to fly, their colleagues each made their own contribution to the Shuttle's development; but with barely twenty-five astronauts left in the office, and faced with the complexity of the Shuttle, they were stretched thin and in danger of being overwhelmed. New blood couldn't arrive soon enough.

NASA, fortunately, was ahead of the game. In the summer of 1978 the first fresh intake of new astronauts since the sixties had arrived in Houston. And many of them were cut from the same cloth as their predecessors.

THIRTY-THREE

Vietnam, 1972

As SOON AS they went feet dry – screaming in over the beach from the Gulf of Tonkin – the shooting started. The squadron boss had told him the sortie would just be a gentle introduction to the war, a chance to dip a toe in the water. Flying down low, escorting a North American RA-5C Vigilante on a reconnaissance mission, pilot Jon McBride gripped the stick of his big McDonnell-Douglas F-4 Phantom and pushed on through the barrage of flak: anti-aircraft artillery, small-arms fire and surface-to-air missiles. He'd never experienced anything like it. For thirty minutes he stayed with the speeding 'Vigi', nudging the Phantom in and out of afterburner to keep up. Clean and unburdened by missiles, the big RA-5C slipped through the air, cruising near supersonic speed in military power, while McBride's fighter gulped fuel. *We damn near got hit fifty times*, he thought as they finally swept out of North Vietnam and back towards the USS *Saratoga* to land. If, that is, he managed to plug into a tanker before the two J79 engines gasped and expired when his tanks ran dry. McBride had been on the ship for just twenty-four hours. It was his first combat mission. 'Skipper,' he told the VF-103 squadron boss on his safe return to the Yankee Station, 'I really appreciate the warm-up . . .'

Getting to Vietnam hadn't been easy. After a career flying jet fighters, McBride was surprised to find himself assigned to fly low, slow Rockwell OV-10 Bronco twin-props to provide close air support for troops on the ground. Right out of test pilot school at Naval Air Station Patuxent River, though, he volunteered to fight in Vietnam, and the Bronco squadron was, at the time, the only route available. But when, before he deployed, the Navy disbanded it, McBride returned to the Phantom and the prospect of a Mediterranean deployment. Then the Admiral called. 'We've just lost a crew in Vietnam,' he told McBride, 'would you like to go?'

Dan Brandenstein was pushed back into his ejection seat as he was catapulted from the deck of the USS *Constellation* on another night-time bombing mission. Brandenstein loved flying the Grumman A-6A Intruder. The jet was no beauty, but she was a tough old bird, capable of flying in foul weather that kept everyone else chained to the deck. The night launch was a bonus: Brandenstein always felt more comfortable knowing that the enemy couldn't see him. As his heavily loaded aircraft accelerated away from the ship, pushed through the dark by 18,000lb of thrust from the two Pratt and Whitney J52 turbojets, it suddenly lost all electrical power. Without hesitation, Brandenstein reached forward to pull a handle to release the ram air turbine, a little propeller that dropped out of the fuselage to generate power from the airflow. Brandenstein and his Bombardier, sitting alongside him, didn't know what had caused their electrical failure, but the RAT seemed to have taken up the slack. They decided to press on with their mission over enemy territory.

You never went round again. That was the rule. But on this occasion, in the cockpit of his A-6 Intruder, Rick Hauck just saw red. Instead of the usually wayward fire that greeted them in the dark skies over Laos, tonight the North Vietnamese gunners were gunning at him and his Bombardier with worryingly accurate radar-laid anti-aircraft artillery. *They're trying to kill me.* After completing his bomb run and climbing out of range, he thought, *we've got two more anti-personnel weapons we can drop; let's go down and throw 'em*. He tipped the A-6 on its wing and carved down towards the triple-A site. Then he switched on the Intruder's bright exterior lights in the hope he could provoke the enemy gun battery into revealing itself. Flying attack missions off the USS *Coral Sea*, Hauck thought he was bulletproof.

McBride, Brandenstein and Hauck applied for astronaut selection in the autumn of 1976, travelling to Houston in the summer of the following year after NASA had filtered a total of over 8,000 applications into a more manageable shortlist of around 200. For a week in that summer of 1977 the candidates were subjected to medicals, psychological tests – *Have you ever wanted to kill yourself? Were you ever abused as a child? If you could be reincarnated what would you come back as?* – formal interviews with a selection board, and the experience of being zipped, fetal, into a small dark fabric ball designed to transfer crew on orbit from a damaged Shuttle to a rescue vehicle, then left without word of when they'd be let out. That would weed out the claustrophobics. And at the end of the day, they were invited to the Pour

House for a beer with Dick Truly, Bob Crippen, Joe Engle and George Abbey. It was another chance to see how people fitted in. At the end of each selection week, in an effort to make the Navy applicants feel welcome, Dick and Cody Truly hosted dinners at their home.

Cody got good at talent-spotting. 'Rick,' she said, pulling Hauck aside before he left at the end of the evening, 'I'll bet you a fifth of Chivas Regal Scotch that you are invited to come down here.'

He grinned. 'How can I lose with that bet?!'

McBride, Brandenstein and Hauck were selected as astronauts in January 1978. They were exactly the kind of men Deke Slayton was hoping to recruit. Once it became operational, NASA planned to fly the Shuttle every couple of weeks. To ensure there were flight-ready crews, Slayton had argued for the creation of an elite cadre of around twenty Shuttle astronauts – a squadron – who could fly in rotation as many as six or seven times a year, growing in experience and skill and, as a result, he thought, reducing the amount of training. It was the only way, he reckoned, of treating the Shuttle like a truly operational vehicle. It didn't work out quite as Slayton had hoped.

When thirty-five new astronaut candidates, all of them interviewed over the summer of 1977, arrived at Johnson for day one of their new job in July 1978, there were plenty more with credentials similar to the three Navy Vietnam veterans. One Air Force pilot had flown 324 classified missions over Cambodia out of Bien Hoa AFB in the little Cessna A-37 attack jet that Joe Engle once flight-tested at Edwards; another had flown top cover for the final evacuation of the US Embassy in Saigon with the fleet's first frontline Grumman F-14A Tomcat squadron; there was an

Air Force tanker pilot who'd listened to reports of Neil Armstrong walking on the moon on *Voice of America* from the cockpit of a KC-135 over Laos; and a black Air Force combat search-and-rescue helicopter pilot who'd been prompted to apply by a NASA advertisement featuring *Star Trek*'s Lt Uhura – actress Nichelle Nichols. But while the astronaut corps had been, traditionally, made up principally of test pilots, NASA was not, this time, only looking for more of the same. Actively encouraging applications from minorities and women, there were, alongside the men with the Right Stuff, physicists, astrophysicists, geologists, astronomers, engineers, surgeons, an oceanographer who'd previously had her heart set on piloting the Alvin sub-mersible to the depths of the ocean, and an Assistant Professor of Civil Engineering who'd been equipping calves with artificial hearts. There was even an academic with a criminal conviction. And a beard. The oldest of them was thirty-nine, the youngest just twenty-six. The starting salary was $11,000, though most of them were no longer listening when, after learning they'd been selected for astronaut training, they were informed of the terms and conditions of government work. They'd have done it for nothing.

The new astronaut candidates, learning the NASA acronym habit early, christened themselves the TFNGs. For public consumption, this stood for 'Thirty-Five New Guys', but in reality they revelled in the earthier appellation 'The Fucking New Guys'. At a stroke, their arrival changed the character of the Astronaut Office. One of the TFNGs noticed that within a month or so, the habit of calling everyone by their surnames was changing. Deke Slayton, labelled a GVA (grizzled veteran astronaut) by the irreverent new arrivals, had got more than a squadron, he had a university.

The TFNGs were welcomed by Chris Kraft, George Abbey and John Young in Room 966 on the ninth floor of Building 1. Young, his eyes darting around the room, evidently not entirely at home standing in front of an audience delivering a speech, nonetheless left them with a few words of sage advice before he wrapped up: 'Don't talk about nothing you know nothing about.'

THIRTY-FOUR

THE COCKPIT SHUDDERED as the Shuttle passed through Mach 1. Young and Crippen shook in their seats a little as the turbulence of passing through the sound barrier layered over the persistent vibration from the big solid rocket boosters.

You're go at throttle up.

Roger, go at throttle up.

With that, the three main engines returned to delivering 100% of their rated thrust. Prior to the point at which the Shuttle went supersonic, the three main engines were throttled back to 65% so as not to exceed the structural limits of the stack in the thick lower air. Past the point of maximum dynamic pressure, as the speed continues to grow, the air begins to offer less resistance until, all being well, it releases its grip completely. A minute after the Shuttle's main engines throttled up out of the thrust bucket, a line reading 'Pc<50' on the screen displaying the back-up flight software began to flash double bright. The pressure inside the boosters' combustion chambers had now fallen below 50 psi, meaning their fuel was spent. The computers initiated the separation of the big solids.

You're go for SRB sep.

Roger on the sep.

Young and Crippen felt a firm thump through the cockpit floor as a series of small rockets propelled the two redundant boosters away from the stack. The ride became smoother. Until, less than thirty seconds later, the master alarm went off.

To the right, on the panel beneath the right-hand CRT display, one of the main engine shutdown lights blinked red. They'd lost a main engine. At this point, a little over two minutes into the ascent, they wouldn't make orbit with an engine down. Young reached forward with his left hand and stabbed at the master caution light on the coaming above the instrument panel. With his right he moved the rotational abort mode controller one notch to the right to 'RTLS', then punched the abort switch next to it. Return to launch site.

Now things got busy.

Crip flipped through the ascent checklist cue cards. With the RTLS abort sequence initiated, a pair of smaller rocket engines mounted in pods above the three main engines ignited to burn off their fuel. Without emptying the tanks, the Orbiter's centre of gravity would sit too far back for her to fly. At the same time, the flight control software increased the angle of their trajectory, in an effort to steal as much height as possible from the power of the two remaining engines.

The Orbiter was flying upside down beneath the massive bullet-shaped external tank, arcing backwards like a high-board diver. The flight profile was mainly to ensure the right loads and thrust vectors. But it also enabled the Orbiter to perform the RTLS abort manoeuvre. In a nominal ascent, the higher they climbed, the flatter their trajectory became, trading rate of climb for acceleration, as they powered towards an orbital speed of 17,500 mph. But with the abort sequence initiated, the change in pitch needed to become more radical. The nose of the Shuttle needed to pitch down through the arc of a circle in a manoeuvre called the powered pitcharound. From arcing elegantly into space, the astronauts needed to pitch the nose of the Shuttle over through 180° so it was facing away from the direction of travel, then use the power from the remaining two engines to slow her progress. Travelling backwards through the sky, her rocket bells and tail leading the way, she was no longer a flying machine but a projectile, only the sparsity of the upper atmosphere preventing her from being ripped apart by dynamic pressure.

Through the cockpit glass the curve of the horizon swung away from view, replaced by the blackness of space as the nose of the Orbiter pitched round.

Although still speeding further away, her nose was now pointing back towards the launch site. And the two remaining main engines, now operating at a maximum 104% of their rated power, began to slow her progress as, nose pitched down, her horizontal speed came to a stand-still. Stalled, she then began to gather forward speed from the thrust of the two big rocket engines as she fell towards 200,000 feet.

Next, Young had to shut down the main engines and jettison the external tank without it smashing straight back against the belly of the Orbiter. At such a high angle of attack, the dynamic pressure was pushing the tank up against the belly of the Orbiter. To ensure a clean separation, Young needed to jettison it as he pitched the nose down at a rate of 3° per second so that the tank was actually shoved away from underneath them – 'punched off' in test pilot speak. Young cut the two remaining main engines and called 'MECO' – main engine cut-off – then applied forward pressure on the stick, to lower the nose. All three main engine lights glowed red. The Orbiter bucked as a series of explosive bolts fired, severing the connection with the external tank. Immediately afterwards, thousands of pounds of propellant remaining in the Orbiter's pipes were dumped. Young, Crippen and the Orbiter were now on their own, gliding to Earth. Young struggled to suppress the Orbiter's yaw as he tried to settle the harassed Shuttle into a stable descent. The worst was over. The engine failure had come late, but they'd been lucky. This time they hadn't gone so far that they couldn't still make it back to the runway at Kennedy.

It worked in the simulator. It was supposed to – no point practising it if it couldn't be done. But it was certainly more than possible to screw it up, and both Young and Crip had augered in often enough as they trained to fly the tough, high-speed semi-aerobatics demanded by the return to launch site abort. If they got the procedures right, though, the sim said they would survive. Whether or not they'd enjoy the same outcome in the event of a real-life engine failure was a subject of great debate within the Astronaut Office.

From the outset, abort procedures had been high on the astronauts' list of priorities. In the Gemini capsule, the crew had ejection seats. For the Apollo missions, the capsule sat beneath a distinctive scaffold escape tower conceived by Max Faget which, in the event of the Saturn V booster failing, pulled the capsule free of the stack and carried it out of harm's way before parachutes in the nose carried it to the ground. In the Shuttle it wasn't possible. The crew compartment and the rest of the Orbiter were one and the same. Ejection seats were a possibility for the two pilots on the flight deck, but the Shuttle would carry crews of up to seven people, half of whom were on the mid-deck below. It was impossible to provide them with a means of rapid escape. They had to find a way of ensuring that the Orbiter itself provided the crew's means of escape. Their vessel was its own lifeboat. They had to stay with her.

When suggested abort procedures first came in from the contractors bidding to build the Shuttle, Fred Haise and Joe Engle were among the astronauts poring through piles of paperwork. Haise chuckled as he thought through the unlikely set of manoeuvres one of the plane builders had imagined the crew might perform to escape disaster. 'You couldn't do that,' he remarked. 'You couldn't go off the booster, go over on your back, twist it around, come back and land.'

Joe Engle looked up. 'It's a piece of cake,' he quipped. 'I did it down at Galveston flying a T-38 down the beach the other night!'

'Bet you couldn't do it in a fog . . .' Haise chipped back.

Once the configuration of the Shuttle was finalized, the crews had to cope with the hand they'd been dealt. Much of the difficulty lay with the characteristics of solid rockets. Being able to throttle the engines had been considered a prerequisite for human spaceflight. When, during the ascent of *Apollo XIII*, one of the Saturn V's massive F-1 engines developed a fault it was shut down immediately, ending any further danger its malfunction might have caused. While two solid fuel boosters strapped to the side of the Titan IIIC rocket earmarked for the MOL programme had a facility to terminate the thrust by blowing the nosecones off the rockets, the Shuttle's SRBs did not. Not only would the system's inclusion have incurred a weight penalty, it was also feared that by triggering it and blowtorching the path ahead of the Orbiter it might only reduce the crew's chances of survival.

Instead of trying to tame the blazing, unstoppable solids, NASA turned the problem on its head and began to regard them as the solution. The big boosters could actually provide the astronauts' means of escape. As long as the boosters were burning the Shuttle would continue to climb to altitude, and, of course, if it had insufficient power to reach orbit, the winged Orbiter was designed to return a crew safely from altitude. It was what it was *for*. In the event of a main engine failure preventing them from reaching space, the big solids themselves would perform the function of Max Faget's escape tower, propelling the crew to an altitude where they had, relatively speaking at least, time and options.

If they were high and fast enough when the main engines quit, there was the possibility of either an abort to orbit – ATO – in which the Shuttle was able to reach a safe orbit, even if it was lower than planned and perhaps meant the mission objectives were scrubbed; or there was the abort once around – AOA – in which the Shuttle, failing to reach orbit, had sufficient height and speed to complete a single revolution of the globe before re-entering to land. But losing the engines too soon would leave Young and Crippen having to pull off the demanding return to launch site – RTLS – manoeuvre to get back to the Cape.

Joe Engle didn't like it at all. What he baulked at was surrendering the Orbiter to the mercy of the remaining main engines. In performing the powered pitcharound necessary to present the rockets in the direction of travel in order to slow down, you had voluntarily turned a good aeroplane into an unflyable, unaerodynamic ballistic missile. You'd relinquished control. Many of the other astronauts weren't even sure the manoeuvre was survivable. *It would be a miracle*, thought Dick Truly.

'*Six* miracles followed by an Act of God,' John Young called it.

It was sporty, for sure, Bob Crippen argued, but he tried to make a case for test-flying the RTLS manoeuvre to prove it actually worked beyond the cockpit of the motion-base simulator. He was on his own, though. Young thought the idea was nothing more than Russian roulette.

'We don't need to practise bleeding,' he said, shutting down the conversation.

In the Pentagon, evidence was beginning to emerge that the Soviets might soon be facing similar dilemmas.

The National Reconnaissance Office had launched its second digital spy satellite from Vandenberg on 14 June 1978, the first since Hans Mark's appointment as the intelligence agency's Director. But when Mark returned home from his office on the third floor of the Pentagon, news of the successful launch of the KH-11 wasn't something he could share with his wife, Marion. Being the Director of an organization that was not even acknowledged to exist could present challenges. Fortunately, he maintained, he'd married someone smarter than he was. Presented with the ominous-looking red telephone that had been installed in their bedroom since the move to Washington, Marion simply teased him by asking, 'Which one of your girlfriends are you calling now?'

Like the KEYHOLE satellites themselves, another crucial addition to the NRO's capabilities was similarly off limits at home. Mark's directorship also covered the Aerospace Data Facility East, a new NRO ground station that opened in Fort Belvoir, Virginia, in 1977 – the site where the images taken by the KH-11 KENNEN satellite were processed after being relayed from space via a network of SDS communications satellites. The top-secret two-storey windowless block in Virginia was designated Area 58.

In 1978, the NRO's increasingly capable spy satellite network recorded evidence of new building work at the Soviets' Baikonur launch site. Not only was a pad that had last been used for the USSR's catastrophic attempts to launch a moon rocket to rival the Saturn V being redeveloped, but a long runway was under construction nearby. Reports on a Soviet Shuttle in respected magazines like *Aviation Week* and *Flight International* had begun to appear with increasing regularity and conviction, and the latest KEYHOLE imagery lent weight to them.

Between October 1977 and September 1978, the Soviets conducted flight-testing of a small spaceplane prototype built by design bureau Mikoyan-Gurevich. Launched from beneath the wing of a Tupolev Tu-95 'Bear' bomber, the little MiG 105.11 'Lapot' was not itself capable of reaching orbit, but was instead a proof of concept. Enjoying much in common with the American lifting bodies flown at Edwards in the sixties, it was also an aerodynamic prototype of the Soviets' own cancelled sixties spaceplane, Spiral. When the time came to make a decision about the configuration of a rival to the US Shuttle, there were powerful voices arguing that, with years of testing and research behind it, Spiral should form the basis of the Soviet Orbiter.

Former Mikoyan engineers produced a proposal for a scaled-up Spiral capable of carrying the same load into orbit as the NASA design, but in the end they were outvoted in favour of apeing the Space Shuttle. 'The Americans,' those working on the programme were told, 'are not dumber.'

It was hoped that by closely following the layout of the Rockwell-built Shuttle, they would save both time and money in achieving their goal. In order to do so, the KGB embarked on a large-scale and systematic campaign to acquire NASA documents and studies by contractors. The CIA would later single out the heatshield and avionics as being of particular interest to the Soviets, concluding that by the end of the decade they were prepared to pay over $140,000 for information on the latter. The company had also done what it could to feed the Russians with mis-information. When it came to the way the Soviet Orbiter looked, though, there was nothing they could do to deflect them from pursuing a machine that, when the Soviets froze their design in the summer of 1978, looked strikingly similar.

Ironically, Gleb Lozino-Lozinskiy, the leader of the Mikoyan team behind the pitch for the 'Big Spiral' develop-ment, would end up as head of the design bureau building the new spacecraft. Forced to explain the decision to follow the Americans so closely, he'd argue, 'if you have the second design of a bicycle and it is a much superior bicycle, it would still look like the original bicycle'. But there was one key decision made by the Russians that, by the end of 1978, was looking as if it might turn out to be a smart one: the Soviet Shuttle would not, unlike the American spacecraft, include a trio of powerful, cutting-edge technology rocket engines in its tail.

THIRTY-FIVE

Pearl River, Mississippi, 1978

IN LATE DECEMBER a massive fire tore through the main engine test stand at NASA's National Space Technology Laboratories on the banks of the Pearl River, Mississippi. For over three minutes the rocket appeared to be running normally until without warning sheets of orange and yellow flame erupted from the machinehead, engulfing the engine and test rig more completely than any previous fire. Such was the damage that it halted further testing of the Space Shuttle main engines for a month while the test stand was rebuilt.

For two years NASA engineers had worked in parallel with their counterparts from Rocketdyne, conducting separate tests at a facility in Santa Susana near LA, to try to build an engine that didn't rip itself to pieces. For the most part the problem lay with the turbopumps needed to deliver fuel to the engine's combustion chamber at the required pressure. No bigger than a Christmas cake, the turbines inside the pump had to cope with reaching 38,000 rpm from a standstill in a fraction of a second, then endure the massive centrifugal forces that followed. They were cracking under the strain. And when they failed, the engine, basted in liquid hydrogen and oxygen, would self-destruct so completely that there was often little left to analyse but splashes of molten metal.

The Space Shuttle, the astronauts had been told, would be as 'reliable as a DC-8' airliner. After hearing more about the challenges faced by rocket engineers, John Young and Bob Crippen didn't come away feeling entirely convinced that they were looking at DC-8-level reliability. Yet the Shuttle main engines were deliberately being tested to destruction. That was the very method through which points of failure – and those aspects of the design that needed fixing – revealed themselves. 'We never learned as much,' said J. R. Thompson, the engineer running the main engine programme at Marshall Space Flight Center, 'as we did during the investigation coming out of a failure.' He was going to keep blowing up the engines, identifying the cause of the explosion, addressing it, then running the engines again until they stopped destroying themselves.

With the airframe itself, however, NASA made a decision to take the opposite approach.

On display in engineer Tom Moser's office was a ceramic eagle given to him by Programme Manager Aaron Cohen, a thank you for saving $100 million from the Orbiter's development programme. He'd earned it first by successfully making the case that the Shuttle did not need the same structural strength as the average commercial airliner or jet fighter, a conclusion that had been endorsed by aviation industry experts. But it still needed to be tested. Initially it seemed sensible to follow the methods aeroplane builders used to test new designs, and so, as was the case for any new commercial airliner design, two extra airframes were planned.

The first was to test for metal fatigue, a problem that had revealed itself through the tragic, puzzling losses of the world's first jet airliner, Britain's elegant De Havilland Comet, in the early 1950s. At the time the cause of the accidents was unknown, but wreckage salvaged from the bottom of the Mediterranean revealed that the aircraft's cabin had exploded with such sudden violence that blue and gold paint from the fuselage was etched on to the outside of the wings. Tests at Farnborough's Royal Aircraft Establishment traced the cause to cracks forming in the corners of the airliner's square passenger windows. In the same way that a spoon bent forward and backward too often will eventually snap, the fuselage of an airliner, subjected to tens of thousands of hours of flying, will weaken.

The Shuttle was different. Designed to fly just one hundred times, fatigue was never going to be a problem. So Moser successfully argued that the airframe for fatigue testing be cancelled.

Then Moser's Structures and Mechanics Division turned its attention to the other test article. Alongside fatigue testing, the aviation industry also provided an airframe specifically to test the ultimate limits of any new design's structural strength. Pressure was applied using hydraulic jacks until something snapped. It was an inherently damaging process, and when a fleet of only four Orbiters was planned, the Shuttle programme could scarcely afford the luxury of bending what was otherwise a perfectly serviceable airframe. In the hope of preserving it for possible conversion into an operational Shuttle, Moser proposed that, instead of actually testing it to destruction, they go only part of the way, then extrapolate their results to pre-predict whether or not it would survive the full loads it was built to withstand. This answer came back from the NASA Chief Engineer's office: 'You guys are crazy as hell. We're not going to accept that.'

Moser reassured them, providing the analysis done by his team to support the argument. Again, experts from commercial planemakers like Boeing and Lockheed were called in to go over the numbers. They were unable to find fault with Moser's suggestion. And so in August 1978, a bare aluminium Orbiter was caged inside a 430-ton steel frame at Lockheed's Palmdale plant and tested to 20% less force than it was designed to cope with. There were no surprises. The test airframe emerged sufficiently unscathed to go on to form the basis of the second Space Shuttle, and the expected cost of building then breaking two perfectly good Orbiters was saved. Moser gratefully accepted the ceramic eagle from his boss.

There was just one cause for concern brought to light by the testing done at Palmdale: a structure in the nose supporting the propellant tanks for the forward reaction control system – the jet thrusters that were to be used to manoeuvre the Shuttle in space – failed. A redesign was approved, but it was a part of the spacecraft that was already a cause for concern elsewhere in the Shuttle programme office.

Fred Haise felt that the stresses that might be endured by the forward fuselage of the Shuttle at launch were not sufficiently well understood. Unlike Tom Moser, though, he found himself on the wrong side of the argument. It wasn't structural loads that worried Haise as much as acoustic stress. The decibel levels experienced by the Orbiter would be worse than those endured by Apollo. To try to replicate them, the test centre at Marshall Space Flight Center had to acquire more powerful acoustic horns. The volume alone – the acoustic vibration – from the engines would be enough to kill anyone unlucky or foolish enough to be standing within 800 feet of a launch. When a decision was made to do without a vibracoustic test article for the front of the Orbiter, Haise told Aaron Cohen that it should be reconsidered. He was anxious, not so much about the external structure of the Orbiter, but about the vulnerability to acoustic stress of the computers and electronics in the lower part of the nose. While Cohen took Haise's objection up the chain, the decision not to test it was to stand. Haise remained uncomfortable about it.

It would turn out that he was right to be. But by the time the problem revealed itself again, Freddo was long gone. And it was nothing so mundane as battles with management that prompted his retirement from the Astronaut Office. It was the sun.

On 15 December 1978, NASA told the President that *Skylab* couldn't be saved. As the prospective date for the Shuttle's first flight moved right, heightened solar activity saw the inevitable demise of the space station move left until a point when two schedules crossed. With that, the Buck Rogers mission that Haise and Jack Lousma had been training for so enthusiastically was gone.

Disappointed, Freddo hung on for a while, until in early 1979 he took a job running space programmes for the Grumman Aerospace Corporation. Concerned that his departure wouldn't disrupt preparations for the early Shuttle flights he suggested that George Abbey move his Pilot, Lousma, into the Commander's seat. Abbey and John Young agreed, and in the reshuffle that followed they assigned

Gordo Fullerton to be Lousma's Pilot, and lined up T. K. Mattingly and Hank Hartsfield as Commander and Pilot of a fourth crew.

There were now veterans of the old Air Force MOL programme sitting in the right-hand seat for all four of the Space Shuttle's planned orbital test flights.

THIRTY-SIX

Palmdale, California, 1979

SHE WAS CHRISTENED *Columbia*. The first Space Shuttle Orbiter built to travel into space was rolled out of Rockwell's Palmdale plant in March 1979. From the front, muscular and squat, her wide, blunt nose had the purposeful appearance of a catfish snout. She'd lost the long white aerodata boom that had extended from in front of the cockpit of *Enterprise* and instead gained a rakish Lone Ranger mask of black thermal tiles around the flight deck window. She and the three Orbiters that followed her were to be named after famous sailing vessels. *Columbia*, an eighteenth-century sloop, was the first American vessel to circumnavigate the globe, her name (after Columbus) considered to be the female embodiment of the United States. *Challenger*, *Discovery* and *Atlantis*, all three sailing ships that had conducted pioneering scientific research, would follow. With the names made public, NASA at least forestalled any possibility of *Star Wars* fans, encouraged by the success of their *Star Trek* counterparts, mounting a letter-writing campaign to get one of them named the *Millennium Falcon*.

As *Columbia* emerged for the first time from the hangar

into the California sun, though, she was like a ship without a keel. She may have looked the part but she was a very long way from being ready to fly. For now that didn't matter. NASA, determined to at least give the appearance that the programme was still forging ahead on schedule, had ordered her premature delivery to Kennedy Space Center, the launch site, where there was a waiting army of Rockwell employees, hired to prepare the Orbiter for flight, who now had the job of finishing her. But if public relations was behind the decision to move her from Palmdale, then *Columbia*'s first flight across country, attached to the back of the 747 carrier aircraft, was to have a calamitous effect.

As *Columbia* was flown over Houston, Deke Slayton took up a T-38 from Ellington Field to get pictures of her in the air. While a photographer in the back seat of the cockpit snapped away, Slayton cast an eye over the apparent wreckage of a heatshield he reckoned *would crumble if you looked at it wrong*. Thousands of tiles were missing. Of the near 31,000 that formed the major part of the heatshield, many had not been attached before the Orbiter was ordered to Kennedy. A short test flight from Palmdale prior to embarking on the transcontinental journey had ripped off nearly all the polystyrene substitutes she'd had stuck on in order that she look the part for her public. She was a sorry sight by the time she arrived at Kennedy. Damaged in transit. Even Crip, one of the two men training to fly her, had to admit *she was in a pretty bad state*.

In the Structures and Mechanics Division at Houston, it wasn't news to Tom Moser. He'd tried to persuade Rockwell to test the structural integrity of the tiles, rather than just their thermal properties. But, squeezed by budgets, they told him, 'We're not analysing those damn tiles like a piece of structure.' In the end, he took the initiative himself, organizing in-house flight tests. He attached tiles, as they would be bonded to the Orbiter, to the airbrakes of T-38 jets to subject them to high dynamic pressures in flight. And they came off. *Oh Christ*, he thought, *that should not have happened*. Nor was there any obvious reason why. Each element of the bonding process was, on its own, known to be strong enough. With Rockwell already applying the tiles to the first orbital Shuttle, the discovery had the potential to be a showstopper. It was exactly what Hans Mark's letter had warned of in the weeks that followed the Shuttle programme's approval in 1972.

Throughout a Shuttle flight, the Orbiter's metal airframe flexes and bends in response to heat and aerodynamic loads. In space, with one side heated by the sun, the other shaded and frozen, the entire structure bananas. The fragility of the silica compound used to safeguard the Shuttle from the heat of re-entry ended any hope of building the thermal protection system out of large, unyielding sheets. Instead, the heatshield became a complex tessellation of 30,757 numbered tiles roughly 6 by 6 inches square and up to 5 inches deep. Each specially cut and shaped tile occupied a unique position on the surface of the Orbiter.

Even assembling the heatshield like a jigsaw puzzle wasn't enough to ensure their integrity, however. The tiles could cope with a deflection of just one-sixteenth of an inch before they shattered like glass, so it was understood from the outset that the brittle silica tiles couldn't be bonded directly on to the aluminium skin of the Orbiter. Instead, to absorb the movement of the airframe beneath them, a layer of Nomex nylon felt, called a strain isolation pad (SIP), sandwiched by adhesive, sat between the tiles and the metal below, allowing the tiles to float above the source of stress. But it turned out to be the padding itself that was causing the problem. To add strength to the nylon felt, it was needled with perpendicular fibres. These caused stress concentrations that acted as a catalyst for failure. Subjected to forces as small as 2lb per square inch, the delicate silica tiles would fracture and tear off. Because of the aggravation from that stitching, the tiles were barely half as strong in practice as they had been on paper.

They would not survive spaceflight.

Engineering boss Max Faget stepped in, suggesting he should lead the search for a solution, but was told that he already had more than enough on his plate. Instead, Tom Moser got handed the job. He knew from the outset that he and his team had their backs to the wall.

NASA was now in the spotlight for all the wrong reasons. By June 1979, with uncertainty over where and when *Skylab* would come down, entrepreneurs were selling beanie hats advertised as *Skylab* early-warning devices. When falling wreckage hit you on the head you could consider yourself warned. Only getting the Shuttle into space offered to turn things round. And even as pieces of the space station began to come down in northern Australia in July, cinema audiences were being offered a glimpse of what an operational Space Shuttle might look like. Or at least that's what many assumed when, in June, *Moonraker*, the new James Bond film, was released.

Inspired by the success of *Star Wars*, Bond's producers decided to launch their super spy into space to thwart villain Hugo Drax's latest variation on the theme of taking over the world. Drax planned to release lethal spores from an orbiting space station, then repopulate the Earth with his own perfect physical specimens. And Jaws, his giant, metal-mouthed henchman. To get to and from his space base Drax employed a fleet of stolen laser-armed Space Shuttles, sporting yellow go-faster stripes and logos seemingly inspired by Rockwell's stylish Saul Bass creation, and launched them from inside the traditional hollow mountain headquarters.

That no Shuttle had yet flown in space perhaps gave the film-makers a licence for some of their wilder speculation, but so much was wrong that, as entertaining as it might have been, the astronauts could only smile and shake their heads at the ridiculousness of it all. In at least one respect, though, it was evident that the film-makers had done a little homework. In the movie, alongside 007's own mission to save the world, a military Space Shuttle is launched from Vandenberg Air Force Base, the site from which the Department of Defense planned to launch its own Space Shuttle missions once the real Orbiter became operational.

Inside the Pentagon, Hans Mark remained both Vandenberg's and the Shuttle's most committed and powerful advocate, but he was no longer having to bang the drum alone. By the summer of 1979, Lieutenant-General Tom Stafford had been in Washington for over a year in his position as Deputy Chief of Staff of the Air Force for Research, Development and Acquisition. It was a job that placed him in the forefront of many of the Air Force's most classified programmes, including the development of the Lockheed F-117 Nighthawk stealth fighter, the Northrop B-2 Spirit stealth bomber and a number of black-world drone programmes at a time when unmanned aeroplane technology was still in its infancy. Not many in the Air Force were particularly enamoured with the idea of the latter. The fondness the Air Force had for pilots was in marked contrast to its feelings about astronauts. But in Stafford, Hans Mark had a high-ranking ally in his effort to change their view. Soon after Stafford's arrival in the Pentagon, the two of them began work on a study into the future of manned military space operations. Submitted to Secretary of Defense Harold Brown in October, they called it 'The Utility of Military Man in Space'.

In *Moonraker*, the Shuttle launched from Vandenberg carries a squad of heavily armed Space Marines who, as the Orbiter approaches the enemy space station, float up out of the payload bay to conduct their assault. The ambitions of Mark and Stafford's study may have been less flamboyant but they were still, ultimately, substantial. As a long-term goal, the strategic merits of a manned military space station were discussed in detail, but there was also a more immediate suggestion. Made possible by the decision, to which Mark himself had contributed, to allow Payload Specialists to fly aboard the Shuttle after months rather than years of dedicated training, Mark created a new cadre of Air Force astronauts known as Manned Spaceflight Engineers, or MSEs.

As *Moonraker* finished its successful run in cinemas in August, the first thirteen MSEs were selected. All had backgrounds in existing Air Force or Navy space programmes, including satellite control from inside the Sunnyvale Blue Cube. Just one was a pilot. And when they first reported for duty they were described by National Reconnaissance Office Program A boss Major-General John Kulpa as 'the future of the Air Force in space'. Mark's hope was that they would act as a conduit, educating NASA about the Air Force in space and vice versa. In the end, though, they found themselves in an uneasy position, neither entirely of the Air Force nor of NASA; not quite fully fledged astronauts. It was a situation made painfully apparent by any comparison with the experience of the TFNGs, the new group of NASA astronauts who'd arrived in Houston the previous year.

As a graduate student hoping to study vulcanism on the deep ocean floor, claiming an aviation world record wasn't something Astronaut Candidate Kathy Sullivan had anticipated. But on 1 July 1979, wearing a full pressure suit in the back seat of one of NASA's big-winged Martin WB-57F reconnaissance aircraft, she and her pilot reached an unofficial women's world record altitude of 63,300 feet during a four-hour flight that took them over Big Bend, Texas.

Through nearly a year of briefings, lectures and training on everything from space physics to 'People and Requirements: It Takes a Bunch to Make Things Work in NASA', the TFNGs felt they were drinking through a firehose. From the spring of 1979, though, George Abbey had begun assigning them real responsibility in support of the crews preparing for the orbital test flights.

Before getting pressure-suit-qualified and flying from Texas through Panama, Peru, Uruguay and the Falkland Islands, sucking food out of toothpaste tubes on high-altitude air-sampling missions in the WB-57F, Kathy Sullivan was assigned to work with Dick Truly alongside Rick Hauck drawing up emergency procedures checklists for the Shuttle's ascent and re-entry.

Many of the new astronauts put in time at SAIL, the Shuttle Avionics Integrated Laboratory. Here, they were charged with the Sisyphean task of testing and debugging the Orbiter's software through every phase of flight using what John Young called the 'Iron Bird'. In truth, she was no bird at all. SAIL was a nervous system without flesh: the computers, wiring and electronics of the Space Shuttle, laid out exactly as they would be in a flying Orbiter, but unsheathed by an airframe. It was repetitive, joyless and often frustrating work. But, as the third long pole alongside the development of the main engines and the heatshield, it was also crucial to the programme's success.

Software used for critical computer systems like air traffic control contained, on average, ten to twelve errors for every thousand lines of code. NASA would reduce that figure to just 0.11 per thousand, but it was laborious, expensive work, each line of code costing twenty times what it cost to develop equivalent systems in commercial aviation. As vital as the work was, to alpha male test pilots who'd flown combat tours in South-East Asia, SAIL could feel like exile.

Inevitably, there was second-guessing within the ranks of the TFNGs about the significance of the assignments they received from George Abbey and John Young. Requiring extended periods away from Johnson, did the air-sampling mission mean Kathy Sullivan would be out of sight and out of mind? Was SAIL just a place to mop up a second tier? None of them knew; and Abbey and Young gave no indication of their thinking. There was one assignment, though, on which the whole TFNG group could agree: Jon McBride had landed a real plum.

In his last assignment before joining NASA, McBride opened the show for the Navy's Blue Angels display team, warming up the crowds in a spectacularly painted red, white and blue

F-4 Phantom. For ten minutes, McBride flew a solo display, carving out a tight routine trailed by dirty smoke from the jet's twin J79 engines, able to keep within the airfield boundaries only because of extensive modifications to the big interceptor. His last display in the jet was in May 1978 in front of his prospective TFNG classmates overhead Ellington Field. He had mixed feelings about handing the spectacular-looking Phantom back to the Navy, but he got to keep hold of his matching red, white and blue 'Captain America' bonedome flying helmet. On joining NASA, the affable, sandy-haired test pilot had his eye on becoming the first West Virginian in space.

After completing his basic training at Johnson, McBride was called in to see Abbey and Young. 'We're passing out new assignments for you guys,' they told him, 'and we'd like you to lead the chase team for recovery of STS-1.' McBride was going to have to join on the returning Space Shuttle at altitude then shadow her descent and approach to the Edwards lake bed. It didn't take long for him to appreciate that he'd been handed what was perhaps the most challenging aerial rendezvous ever attempted.

'How are we going to do this?' McBride asked as he mulled over the trajectories with Dick Gray, the NASA staff pilot working alongside him on the chase assignment. 'It's never been done before; nobody has ever tried anything like we have to try . . .' It was two weeks before the two of them even got airborne. Using flight profiles drawn up by a Houston aerodynamicist, they identified the best point to try to pick up the diving Orbiter as wherever she was at 40,000 feet, just as the speed dropped below the speed of sound. So fine were the margins, they discovered, that if the chasing T-38s missed their mark by more than three or four seconds, early or late, Young and Crippen would be descending into Edwards on their own. Not only that, in order to ensure that their sleek little jets didn't slip away from the blunt-nosed Orbiter, McBride and Gray had to find a way to inhibit their passage through the sky. Their solution: to lower the T-38's landing gear at near supersonic speed. That was way beyond the gear's design specs and would require help from Northrop to beef up the undercarriage, but it did the trick.

Alongside modifications to their jets, they also needed to develop infallible procedures. If they were to successfully arrive alongside the returning Orbiter, McBride's chase pilots had to be able to roll out of a 360° circle in exactly the right parcel of sky at the right time. They measured their turns in minutes. A perfect three-minute circle would deliver them to the right place, but if the Shuttle was then going to be five or ten seconds early or late, they needed to be able to extend or crank up their turn to accommodate it. Up in thinner air at 40,000 feet the dart-like T-38 didn't have the wing area or thrust for really hard turns. If they needed to lean into the turn to pick up an early arrival, there was going to be barely half a g of flexibility with which to do so.

Lastly, they had to make sure they could perform the manoeuvre in three different locations: Edwards, the White Sands El Paso diversion strip, and the runway at Kennedy Space Center, which would be used in the event of a return to launch site abort. All three airfields were at different elevations. McBride and Gray pencilled in the basics of the plan then travelled to the Air Force bases at Vandenberg, Edwards, Holloman, New Mexico, and Patrick in Florida to work out the detail of the timings and distances and radio calls with the Air Force flight controllers who'd be directing the one-shot manoeuvre from behind their radar displays.

McBride was just relieved that he'd been handed the assignment so far in advance of the first flight. Even six months, he thought, wouldn't have been enough to perfect such a complex task loaded with so many variables.

While McBride wrestled with the challenge, fellow naval aviators Dan Brandenstein and Rick Hauck were assigned to support crews for the Shuttle's first flight. There was no debate over where that seemed to place them in the pecking order. Working right on the frontline alongside Young, Crippen, Engle and Truly, *it was*, thought Hauck, *the best job any of us could have had*. While Hauck was first dispatched by John Young to Hamilton Standard in Windsor Locks, Connecticut, to report back on difficulties with the systems used for cooling the Shuttle on orbit, Brandenstein was given the job of following the development of the trouble-some heatshield.

Every Monday morning, the astronauts reported back on their particular area of responsibility at the weekly pilots' meeting in Building 4. Being the tile guy meant Dan Brandenstein was having to get used to reporting bad news.

Tom Moser's options didn't look good. There were tile compounds available that were strong enough to survive the stress concentrations that were breaking them, but they were also heavier. Using them would impose an unaccept-able weight penalty. Moser's Structures and Mechanics Division also explored the possibility of removing all the tiles, covering the SIP felt with a thin aluminium foil skin, then bonding the tiles on to the new, smooth surface. Again, it would add weight. Worse, they quickly realized the foil would expand and contract in response to heat and cold and so might only make the problem worse. Both solutions would also mean taking apart the intricate jigsaw that made up the heatshield then putting it together again from scratch. Moser was acutely aware that unless the problem was fixed, the Shuttle was going nowhere. As divisional head, it was on him. And there were important people now banking on seeing the Orbiter fly.

THIRTY-SEVEN

Vienna, Austria, 1979

SURROUNDED BY THE baroque grandeur of Vienna's Hofburg Palace, President Jimmy Carter and a visibly ailing Leonid Brezhnev signed the SALT II Strategic Arms Limitation Treaty in June 1979. Although hailed by Carter as 'a victory in the battle for peace' there were many who viewed the agreement to limit the number of intercontinental ballistic missiles, or ICBMs, operated by each country to 2,250 with more ambivalence. Hans Mark was one of them, but not because he regarded the reduction as inadequate.

The previous month, Mark had accepted the position of Secretary of the Air Force, and in doing so gave up day-to-day control of the National Reconnaissance Office and Air Force space programmes, but assumed greater influence for the overall Air Force budget. At the top of his list of priorities were the modernization of the country's strategic nuclear deterrent forces, the enhancement of the military airlift capability, and improving the ability of the Air Force to conduct operations in space. Mark worried that SALT II would restrict his ability to pursue the first two on the list, but when it came to his commitment to space, and with it the Space Shuttle, the SALT II negotiations were going to provide an unexpected bonus.

Houston boss Chris Kraft used an aviation analogy to describe the state of the Shuttle programme as NASA looked forward to the first flight: it was behind the power curve. Since the day the programme was approved they'd been underfunded, pushing what he called a bow wave amounting to a 10% shortfall on what they actually needed to get the Shuttle operational. After years of robbing Peter to pay Paul, NASA needed fresh funds. But, faced with continuing challenges with the main engines, the thermal protection system and software, and believing that further funding from Washington would not be forthcoming, the Shuttle programme's managers, in desperation, began to consider the possibility of turning the Shuttle into a research project. By allowing the programme to eat itself – by diverting funds allocated to getting the Shuttle operational to just getting it into space – they felt they might at least be able to get her off the pad.

Hans Mark learned in August from friends at NASA that the Shuttle needed more money and time. For the first time, the space agency was planning to submit a supplementary budget request to Congress. Mark, concerned that any delay would give the Air Force the pretext they needed to abandon their commitment to the Shuttle, and with it the Vandenberg launch site, threw his weight behind NASA's request. In November, he was invited to attend a meeting with the President in the White House Cabinet Room to discuss the programme's status. Sitting two places to Jimmy Carter's left, Mark spoke briefly to outline the Shuttle's importance to national security. The President quizzed

NASA's representatives about the difficulties with the engines and heatshield. His questions, though, were no longer to help him make a decision, but to allow him to justify it. Before convening the meeting, Carter had already made up his mind.

Earlier in the year, shortly after his return from the SALT II talks in Vienna, the President had called NASA Administrator Robert Frosch in to a meeting at the White House. 'I want to tell you how wonderful that Space Shuttle is,' he told Frosch. Acutely aware of the programme's struggles, Frosch could feel himself tighten. Carter went on to relate his discussions with the Soviets and how he'd enthusiastically endorsed the capabilities of the new space-craft being developed by NASA. 'It carried the day,' said the President. Frosch left with a sinking feeling, uncertain how he could possibly reconcile expectation and reality. Days later he returned to the White House to tell the President that the Shuttle programme was in trouble, that they had insufficient funds to meet the launch date, and that there were possibly insurmountable problems with the tiles. Carter listened, then he asked, 'How much do you need?' Surprised, Frosch estimated a figure of $600 million now, and another $400 million the following year.

'You'll get it,' Carter told him.

Despite some opposition on Capitol Hill, NASA's budget supplement was approved. And the Air Force, having always resisted any direct contribution to the Shuttle's develop-ment, paid for it. Hans Mark was only too happy to see the re-engining of the Air Force Boeing KC-135 Stratotanker air refuelling fleet delayed in favour of the Shuttle. November 1979 would be the final financial wobble. The money to get

Columbia into space was now assured. If the technical challenges could be overcome.

Materials engineer Glenn Ecord was working in the Building 13 laboratory at Houston over the weekend. A member of Tom Moser's thermal protection team, he'd been asked to consider the problem of strengthening and repairing the tiles. Exploring different ideas, one appeared to show promise. By mixing a powdered silica known as TEOS – extremely fine sand, essentially – into water then stabilizing it with ammonia, Ecord was able to paint the suspension on to the tiles with a brush. Once the water evaporated, the powder – fine as talc – packed itself in among the fibres and set like cement. Ecord then baked it in an oven at 350°F for three hours, densifying – as the strengthening process was christened – any part of the tile to which it was applied. Ecord took the idea to Moser on Monday.

This looks good, thought the Structures and Mechanics boss. For the first time in months, Moser wondered whether there was a solution to the tile problem on the horizon. 'Let's make sure it's right,' he told Ecord. Within a couple of weeks, it was clear that the elegantly simple idea could work. Armed with test data drawn up using the facilities inside Building 13, Moser was able to convince the programme management that his team had cracked it. Ecord's breakthrough meant broken and damaged tiles could quickly be repaired, while Ludox, a densification process developed in parallel by Rockwell, allowed the TPS team to double the strength of the tile where it was attached to the stitched Nomex felt below. It was enough to prevent the fracturing that had Moser worrying they were going to have to start all over again, at vast expense.

And while, initially, Aaron Cohen wasn't convinced that densification was going to work, subjecting the treated tiles to real-life aerodynamic loads in wind tunnels and in flight using NASA's F-104 and F-15 jet fighters confirmed that it did. There was swift approval for the proposal from NASA HQ in Washington. But if Ecord's first tile test in the Houston lab represented tile one, it still left another 30,756 to go. The workforce at Kennedy still faced a monumental task to prepare *Columbia*'s heatshield for flight.

When *Columbia* first arrived at Kennedy Space Center from Palmdale barely 1,500 of the silica tiles were ready for flight, most of them applied to the easier-to-work-with flat underside of the Shuttle. To complete the work on the spacecraft's heatshield, the Orbiter Processing Facility at KSC, a high-bay hangar designed for a workforce of 200, became home to combined teams from NASA and Rockwell totalling around *3,000* people. New employees, drafted in to provide the necessary manpower, were given a chance to practise the process of installing the tiles on the aluminium skin of an old Douglas DC-3 piston-engined airliner that hung inside the facility. Enduring a scramble for car-parking, inadequate catering and long lines for the loos, the tile installers worked in three shifts, seven days a week, to complete the work, and yet by the summer of 1979 they were installing barely 200 of the 600 tiles per week called for by the schedule. By the end of the year, with the additional requirement to remove vulnerable tiles in favour of individually fashioned, densified replacements, there were days and weeks when they were removing more tiles than they were fitting. Progress was so slow that, on average, each worker was installing just 1.3 tiles per week.

Every morning at 7.30, Tom Moser met with Orbiter

Programme Manager Aaron Cohen and Houston Director Chris Kraft to review progress. An essential part of the routine was the phone call to the Cape. Each morning, the question was the same: 'How many tiles did you take off and how many tiles did you put on?' Cohen's wife had begun telling friends that her husband's hair was going grey in inverse proportion to the rate at which the tiles were being fixed to *Columbia*.

To test the strength of the tiles' bonds to the airframe, Moser's team had designed a device which, using suction cups, pulled on them in place on the Orbiter. It seemed straightforward, but, from NASA HQ, Manned Spaceflight boss John Yardley saw an immediate problem with it. 'I got a question for you guys,' he said. 'When you pull on that tile, how do you know you didn't decrease the strength . . . so it's not going to be as strong as it was before you pulled on it?'

'Good question.'

In response, the Structures and Mechanics Division spent a weekend using a sensitive microphone to record the sounds of tiles as they were pulled, analysing just how much noise they could get away with before the act of testing it caused it to fail. They predicted that 5% of tiles tested would fail the pull test. Down at the Cape, as large numbers of tiles were put through the time-consuming new test, with workers listening in every case for the tell-tale sound of failure, the figure turned out to be 15%. The near 31,000 tiles that covered the Orbiter had been plenty. But inside the Orbiter Processing Facility, with some tiles needing to be fixed multiple times, the overall number installed rose to more than 75,000. Just one thing was certain: one way or another, they were going to have to put on more tiles than they took off. It was a job that would take 670,000 man hours to complete.

Such was the anxiety about the heatshield, however, even post-densification, that in January 1980, aerospace company Martin-Marietta received a contract from Houston to develop a tile repair kit that John Young and Bob Crippen could use on orbit. The reaction to that news from Kenny Kleinknecht, the engineer Houston had sent to the Cape to oversee *Columbia*'s preparation for flight, was succinct: 'Well, it would be fine to have that piece of equipment, but if we think we might need it, we really shouldn't be going.'

THIRTY-EIGHT

Vandenberg Air Force Base, California, 1980

WHILE NASA STRUGGLED to return American astronauts to space, the Air Force unmanned space programme was thriving. The first year of the new decade saw Department of Defense spending on space exceed that of NASA for the first time since 1960. And on 7 February, a third KH-11 KENNEN reconnaissance satellite was launched from Vandenberg. Following the de-orbit of the first of the big digital spy satellites a year earlier after 770 days on station, the spy agency once again had two of the big reconnaissance birds operational. On essentially the same orbit, they were separated by a period of around twelve hours. Between them, every day, they could take advantage of the long shadows at dawn and dusk to give greater depth and perspective to their images. They were about to prove their worth.

In the spring of 1979, a small, chaotic group of Iranian Marxist revolutionaries had managed to seize the US Embassy in Tehran. They were quickly thrown out by supporters of the country's new leader, the radical cleric Ayatollah Khomeini. In the aftermath of the attack, the

CIA's Iran Branch chief in Langley contacted the head of the Tehran Station to reassure him that another incursion was unlikely. 'The only thing that could trigger an attack,' he said, 'would be if the Shah was let into the United States – and no one in this town is stupid enough to do that.'

On 21 October, the exiled Shah of Iran was admitted to a New York hospital for cancer treatment. Two weeks later, 3,000 angry supporters of the Ayatollah swarmed over the walls of the US Embassy in the Iranian capital, overpowering the Marine guards to take hostage over sixty embassy staff. The sight of American diplomats blindfolded and hand-cuffed while crowds outside burned the Stars and Stripes and chanted for the Shah's return shocked America. Before the end of the month, Ayatollah Khomeini had released all the women and African-American hostages, but a warning was to follow: if America attacked his country, the remaining fifty-three hostages would die 'on the spot'.

Work on a rescue attempt, however, had already begun. The foundation upon which the US Army's Delta Force planned their operation was overhead photography from the two KH-11 satellites. While a CIA agent on the ground was able to supply detail about what was going on inside the embassy, augmented by increased signals intelligence monitoring Iranian communications, no military option was possible without detailed information about the physical layout of the compound and its surroundings. The KENNEN images were able to reveal changing daily patterns of activity with a degree of fidelity that had a member of the ninety-three-strong Special Forces assault team talking about the tiles on the roof and the grass in the gardens. More importantly, the pictures revealed that large numbers of poles had been erected around the 27-acre grounds of the embassy to prevent the possibility of any helicopter being able to put down. At the same time they showed that the nearby Amjadeh football stadium was clear of obstacles, making it a potential alternative landing zone.

With sanctions, UN diplomacy and embargoes all failing to make any progress in resolving the crisis, on 11 April 1980, the President authorized Operation EAGLE CLAW to rescue the hostages. What unfolded was a catastrophe.

Already aborted because mechanical problems had reduced a planned force of eight Sikorsky RH-53D Sea Stallion helicopters to just five, the operation quickly turned from failure to tragedy at Desert One, a forward operating base 200 miles south-east of Tehran. When the main rotor of one of the huge Navy Sea Stallion choppers sliced into the port wing of a fuel-carrying Lockheed EC-130 transport aircraft, Desert One was immediately turned into a fuel-fired inferno; eight servicemen were killed in the blaze that ensued. When the Iranians trawled through the wreckage and hardware left behind at the site, they found the cockpits of the abandoned helicopters stuffed with KEYHOLE imagery of downtown Tehran.

It was both a tragedy and a national humiliation for America. And a personal one for Jimmy Carter from which his Presidency would not recover. In the mood of gloom that swept over the country in the wake of the Desert One disaster, naysayers were predicting that the Space Shuttle programme offered further heartbreak just around the corner. On news-stands, the April edition of *Washington Monthly* magazine carried a picture of the Shuttle on its cover alongside the headline 'BEAM US OUT OF THIS DEATHTRAP, SCOTTY!'.

*   *   *

STS-1 Commander John Young was under no illusions whatsoever about the dangers involved in riding the Shuttle's first flight. 'Anyone,' he said, 'who sits on top of the largest hydrogen-oxygen fuelled system in the world, knowing they're going to light the bottom, and doesn't get a little worried, does not fully understand the situation.' For good measure, he also pointed out that the last time he could think of when anyone had climbed aboard a flying machine full of hydrogen was the *Hindenburg*, and that, he remarked, 'didn't work so good'. His wife, Susy, didn't want him to fly. When Young had been training for *Apollo XVI*, she'd become aware of a formal NASA risk analysis that rated his crew's chances of surviving the mission as low as 20%. They weren't numbers that the management in Building 1 chose to share with the Astronaut Office, but Susy shared them with her husband. Young could only counter by telling her it was his job. Asked later whether it was a profession that needed daring and courage, he smiled. 'Daring and courage? I hope you don't need any daring and courage because I don't have any of that.' Instead he just kept picking away, firing off memos about the things he *could* fix; those that were out of his control he brushed aside with self-effacing good humour. Asked during one meeting about what kind of cockpit display he favoured while performing the difficult and dangerous return-to-launch-site abort manoeuvre, he maintained that he was the wrong person to ask. 'If we're doing an RTLS,' he said, 'I'll have my hands over my eyes and be going "AAAAGGGHHH!"'

Columbia's first flight was scheduled for March 1981.

THIRTY-NINE

Houston, 1980

MOL VETERAN DON Peterson was sitting in the cockpit of the Shuttle Main Simulator when it occurred to him that there was a procedure no one else had yet tried. As he sat with his feet on the rudder pedals, looking at the cockpit dials and displays ahead, he wondered: could the Shuttle ascent be flown manually? *It's always been done automatically*, he thought, managed by the computers; *question is, could a crew do that?*

The Shuttle ascent profile used something called Lambert Steering. Climbing at an angle, some of the thrust is used to fight gravity, the rest for acceleration. As the stack ascends and fuel is burned it gets progressively lighter and faster, while, as centrifugal force overcomes the effect of gravity, the angle of the climb flattens into a horizontal orbit.

Along the right ascent trajectory, by multiplying the Shuttle's acceleration and pitch angle, Peterson realized you should get a constant value. For a crew forced to do the calculations in their heads in real time, it appeared to offer a simple enough way of flying a safe launch. And it worked. Peterson successfully demonstrated it in the sim. *It wasn't a perfect orbit*, he had to admit, *but it got you there.*

Peterson's manual ascent crossed another possibility off the list. Every conceivable contingency, no matter how big, small or unlikely, was being explored. On one occasion the SimSups even climbed into the cockpit of the simulator, threw a rubber chicken on to the lap of the Commander and told the Pilot sitting next to him that he'd lost the man in the left-hand seat to a birdstrike. 'You're on your own,' they told the surviving pilot. Throughout 1980, as the first flight approached, it could sometimes seem that, between them, the astronauts and SimSups were becoming increasingly imaginative as they considered potential flight safety issues. But with STS-1 scheduled for the spring of 1981, perhaps that was as it should be. By now, they had to count on the basics being right.

Joe Engle and Dick Truly climbed into the sim, took their seats, and relaxed as the hydraulic rams beneath the cockpit tipped them on to their backs. Another day, another ascent simulation. The crews were now so used to the SimSups throwing failures at them that maybe it did take a little imagination to keep them on their toes. Dialling in the failure of one of the four computers was pretty standard. The SimSups could now hear the familiarity in the astronauts' voices as they responded to pretty much whatever was thrown at them.

As the dials and displays recorded a nominal ascent, suddenly all three of the display screens on the instrument panel blinked off, their flight information replaced in every case by nothing but a big 'X' against a dead black background. All four main computers had failed.

Aw, shoot, thought Truly, *the simulator's crapped out*. It happened often enough, and there was no way the SimSups would pull the plug on *all* of the computers. It was pointless: without them, the Shuttle was out of control; there was nothing more the crew could do. The trouble was, the sim *wasn't* over. The simulator was still working perfectly. It was the software. Using real flight software and a set of four of the same AP-101 processors as the Orbiter, the computer systems were set up exactly as they were in *Columbia*. A flight software glitch that, if it had first revealed itself during a mission, would have caused the loss of the vehicle, had chosen to make itself known just months before they lit the fires at the Cape.

The SimSups froze the simulation and gathered the data so that the teams debugging the Orbiter's flight control software could try to reconstruct the problem, then iron it out.

That, thought a chastened Dick Truly, *was pretty serious*.

A decision was taken to carry a sixth GPC (general purpose computer) on board *Columbia* that, if all else failed, could be installed on orbit by the astronauts. Loaded with the PASS software – the primary avionics software system – needed for the Orbiter's return to Earth, it got dubbed 're-entry in a suitcase'. The computers, though, weren't the only critical system throwing up surprises.

In May 1980, a series of tests conducted in the Naval Surface Weapons Center shock tunnel reinforced concerns about some test results that had first surfaced over a year earlier. A 1979 review conducted by the Air Force Flight Dynamics Laboratory into potential heating issues revealed that temperatures experienced by the two orbital manoeuvring system pods mounted on the sides of the fuselage beneath the vertical tail might be higher than anticipated. Further tests by the Air Force supported the initial findings. Then the Navy weighed in. Their data suggested that under the dynamic and acoustic loads experienced on launch, the pods would flex substantially more than expected; more so than some of the bigger 8-by-8-inch silica tiles protecting them could cope with, without cracking and separating, and leaving the pods unprotected on re-entry. While the thin, undensified tiles covering the OMS pods had already survived a simulated thirty-one missions in the acoustic chamber at Marshall, it hadn't been an area included in the flight-testing conducted by Tom Moser's team earlier in the year. Rather than remove them, the decision was taken to carefully dice them into nine equal parts *in situ*. If they were already in pieces, the effect of any deflection of the airframe beneath was negated. It was further evidence, however, that when it came to the tiles, there didn't seem to be a point at which the programme management could breathe out and regard the issue as finished. In August, NASA went ahead and told Martin-Marietta to build and deliver the on-orbit tile repair kit that earlier in the year they'd been contracted to design.

'We have confidence in the Orbiter thermal protection system,' Houston Director Chris Kraft had said when first considering the possibility of in-flight heatshield repair, 'but at the same time we think it prudent to be prepared to make in-orbit tile repairs.'

No matter how hard they tried, NASA couldn't seem to move the tiles off their to-do list.

None of this was news to Hans Mark. As Director of Ames in 1972 he'd been a signatory to the letter highlighting the potential problems of bonding the tiles to the airframe. Responsible for much of the early effort to develop and test the tiles, he'd followed the heatshield's chequered progress since with interest. And in fighting the Shuttle's corner from the Pentagon the previous year he was up to date with the most recent efforts to get to grips with it. At Ames, Mark had known John Yardley, the NASA official leading the Shuttle programme in Washington, since Yardley had first been drafted in from McDonnell-Douglas. With Mark's own move to Washington their paths had continued to cross. They had recently met to formally discuss NASA–Air Force cooperation in space. Now, though, as the two of them talked about the continuing challenges posed by the thermal protection system and the internal arguments that had led to NASA's reluctant decision to build the on-orbit tile repair kit, Yardley was struck by the level of interest from the Air Force Secretary.

It was clear to Mark that nobody at NASA wanted to order Young or Crippen to suit up and leave the safety of *Columbia*'s pressurized cabin to inspect possible tile damage. Aware of the resources DoD might bring to bear in the event of fears for the integrity of the heatshield he made clear that the Air Force was ready and able to help. 'I think,' he told Yardley reassuringly, 'we could work something out . . .'

FORTY

Fort Belvoir, Virginia, 1980

KEN YOUNG PULLED up to the Fort Belvoir guardhouse, showed his driver's license and recited his social security number. It was his first visit to Aerospace Data Facility East, the NRO's ground station on the banks of the Potomac river, and the the final part of his induction into the black world.

As head of the JSC's Flight Planning Branch Young had already had Secret security clearance, but, when he was called into the Shuttle Program office by his boss, Mission Planning and Analysis Division Chief, Ed Lineberry, and George Abbey's assistant, Jay Honeycutt, in early 1980, he was told it wouldn't be enough. For the job they wanted him to do he required a top Top Secret compartmental ticket and it was going to take six months of background checks by the Office of Personnel Management to get it.

NASA was going to try to capture on orbit images of the Shuttle using the NRO's spy satellites. Fewer than fifteen people at JSC were privy to what became known by Lineberry and Young as 'The Plan'. The two of them would have sole responsibility for planning the conjunctions between the two orbiting spacecraft. Speed, timing, distance, angles, orientation and lighting all needed to be considered. And they represented the only direct point of contact between Houston, the NRO and the Air Force personnel operating the Keyhole satellites out of the Blue Cube.

Lineberry and Young met their DoD counterparts inside a secure room at JSC, impressed with their competence and, in the case of Tish Vajta-Williams, a particularly sharp-minded Lockheed engineer working with the NRO, her strikingly good looks. Such was the concern over security that at times it almost felt as if the DoD liaison team were talking in tongues. Different scenarios and project compartments were assigned specific codenames like LAMPPOST, WAGON WHEEL, FIREFLY and WATER GLASS. All had to be memorized by Lineberry and Young. Taking notes was forbidden, so too even saying 'NRO' out loud. Instead in their hidden workroom in the Mission Control Center there was a single Hewlett-Packard desktop computer and a DoD encryption machine that came with someone to operate it.

For Young, the insight into the classified, DoD space program was eye-opening. While he was impressed by the quality of the people, he was surprised that Air Force mission planning was still mostly a pencil and ruler affair. While there were computers to run the data, the timelining was being done on what looked like big rolls of butcher paper. But if the Blue Cube was spartan, the same could not be said of Area 58, the NRO's Fort Belvoir facility. Even to an engineer who'd cut his teeth on NASA's moon missions, the sophistication of the computer, photographic and communication equipment inside the building's plain exterior was a revelation. As Tish and her NRO colleague showed him round, Young wondered at what was clearly *billions* of dollars worth of kit. Such was the importance to national security of the job they were doing, his hosts told him, that the Environmental Protection Agency had been forced to turn a blind eye to the large volumes of toxic silver salts – old developing fluid – the NRO was flushing into the Potomac river. At least until a biological treatment for dealing with the effluent could be developed. And it was here that any photographs of the Shuttle captured by the satellites would be processed. Here, that any evidence of damage to *Columbia*'s heatshield might first be revealed. But only if Ken Young and Ed Lineberry got their sums right.

Among the handful of people at JSC briefed on the NRO's involvement were the Primary and Reserve crews. Invited into the vault, the four astronauts listened intently as The Plan was described in detail. After considering what was said, John Young asked: 'Well, what do we have to do in all this?'

'Nothing really,' replied his namesake, 'just perform the manoeuvres and orientations we send up. And don't ask anything on the open loop if some attitude change seems weird or unconventional.'

'OK,' replied *Columbia*'s Commander, 'that's easy.' On the face of it, it was.

While the program continued to pursue every opportunity to bolster its confidence in a safe re-entry, John Young's greater concern was with the ascent. If they got into space in one piece he thought that, relatively speaking, coming home would be, as he put it, 'a piece of cake'. Through the summer of 1980 there was better news from Rocketdyne and Marshall Space Flight Center. Over a month in May and June, each of *Columbia*'s three main engines performed a successful 520-second static burn on the test stands at the

National Space Technology Laboratories in Bay St Louis, Mississippi. Then in July, a pre-burner fire was quickly followed up with an exploding oxygen turbopump. It appeared that, despite the progress being made, Young, Crippen, Engle and Truly had good reason to let the possibility of losing one or more of the main engines continue to dominate their abort planning.

John Young and Bob Crippen climbed down from the motion-based sim into the cool air of Building 5. 'Even a blind squirrel can find a nut now and then,' Young told the SimSups, his way of letting them know they'd run a good session. Even if it had meant that he and Crip had been forced, after losing main engines beyond the point at which they could make it back to Kennedy, to fly the Orbiter across the Atlantic only to pull their ejection-seat handles and abandon her off the coast of North Africa.

Despite NASA's best efforts, the combination of abort profiles – return to launch site, abort to orbit and abort once around – didn't quite cover everything. There remained a period between the call of 'Negative return' and 'Press to MECO' (main engine cut-off) in which the only option was to fly the Orbiter down to a safe height and bail out. In April, Crip and Young had been forced to paddle around in rafts in the Building 29 water tank, just to make sure they were ready for it. The whole idea rankled with Young. Yuri Gagarin had ejected from his capsule after re-entry; there had even been a plan mooted by North American to boost an X-15 rocket plane into orbit before, after returning to a safe altitude, the pilot ejected into the Pacific. That had been back in the fifties, though, at the genesis of the space race when just getting into space and back was enough. The Shuttle was supposed to be different. That was the whole point.

Back in the Astronaut Office after the simulator session, Young shared his frustration with Crip, Joe Engle and Dick Truly. 'This is stupid,' he said, 'to throw away a good airplane like this. We could probably land this thing.' The trouble with that was that it wasn't possible to bring up the guidance systems that might provide a heading to an airfield on the primary software. And it wasn't possible to dump the primary software without losing the flight control system that carried the full re-entry software.

It was Dick Truly, whose familiarity with the computer systems had seen some label him, along with Crip, 'Mr Software', who figured out that there might be a way round it. While Young and Crippen were flat out preparing for STS-1, he and Engle had time to pursue it. They were looking for anything they could fix or improve. And this one caught their imagination.

They returned to Building 5 and asked the SimSups to give them exactly the same conditions as Young and Crip had had. As soon as the engine failed, Truly, leaving Engle with the primary computer and flight control, brought up the back-up computer and displays on his side of the cockpit. While Engle flew the Orbiter from the Commander's seat, Truly used the information from the back-up computer to navigate, passing instructions to Engle to gain or lose energy, or head left or right, in order to make it in to an airfield on the other side of the Atlantic. By dividing the work and making use of two computers to run different software simultaneously – which Truly was first to realize might offer a solution – they were able to land the Shuttle rather than abandon her to the sea.

Of course, it needed an acronym, and Truly and Engle's

new transatlantic abort landing became TAL. And although it was too late to programme the new procedure into the PASS software for STS-1, Young and Crippen trained in the sim to fly the abort manually as Engle and Truly had demonstrated was possible. In time, potential TAL airfields were added in France, the UK, Morocco, the Gambia, Senegal, the Azores, Nigeria, and even Easter Island for launches out of Vandenberg.

For the first Shuttle flight, though, Rota Naval Station, a US base in Spain near the Straits of Gibraltar, became the designated diversion in the event that Young and Crippen had to abort across the Atlantic. A NASA support crew was flown in days before the launch. The only thing missing if *Columbia* were to arrive in Europe would be a mid-air welcome from Jon McBride's Chase Air Force.

TFNG recruit Kathy Sullivan used to tease her colleague that the whole Chase Air Force assignment was just the world's biggest boondoggle. And in truth, McBride had enjoyed every minute of it as he, Dick Gray and the two back-up crews punched holes in the sky above Florida, California and New Mexico in their T-38s. In the back seat of his own jet, McBride carried George 'Pinky' Nelson, another one of the 1978 astronaut intake, charged with taking stills of *Columbia* with a top-of-the-range Hasselblad camera after they made their rendezvous. And on that score, McBride was feeling confident. He and his crews had practised their three-minute circles at 40,000 feet till, McBride reckoned, *we could do it in our sleep*. They'd worked out their calls with Young and Crippen, who chopped a couple in an effort to keep the radio traffic to a minimum. McBride's T-38 had been specially calibrated so that he could pass on accurate speed and altitude to the Shuttle crew. They'd practised on each other and with the Gulfstream Shuttle Training Aircraft, but in search of something that might make the same sort of dramatic entrance as the Orbiter, McBride called up Beale Air Force Base in northern California, home to the USAF's Lockheed SR-71 Blackbird fleet. Would they, he asked, be at all interested in flying a Shuttle profile? The Blackbirds of the 9th Strategic Reconnaissance Wing were already collaborating with NASA to help them evaluate the Shuttle's radios and navigation aids; they proved only too willing to play.

After a visit to Beale to plan the sortie with the Blackbird unit, McBride would lead his flight of four Talons into position in the skies over the California high desert. Under the direction of the flight controllers at Vandenberg and Edwards, he completed his three-minute circle and rolled out abeam the starboard wing of the big spyplane.

'We're aboard,' he reported over the radio.

The sleek black jet didn't hang around for long, though. At just Mach 0.9 and 40,000 feet, trying not to outrun a pair of jet trainers dragging their landing gear, the SR-71 was way outside its comfort zone. Less than thirty seconds after reaching the agreed IP (initial point) at Shuttle speed and height, the Blackbird pilot poured on the coals and accelerated away, just giving a smiling McBride time to snap off a few pictures. The Chase team were ready.

The striking-looking spyplane might have provided a flavour of what was to come, but before the Chase Air Force had any chance of meeting the Shuttle on her descent into Edwards, NASA still had to put to rest any lingering concerns about the Orbiter's ability to return safely to Earth, and what they might do if they thought she could not.

FORTY-ONE

Houston, 1980

IN AUGUST, NASA announced that John Young and Bob Crippen would carry the on-orbit tile repair kit developed by Martin-Marietta with them on *Columbia*'s first flight.

It consisted of a manned manoeuvring unit – a 300lb jetpack powered by twenty-four little nitrogen thrusters first tested on board *Skylab*. Worn by a spacesuited astronaut, it provided a means to fly around the Shuttle to inspect the whole of the spacecraft's exterior. Combined with a work restraint station – a kind of window cleaner's cradle – that attached itself to the side of the Shuttle's fuselage, the jetpack allowed an astronaut to use a caulking gun containing a heat-resistant silicone-rubber paste to plug small gaps in the heatshield. To repair any larger areas of tile loss, 160 6-by-6 hardened blocks of the same basic material were to be stored in *Columbia*'s cargo bay.

Tested in Denver inside Martin-Marietta's zero-g simulator and on board NASA's KC-135 Vomit Comet, it seemed to offer promise, and Bob Crippen, the astronaut with prime responsibility for conducting any emergency spacewalks during the first flight, was told to train for the job. But in

Denver, strapped into the simulator that replicated the response in space of the jetpack to his control inputs, he struggled to stay in position for long enough to get any useful work done. Every time he applied even the smallest amount of pressure to the Shuttle he just pushed himself away from the vehicle. He quickly came to the conclusion that if he attempted to do any work on the bottom of the Orbiter, he was going to tear up more tiles than he could ever repair.

'This is not going to work,' he told John Young. 'We gotta make sure they're going to stick on.'

Looking at the prospect of hauling a tile repair system into orbit that the engineers argued wasn't necessary and which Crip said wouldn't work, Young put his foot down. He didn't want the jetpack and tile repair kit on board *Columbia*.

Tom Moser was certain the crew didn't need it. The tiles *were* going to work. Whenever there had been a fresh alarm over the heatshield's integrity it had been addressed. He was confident the system was robust enough to cope with all that was being asked of it – and even one or two things it wasn't. Down at the Cape, one of the engineers working on *Columbia* had fixed tiles to the front of a race car then driven at speed through a flock of chickens in the hope of gaining an insight into whether or not they would survive an in-flight birdstrike. The data he gained was, one suspects, of greater curiosity value than usefulness.

Dottie Lee, first drafted in by Max Faget to work on the earliest designs for the Shuttle in 1969, travelled to Ames to make a final presentation to Walt Williams, NASA's Chief

Engineer, and a dedicated oversight committee of Center Directors, industry experts and academics known as the Aeronautical Space Advisory Panel. Throughout the day, representatives of the different disciplines contributing to the Shuttle had given their status reports to the committee. Lee was the last speaker of the day. For forty-five minutes she talked the group through the nuances of heat transfer and aerothermodynamics, and of the confidence there was in the Orbiter's heatshield. As Lee packed up at the end of her presentation, one of the members of the panel came over to her and said: 'Based on what you've told us today, we're going to recommend to the President that we launch the Shuttle.' The thought was enough to send a chill down her spine.

A week later, a fifty-four-hour simulation of the entire STS-1 mission was carried out. *Columbia*, still inside the hangar at Kennedy – and now requiring fewer than 1,000 tiles still to be attached – was networked to both Launch Control and the Mission Control room in Houston. While Young and Crippen shared the duration of the simulation inside the Shuttle's cockpit with their back-up crew, Engle and Truly, the three teams of flight controllers in Houston worked the same shifts as they would during the mission. This was the second of a series of three full-duration, integrated simulations, and there was now a palpable, growing sense of anticipation around the first flight. After over a decade of development and delays, in the late summer of 1980 the Shuttle programme was coming together.

FORTY-TWO

Kennedy Space Center, Florida, 1980

MERRITT ISLAND, CAPE Canaveral, the home of Kennedy Space Center, wasn't, when NASA first acquired it, much of an island at all. An 88,000-acre swamp capillaried with grey-green water that was called home by alligators, manatees, dolphins, storks, wild pigs, tortoises and plagues of salt-marsh mosquitoes, it was as liquid as it was solid. The job of making it ready for service as a spaceport was handed to the Army's Corps of Engineers. Fifteen thousand individual tracts of privately owned land were bought up, then dredged, drained, squeezed and packed with sand before construction work began on launch pads, firing rooms and processing facilities. At the end of September 1965, the workforce moved into their new buildings: a tableau of tan, beige and grey concrete modernism that, in a landscape devoid of trees, seemed to sit on top of the landscape rather than truly being part of it.

At the heart of the complex was the Vehicle Assembly Building – the VAB. Founded on a hive of thousands of piles driven 160 feet down through the soft ground until they found the bedrock below, the VAB was, at the time of its completion in the mid-sixties, the world's largest building. A third bigger than the Great Pyramid and enclosing a greater volume than either the Pentagon or nearly four Empire State Buildings, the giant iron-lattice cathedral seemed utterly alien on Merritt Island; a vast, pale, corrugated box standing over fifty storeys high that dominated the view from any direction. The scale of the VAB was to accommodate the towering Saturn V moon rockets, assembled vertically within a building that was large enough, on hot, humid days, to generate rain showers from beneath its ceiling, then stacked on top of a 2,700-ton caterpillar-tracked flatbed transporter that carried it to the launch pad 3 miles away, where the Cape met the Atlantic.

Now, both VAB and transporter had been re-engineered and repurposed for the Space Shuttle. And on 5 November, the Kennedy workforce bolted the two solid rocket boosters, each of their steel cases packed with 1.1 million pounds of propellant, to the flanks of the white-painted external tank. Inside the Orbiter Processing Facility, *Columbia*'s three main engines, test-fired and recertified by Rocketdyne, were carefully manoeuvred back into the waiting engine bays by specially adapted forklift trucks and reinstalled ready for flight. On the 24th, in front of cheering crowds, *Columbia*, complete save for a handful of outstanding tiles, was towed outside for the first time since she'd arrived at the Cape on the back of a 747 in March 1979 looking, as one witness had put it, 'like something the cat dragged in'. A small crowd of the Kennedy workforce cheered her progress across the few hundred yards to the huge VAB where she'd be prepared for launch.

It was this part of the process that Bob Crippen wasn't so keen on witnessing. The prospect of seeing 85 tons of

Space Shuttle that had taken a decade and billions of dollars to create lifted from the ground then swung nose up to dangle vertically from beneath the roof of the VAB just made him too nervous. And yet after being winched up from the ground by a pair of cranes on 25 November, she hung there overnight, suspended like a shot pheasant, high above the concrete floor until, the following day, she was finally mated to the external tank and solid rocket boosters. The first live Space Shuttle stack had been assembled. Nine days later, they powered her up.

While Crip's anxiety about seeing her might now have subsided, John Young said no to an invitation to go and look at his spacecraft for entirely different reasons. When offered the opportunity to climb up to a walkway near the roof of the VAB to look down on *Columbia* from above, the STS-1 Commander declined.

'I'm afraid of heights,' he quipped.

Perhaps it was being trussed up in a suit and tie that contributed to John Young's apparent awkwardness in front of the microphones and cameras in a press conference. Off duty, at home, he was more likely to be wearing a Stetson, denim jacket and jeans. That, somehow, seemed a more appropriate image to accompany the pithy cowboy rhythms of his voice. It was said of Young that he had the Right Stuff before we all knew what the Right Stuff was.

As the date of the first flight drew near, media interest in him and Crip grew stronger, and press conferences became more frequent. This ensured that Young had plenty of opportunities to share his own, distinctive reflections on what lay ahead. He may not have enjoyed the hoopla much, but, like batting away 20% odds of survival, he did it because it came with the job. What's more, he was good at it. Young himself professed bemusement. He maintained that people found him funny when he was trying to be perfectly serious, but he had a gift for a well-timed, bone-dry remark delivered, it seemed, to some arbitrary point in the middle distance. It often saw Crip reduced to playing the role of straight man, laughing as hard as the rest of the room. Uninformed or trivial questions were dealt with patiently, but often not without an edge.

During the Apollo years, a reporter once asked Young why all the vital burns and manoeuvres took place on the moon's far side, out of reach of radio contact from Earth. 'Is it an insurmountable problem,' he wanted to know, 'to have these manoeuvres take place on the front side?'

'It's an insurmountable problem, alright,' Young replied. 'You ought to talk to that guy Kepler.'

When discussing the Shuttle, he'd lost none of his technique. Will boredom, he was asked, be a problem in future spaceflights?

'Gee, I hope so . . .'

Will there ever be a passenger Shuttle?

'If the US government want to, we could do it.'

But is it likely?

'Governments are funny critters . . .'

A question from a European journalist was more pointed: 'It's still not entirely clear to me whether you can abort with the ejection seats during the burn of the solids . . .'

Young paused, apparently considering his answer, before putting the microphone to his lips, and saying, 'You just pull the lil' handle . . .'

Young looked heroic, the journalist a little foolish. To those watching, it no longer mattered that neither he nor

Crip thought that pulling the 'lil' handle' to eject into the 750-foot-long exhaust plume of the SRBs offered much chance of survival; a question about danger and mortality had been deftly avoided.

Asked what he'd like his epitaph to be, Young was barely able to answer before beginning to chuckle at the thought of it. 'Good grief,' he attempted, 'I'd not like to see any epitaph. I have no idea. I worry about a lot of things, but I'm not really worried about that yet. You're really pushin' me, aren't ya?' His broad grin threatening to get the better of him again: 'Shoot . . .' He tried one more time – 'I'd feel sorry for the guy who wrote it' – before he gave up, laughing infectiously at the very idea of something so absurd.

Low cloud and chill air hung over the Cape when, at dawn on 29 December, the huge doors of the Vehicle Assembly Building rolled open – a process that took forty-five minutes to complete. Around 8 am, the Shuttle stack emerged from inside the sanctuary of the high bay, rolling slowly on top of the crawler towards the pad along a track laid deep with crushed rock and river gravel – a 3½-mile journey on the back of a 5,500-horsepower machine as big as a baseball diamond that took ten hours to complete. By 6.30 that evening, as the sun went down behind the VAB, *Columbia* took her place on Pad 39A, casting shadows that reached out towards the Atlantic.

It was a bittersweet moment for Hans Mark. Jimmy Carter had been defeated in the November election, and Ronald Reagan was due to be inaugurated as President in January 1981. Mark knew that as eager as he was to cement the Shuttle's future with the Air Force and ensure the completion of the Vandenberg launch site, his days in the Pentagon were numbered, his influence within the Air Force already gone. He could only hope that the new administration would push on with STS-1, the Shuttle programme and the military's West Coast spaceport without delay. At the same time he knew that if he could engineer a return to NASA as either Administrator or Deputy Administrator, then he might bring his experience of NASA–Air Force relations to bear on the programme from the other side of the fence. Influential friends and supporters championed his cause.

Early conversations appeared promising, but, as he packed up his Pentagon office to make way for the incoming Secretary of the Air Force at the beginning of February 1981, nothing was certain.

FORTY-THREE

Kennedy Space Center, 1981

THERE WERE THREE different ways to get back to the ground
from the top of the launch tower. You could fly the Shuttle
into space and back, take the lift down the launch tower to
the pad, or use the zip wire. Earlier in the year, Young and
Crippen had asked to try the flying fox-style escape wire
that, in the event of an emergency on the pad, was there to
spirit them from the top of the launch tower to a waiting
M113 tracked armoured personnel carrier in which they
could then drive themselves to safety. Both astronauts liked
the idea of a 54 mph fairground ride to the ground, but
while they were allowed to climb into the basket in the
pressure suits, no one was prepared to let such a precious
cargo try it for fun. On 19 March 1981, after completing the
terminal countdown demonstration test, they took the lift.

Young and Crippen were in high spirits. The countdown
test, a dry run of the launch, was the last major test before
they flew. There was less than three weeks to go. After the
countdown ended at 7.24am, the two astronauts climbed
out of their ejection seats and crawled back out of the
Columbia's cabin. They spoke to engineers in the White

Room at the top of the gantry, before heading back to the crew quarters in the Operations & Checkout building. The Shuttle systems were powered down at around 8.30.

'Hey,' Crippen said, 'we're getting pretty close to flight.'

While, in press conferences, the experienced John Young had the air of a man who'd been there, done that, some journalists detected a quality in Crippen that they described as almost akin to yearning. After fifteen years of service as an astronaut, Crip was on the cusp of earning his long-cherished astronaut wings.

Then the bad news reached the astronauts: there had been an accident on the pad.

After a twenty-second test-firing of *Columbia*'s three main engines on 20 February, there'd been a concern that nitrogen, used to purge the aft compartment of the Orbiter as a precaution against fire during launch, was leaking and that it might find its way into the cockpit. Instead of halting the flow of nitrogen at the end of the countdown test, a decision was taken to extend the duration of the purge to give ground crew an opportunity to pinpoint any leaks. News of the change in procedure failed to reach Rockwell technicians John Bjornstad, Forrest Cole and William Wolford. The three men checked in to the monitor station at the 130-foot level of the launch tower at 9.15. They made their way past a bundle of hydraulic lines into the vehicle near the main engines. Six minutes later, another Rockwell contractor, Jimmy Harper, discovered Bjornstad and Wolford unconscious inside the aft compartment. Harper made to help but then also collapsed. He was fortunate to fall backwards and away from the concentration of gas inside the Shuttle. As he fell he was spotted by two colleagues, one of whom, Nick Mullon, managed to drag Wolford free of *Columbia* while the other rushed to get help. On his return, the two of them went back in to haul Bjornstad out, after which Mullon too lost consciousness. It wasn't until fire and rescue crews arrived nearly a quarter of an hour after the three men first entered the compartment that Forrest Cole was finally pulled free.

Nitrogen is not in itself toxic. It makes up nearly 80% of the Earth's atmosphere. But in a pure nitrogen environment, without the oxygen that makes up over 20% of the rest of the atmosphere, loss of consciousness comes quickly. The USAF Flight Surgeon's Guide explains that at an altitude of over 43,000 feet, where the oxygen concentration is equivalent to just 3.6%, a pilot will last between nine and twelve seconds before experiencing symptoms of hypoxia. At those concentrations, a loss of consciousness is likely to occur within forty seconds, death in minutes. Breathing pure nitrogen is so lethal – and yet doesn't produce a feeling of suffocation – that it's been argued it should be considered as a humane method for executing prisoners on death row.

While rescue crews had pulled all five victims clear by 9.28, John Bjornstad, a father of three, died shortly after noon. Forrest Cole would die thirteen days after the accident. Years later, Nick Mullon too was killed by complications from the nitrogen asphyxiation.

Their tragic deaths, the first in the US space programme since the 1960s, were due to faulty procedure and miscommunication rather than any failing in *Columbia*. But they served as a powerful reminder that it was often the things you hadn't planned for – the unknown unknowns – that were most likely to get you.

* * *

Houston Director Chris Kraft knew it. He'd been there throughout NASA's two decades of triumph, tragedy and near misses. And despite the terrible, avoidable deaths of Bjornstad and Cole, Kraft felt they were running out of things they could meaningfully still test before lighting the fires.

After the destruction and delay that had bedevilled the development of the main engines, the massive fire in December 1978 had marked a turning point. By early 1981, having completed nearly 100,000 seconds of operation on the test stands – equivalent to nearly sixty-four launches – the main engines were flight-qualified and man-rated. The final flight software was released in February and loaded on to *Columbia*'s mass-memory device the following month. There were still quirks, including a sub-system that displayed 'off' when it should have read 'on'. To fix it would have meant revalidating the software, and there wasn't time. Instead they flew with a book of programme notes detailing known flaws in the software.

The crew were similarly well prepared. Young and Crippen had each shot in the region of 1,500 landing approaches in the Gulfstream training aircraft. They'd spent well over 1,000 hours in the simulators and many more hours in the classroom, familiarizing themselves with the flight data file – a tower of paper 51½ inches high and weighing 210lb. The delays to the programme had not been without benefit. John Young maintained that, as a result of the extra time, he was 'one hundred and thirty per cent trained' for the mission.

Over the first three months of 1981, the preparations for flight had gathered pace. The auxiliary power units – APUs

– powering the Orbiter's flight controls were loaded wth hydrazine and powered up; *Columbia* was disconnected from the umbilicals that provided for her from the launch tower in order to test her internal systems. A mission verification test to check the readiness of Mission Control and the global network of tracking stations culminated in the arrival of T-38 jets, playing the part of the Shuttle, escorted by Jon McBride's Chase team, at Edwards and the White Sands diversion strip. One T-38 landed at Edwards, and taxied past a waiting mock-up of the Shuttle, parked up on the lake bed at the point where *Columbia*'s landing roll was expected to end. As the T-38 returned to the hangar, the recovery convoy from Kennedy rehearsed their role on the facsimile Orbiter. On Pad 39A, the external tank was pumped full of 384,000 gallons of liquid hydrogen, followed by 140,000 gallons of liquid oxygen, then emptied again, revealing a weakness in the foam heat insulation that saw chunks fall off and engineers at contractor Martin-Marietta working round the clock to fix it and apply a solution.

The last of the Orbiter's thermal protection tiles were installed as *Columbia* waited on the pad.

Yet while it had been determined that the Shuttle was good to go, there was still the odd dissenting voice. Inside NASA's Langley Research Center, a small but influential pocket of opposition was led by an engineer named John Houbolt. And Houbolt had form. His forceful opinions had once turned him into something of a pariah inside NASA. The trouble is, on that occasion he'd been right.

Houston's original plan to put men on the moon was for all three astronauts somehow to land the upper stages of their Saturn V rocket on the lunar surface, before blasting off again at the end of their exploration to return home. It was realized early on that this approach, labelled 'Direct Ascent', put the mission beyond the limits of what a single Saturn V rocket might carry. At Marshall, Wernher von Braun proposed getting round that by launching a series of Saturns, combining their payloads in Earth orbit then continuing the mission to the moon. But working out of Langley Research Center, John Houbolt had an epiphany. The way to do it was not to use the heavy, complicated Command Module as a lunar lander, but instead to take a hollowed-out, lightweight vehicle dedicated to taking the astronauts to the surface and back to dock with the Command Module waiting in lunar orbit. His rough calculations suggested that, if adopted, it might be possible nearly to halve the weight of the Saturn V rocket's upper stages. It was a technique called lunar orbit rendezvous (LOR), and it scared the life out of everyone working in Houston. If, for any reason, the rendezvous around the moon failed, the moonwalkers would simply be left circling like a broken satellite, unable to come home, talking to Houston only until their fuel cells expired.

Such was Houbolt's conviction, though, that, abandoning any pretence of protocol, he bypassed the normal chains of command and wrote directly to NASA's Deputy Administrator. While admitting that his nine-page letter was 'unorthodox', he went on to blast his superiors, Houston's programme managers, claim he was 'a voice crying in the wilderness', and suggest that failing to adopt LOR put NASA's very existence at risk. The reaction to the letter in Houston was furious; Max Faget immediately and loudly dismissed Houbolt's numbers as lies. But as the overwhelming challenges of landing vertically on the moon in a rocket which even after shedding its first and second stages was still similar in size to the one that had boosted John Glenn into orbit in 1962 – then launching again – began to bite, Faget and Center Director Bob Gilruth took another look at Houbolt's numbers. And saw some merit in them. By April 1963, the decision to pursue lunar rendezvous was confirmed. And in July 1969, after Neil Armstrong had successfully piloted *Apollo XI*'s bug-like Lunar Module to a safe landing on the moon, Wernher von Braun turned to Houbolt, gave a thumbs up, and admitted, 'John, it worked beautifully.'

Now, Langley, driven once more by Houbolt's certainty, expressed their concern about the ability of the tiles to stand up to the fierce, complex aerodynamic loads at the point where a thick steel A-frame strut attached the nose of the Orbiter to the external tank. The men from Langley thought it best to conduct further wind tunnel testing of the area. At Houston, they weren't so sure. *We've got it covered*, thought Tom Moser; *we understand all the loads on it and we've got enough margin*. In truth, he'd kind of had it with Langley by now. He'd spent a year or more being challenged on every aspect of the thermal protection system, answering what he regarded as sometimes ridiculous questions from the theoreticians in Virginia while he and his team worked on solving real-world problems. He wasn't sure they were contributing a damn thing. *Research Center*, he thought, the clue was in the name; his job was to *design it, build it, fly it*. But the red flags kept coming up, and so, with weeks to go, he, Aaron Cohen and Chris Kraft agreed: 'OK, we'll do the test.' *To satisfy the concern*, they told themselves.

When he found out what was planned, John Yardley demanded to see all three of them at 7.30 the next morning.

All three were down at the Cape. Yardley was livid, tearing strips off them for agreeing to perform a test none of them even thought was necessary and, in doing so, putting at risk *Columbia*'s first flight. Moser and Cohen said nothing, unable to bring themselves to speak up in support of running the test. It was Kraft who took the bullet. 'They didn't want to do it,' he told Yardley; 'I agreed to do it.'

'John, it's OK,' they tried to reassure him. And it was. They performed the wind tunnel test Langley had asked for and, as they had expected, the thermal protection system came through. But Kraft was determined not to get tripped up again. It was time, as he put it, 'to light the candle'. After repairs to the damaged external tank were completed, the launch was set for 7am on 10 April.

With the test on the tiles around the forward strut behind them, Kraft phoned Langley Center Director Don Hearth. 'Don,' he asked, 'do your guys have any issues? We think we answered your questions; we're comfortable; we're ready to fly.'

'Let me get back to you tomorrow; I'll just check with everybody.'

The next day, Tom Moser was in Kraft's office in Building 1 when Hearth called back. It was exactly the response they were hoping for. 'We're good,' Hearth told them. 'We don't have any issues anywhere.'

As the flight approached, Hans Mark found himself in a perplexing position. On the one hand, he'd got what he wanted: on 11 March he'd learned from the White House that the President wanted to nominate him as Deputy Administrator of NASA. On the other, he was not scheduled to take up the post until two weeks after *Columbia* flew. There had yet to be any kind of announcement about his appointment and so there was no formal reason, so far as anyone else knew, for him to be involved in the Shuttle's launch. But Mark wanted to be there, following the Shuttle's historic first flight from where, as the recently appointed Director of Ames, he'd watched Neil Armstrong land an earlier spacecraft called *Columbia* on the moon. Mark phoned Chris Kraft in Houston and asked whether he'd be welcome in Mission Control. Kraft, who already knew of Mark's imminent return to NASA, had no hesitation in saying yes.

While at Houston, Kennedy and Edwards the final preparations were made for the first flight, inside the Blue Cube at Sunnyvale AFS, the Air Force was also getting ready. Much of the groundwork had been done on Mark's watch at the Pentagon.

In the summer of 1978, Program 'S', a Shuttle Operations Office, designated VOS, was set up to plan and coordinate bluesuit support for the orbital test flights. While some of the baseline support for the Shuttle from the Satellite Test Center had first been agreed back in 1973, this had evolved over the years that followed. Ultimately, the Air Force was eager to move command and control of its space programme away from Los Angeles to a more secure, remote location. As Air Force Secretary, Hans Mark had endorsed the move as long as a new Consolidated Space Operations Center included a mission control centre for the Shuttle. And while, in October 1979, it was decided to build the new facility in Colorado Springs, until it was ready, the Air Force would be working hand in glove with NASA. A small cadre of Air Force personnel was assigned to Houston to learn their trade as flight controllers alongside their NASA counterparts. At the same time, the Air Force paid for the creation of a secure, ring-fenced presence inside Houston from where it could support what it called 'controlled mode' – that is, classified – Department of Defense Shuttle missions. The traffic was not all one way, however: NASA also still needed the Air Force.

During *Columbia*'s first flight, Young and Crippen would be in contact with the ground for barely 25% of their time on orbit. A network of NASA ground stations capable of relaying voice transmissions between the crew and CapCom in Mission Control, Houston, was dotted around the world in Ascension Island, Chile, Bermuda, Botswana, California, Senegal, Guam, Hawaii, Spain, Florida, Australia, Ecuador and New Mexico. There was a gap, though: NASA had nothing between Botswana and Yarragadee in Western Australia. But the Air Force did. The Indian Ocean tracking station on Mahe in the Seychelles had weathered independence from Great Britain, coups, and the extortionate demands of the Seychelles government over lease negotiations, to remain a valuable part of the Air Force satellite tracking network. In early January 1979, modifications that would allow it to support the Shuttle were completed. Using a four-strong constellation of Air Force communications relay satellites, conversations with *Columbia* would bounce to and from orbit between Mahe and Houston via the Blue Cube.

While the men and women of Program 'S' looked forward to making their contribution, elsewhere in the Blue Cube the Special Projects Office continued their vigil of the Soviet Union.

In August 1980, at a rocket testing facility on the Khimka River, north-west of Moscow, design bureau KB Energomash conducted the first test-firing of their RD-170 engine, designed to power the Soviet Shuttle's strap-on boosters. And in December of the same year, under the direction of the USSR's second man in space, Cosmonaut Gherman Titov, a K65M-RB5 rocket was launched from pad LC-107 at the Volgograd Cosmodrome near Stalingrad in the direction of Lake Balkhash to the east. On top was a half-scale model of the Spiral spaceplane designated BOR-4, built to test the aerodynamic characteristics of the design in sub-orbital flight before using it to test the thermal protection system under development for the Soviet Orbiter.

An American National Intelligence estimate expected a Soviet vehicle capable of carrying a 20-ton payload to be ready to fly in the mid-eighties. The Russians called it *Buran*, after the violent blizzards that swept across their Baikonur launch site in winter. For now, though, it was the Americans enjoying centre stage.

Then John Houbolt came out of the woodwork again, pulling the kind of stunt that had so ruffled feathers in 1962.

FORTY-FOUR

Houston, 1981

THE HEATSHIELD WAS safe to fly. The issue of the tiles had been officially put to bed. Until John Houbolt, without any formal authority from Langley, wrote a letter directly to Houston Center Director Chris Kraft. Houbolt had enlisted support from an eminent structural engineer from Stanford University as a co-signatory. 'We implore you,' their letter began, 'not to fly the Space Shuttle Orbiter.' The pair went on to expound their theory that the tiles were going to come off. They may not, the letter explained, come off on ascent, but by the time the Orbiter was ready to come home NASA was going to be embarrassed by major tile loss, caused by the combined effects of vibration and dynamic pressure on the bond between the tile and the strain isolation pad beneath. They were going to lose the vehicle. As a fix, the pair suggested that the Orbiter be cocooned in a steel net to prevent the tiles coming off during flight.

Kraft was furious. He wasn't sure what more he and his engineers could have done to demonstrate the robustness of the system. It had all been said and done. But Houbolt's track record – and his ability to make waves – meant he was impossible to dismiss out of hand. At least not now he'd got the attention of NASA's senior management in Washington. Once again, Tom Moser felt like he was being forced to prove his innocence, and demonstrate to jittery agency administrators that, unlike in 1962, Houbolt was wrong. Other than that, though, the men in Houston did nothing. As far as Moser was concerned, Houbolt was just grandstanding.

At the end of March, Young, Crippen, Engle and Truly moved into quarantine, living in a caravan of trailers lined up inside Building 5 at the Johnson Space Center in Houston, one for each astronaut. In the two weeks they spent removed from the day-to-day, only the astronauts' wives were allowed to visit them without wearing surgical masks. The primary crew catching a bad head cold had been Engle and Truly's last chance to take the helm for the first flight, but after months of good-naturedly encouraging Young and Crippen to go skiing or enjoy contact sports they knew there'd be no last-minute changes now. Certainly minimizing the possibility of the mission being affected by illness was important; even a cold, if it affected the astronauts' ability to equalize their ears, would be enough to have them removed from the flight. But quarantine was also an opportunity for them to focus on the flight without distraction. As the date drew near, there was a tendency for anyone plagued by any concerns either to unburden themselves or explain them to the crew. Most of it didn't need to go anywhere near them. Instead, members of the support crews like Hank Hartsfield, Dan Brandenstein and Rick Hauck acted as surrogates, hearing people out, deciding what might need passing on. When anything did get past the gatekeepers it was unlikely to be good news. It was while Young and Crippen waited in quarantine that they learned that a problem with *Columbia's* reaction control system was not, as the engineer conveying the news put it, 100% corrected.

In the weeks since the final flight software had been delivered, time in the simulator had revealed that the RCS, an arsenal of thirty-eight small rocket thrusters mounted in the nose and tail used to control the attitude of the Shuttle on orbit and re-entry, had a tendency to fire uncommanded, fast lapping up the precious, limited amounts of propellant carried on board.

Well, wasn't that nice? Young reflected as he digested the news. *Always fun to hear about a potentially dangerous problem before you launch. Columbia's* Commander was particularly concerned about hypersonic re-entry when, in the thin air, the aeroplane-style aerodynamic flight control surfaces – the rudder and elevons – were of no use. Much of the uncertainty around the Orbiter's anticipated performance was in this area of the envelope. They were dealing with chaos theory. Tiny variations had exponential effects, capable of unfolding more quickly than a human pilot could possibly react to. Small models of the Orbiter had been tested in the shock tunnels at representatively high Mach numbers for barely seconds at a time. Data captured from that was then scaled up, but the results weren't perfect, only the best that the supercomputer at Ames Research Center could produce. Scaling ratios had a substantial but imperfect effect on the predictions; interference from shock-waves interacting with the walls of the wind tunnel further muddied the water. The hypersonic region was, as Hank Hartsfield, who'd been working on the Shuttle's flight control system, pointed out, 'an uncertain world'.

One clear indication that did emerge, however, was that as *Columbia* sliced down through the upper atmosphere at over fifteen times the speed of sound, she was unstable in yaw; her back end had a tendency to swing out. Unchecked it would rip her apart. The rear reaction control jets mounted beneath her tail really did need to work as advertised.

Unwelcome news of possible disaster aside, quarantine could get dull. The four astronauts broke up their time with flights in the T-38s, and sim sessions, but slipped into a routine, ending each day with dinner, time in a nearby gym and conversation before turning in. At least the two weeks they spent in the trailers was made easier by a generous supply of beer from Crip's mother, happy to share supplies from Crippen's – her Porter, Texas beer joint – with her son and his friends; pleased to be able to contribute something to his success.

On 5 April, just before midnight, the final countdown began.

PART FOUR

Ignition!

'It is this, the unspeakable danger – and the term "danger" is itself taboo among the pilots – that has always given the phrase the first flight such a righteous aura among test pilots.'

Tom Wolfe

FORTY-FIVE

Houston, 1981

BOB CRIPPEN'S BEATEN-UP old red pick-up truck had become something of a trademark; the kind of vehicle driven by a guy who didn't have to try too hard to make an impression. Astronaut Joe Allen's car just drew heat from his colleagues for being an old rustbucket.

After joining NASA as a scientist astronaut in 1967, Allen had spent three years seconded to the agency's headquarters. He took the car with him but the north-eastern winters hadn't been kind to it. Now, back in Houston, he'd been assigned to the support crews, and was lead CapCom for the re-entry phase, with Rick Hauck as his back-up, for STS-1.

On 8 April, two days out from launch, George Abbey collared him. 'Joe,' he said, 'we're going to take John and Crip out to the airplanes. Come with me.'

'Fine,' Allen said. 'Why do you need me?'

'My car won't start.'

'Oh, terrific,' Allen replied, thinking that his corroding jalopy was not the way the crew of America's new spacecraft ought to be arriving at Ellington Field to begin their journey to the Cape to fly her.

The four of them crossed the parking lot to Allen's car. Although just seven years separated them, Young and Crippen already looked generations apart. At fifty, Young wore a gleaming white polo-neck beneath his NASA flight-suit, adorned with more mission patches than any other astronaut. He looked smart. By contrast, Crippen seemed relaxed, bare-chested beneath the open V-neck of his flying overalls – a picture of a frontline fighter pilot were it not for the blue fabric and NASA mission patches.

Settled in the right-hand seat, Young reached into his pocket, pulled out some cash and tried to hand it to Allen, sitting next to him. Allen had recently joined Young and Crippen for lunch and Young – with good reason to be pre-occupied – hadn't brought money with him. Allen had been happy to pick up his tab and had thought no more of it.

'Come on, John,' Allen said, urging him to put the money away.

'No,' Young insisted, 'you don't go fly these things when you got debts.'

Allen took the money and Young was reassured.

'All my debts are paid,' he said, and returned to his own thoughts.

Then Allen realized to his horror that he could see TV antennas outside broadcast trucks camped out at Ellington waiting for Young and Crip's arrival. He cringed at the thought of his wife's reaction to seeing their sad-looking car on national TV. She wouldn't talk to him for days. He quickly came up with a plan. He announced he'd got a favour to ask. 'This sounds a little peculiar,' he admitted, 'but I can't really . . . can I drop you at the parachute shack?'

George Abbey had said the pilots' chutes were already waiting for them in the cockpit of the T-38. Strike one.

'I just need to drop you,' Allen tried again, sounding awkward. 'You just walk out—'

'Oh, I get it,' Young smiled. 'It's the car.'

'Yes.'

'Be happy to.'

So rather than pull up next to their jet, Young and Crippen walked through the hangar to emerge in front of the cameras and flashbulbs on their own, while Joe Allen sloped away back to Houston. The next time he expected to speak to them they'd be in space. The thought of what lay ahead sent a chill through him.

The day after Young and Crippen left Ellington Field for the Cape, Hans Mark arrived in Houston. Not only had Chris Kraft welcomed him into Mission Control, he'd found him a spot in the control room itself behind the programme office console alongside his old friends Programme Manager Aaron Cohen and his deputy, Milton Silvera. Mark was determined not to miss any part of the flight. Instead of booking himself into a hotel, he set up a cot in the infirmary behind Mission Control Center in Building 30. He didn't anticipate needing much sleep.

At the Cape, over 40,000 people had gathered to watch *Columbia* reignite America's manned space programme. Many thousands more were camped up and down Space Coast to catch a view of the launch. Since last sending astronauts to space, NASA had sent the *Viking* landers to Mars, and the *Pioneer* and *Voyager* probes had beamed home extraordinary images of Jupiter and Saturn, but nothing captured the people's imagination like sending men into space. Alongside members of the public, nearly 3,000 journalists were at Kennedy to cover the launch. It was time to give the world what it was waiting for.

There was a knock on the door. 'Hey, Crip! Time to get up.'

He glanced at his watch. 2.45am. The STS-1 Pilot climbed out of bed, showered, shaved, and pulled on his flightsuit.

Out on the pad, as he and John Young got themselves ready, *Columbia*'s external tank was being pumped full of the liquid hydrogen and oxygen that would propel them into orbit.

The two astronauts had a brief fifteen-minute physical, not much more than a temperature and blood pressure check and a peek in their ears – nothing compared to the pair of full-service medicals both had endured earlier in the month. 'OK,' the Flight Surgeon told them, 'ready to go fly.'

First there was breakfast. Sitting together on the top table in front of heavy curtains, behind flowers and cake decorated with the STS-1 mission patch, the two of them laboured through steak and eggs, coffee and orange juice, served by waiting staff wearing surgical masks. The astronauts' faces betrayed the early hour of the morning. There were occasional smiles from Crippen, but, to his right, Young seemed lost in his thoughts, staring across the room as he ate.

Before being helped into their pressure suits, both astronauts were hooked up with sensors to ECGs to record their vital signs. As they climbed into the bulky tan-coloured suits, acquired from the Air Force, who used them to protect high-flying SR-71 spyplane pilots, Young was relieved to see that the Stars and Stripes on the suit's left shoulder had been scaled up. Even Crip had had to agree that the first flags

– 344 –

they saw were almost comically small. Barely a week before the flight, Young had been trying to get the flags changed. 'It has to be a small flag,' he was told by someone in public affairs, 'to balance out the Space Shuttle programme' patch. Young was quick to point out that the United States was bigger than the Space Shuttle programme. The flags were traded up for something more substantial.

Suited up, they sat in a pair of brown vinyl armchairs, a computer tower between them tracing their vital signs on to reams of paper, while fans blew cool air into the pressure suits through umbilical tubes. Half squadron ready room, half hospital, the scene suggested that the astronauts were both undergoing some kind of exotic medical procedure – a blood transfusion or dialysis.

Young flipped through his flight reference cue cards. He was nervous. But he also appreciated that it was right to be apprehensive. That came from a thorough understanding of what he and Crip were about to do. *Some of the potential emergencies*, he thought, *make* Apollo XIII *look like a Sunday School picnic*. He had to remind himself that as a result of both their training and procedures they were ready; *prepared*, he thought, *to deal with anything short of Armageddon*. He also had faith in the vehicle. NASA, as he put it, *had engineered the heck out of it*.

Crip's biggest fear was of simply making a mistake. After the recent humiliation of the Iranian Embassy crisis and then, less than two weeks ago, an assassination attempt on the new President, he felt the country was desperate for good news. *Don't screw up, Crip*, he told himself. *If there's a problem, we can handle it.*

FORTY-SIX

Kennedy Space Center, 1981

AT NEAR 4.30AM they were helped up from their chairs. Each picked up a portable cooling unit and was led by George Abbey to a lift down to the exit. Waiting for them outside in the dark, in front of lines of waving Kennedy employees, was the familiar sight of an Airstream recreational vehicle, the 'Astrovan', reporting for duty for the first time since 1975. Young and Crippen thanked the crowd, climbed into the back with Abbey, and drove off. Before continuing on to the pad, they stopped at Launch Control to drop off the Flight Operations boss. He shook them both by the hand and wished them good luck.

As Young crawled on all fours through the hatch into *Columbia*'s mid-deck, Crip smiled and shook hands with pad-workers in masks, each sporting either 'NASA' or 'Rockwell' on the backs of their white overalls. Young clambered awkwardly in his pressure suit over the rear bulkhead then through the companionway to the flight deck. With the Shuttle on her tail pointing skywards, his seat was now directly above him. He pulled himself up to the cockpit using handholds before dumping himself into his ejection seat. He was enveloped on all sides by grey panels carrying dials, displays, and over 2,000 different switches recessed or caged so that a stray foot or elbow in space couldn't flip them unintentionally. Patched wherever they would fit were checklists and cue cards, mounted on Velcro. Crip followed him and took his place in the right-hand seat before both astronauts were helped to fasten their harnesses, clip into the ankle restraints that would protect their flailing legs in the event of an ejection, and lock their helmets on to the metal hoops circling their necks. Twelve different communication and life support fittings connected them to their spacecraft.

They began running through comms and systems checks as the count continued. At T minus one hour and ten minutes, with the two astronauts settled, the closeout crew retreated back to the White Room, pushed shut the side hatch and secured it. That was a battle Young had lost. He thought an outward-opening hatch on a spacecraft made as much sense as an inward-opening door on a submarine. *Sooner or later*, he thought, *it was going to get you.*

At the scheduled twenty-minute hold, the computers transitioned from pre-launch to flight mode. At the same point, the back-up computer, which until this point had been inactive, came on line. But something was wrong. An error message flashed up on Crippen's display.

Inside Firing Room 1 and Mission Control in Houston, the flight controllers' console lit up with the same information.

'BFS didn't follow the PASS,' Crippen reported from the cockpit. For reasons unknown, the back-up flight system, the emergency computer, had failed to synch with the four primary flight computers.

TFNG astronaut Kathy Sullivan knew as soon as she heard Crip's radio call that *Columbia* was staying put. She was assigned to ABC News alongside *Apollo XVII* Commander Gene Cernan. Clearly the network regarded the last man on the moon as the box office draw, using him for their TV coverage while relegating Sullivan to support the radio coverage. Cernan, though, was now an ex-astronaut; Sullivan, by contrast, had been training to fly aboard the Shuttle for over two years. She listened in as the launch team discussed the situation. 'This ain't going today,' she told the radio team. Meanwhile, Cernan, in the absence of any announcement from NASA, was being called on to fill the airtime. Asked to provide increasingly speculative answers to what might have gone wrong, he was struggling; *getting further off the ranch*, Sullivan worried. She pulled at her producer's sleeve to get his attention. 'Does it matter that he's not giving correct answers?' she asked.

The confusion wasn't limited to those trying to follow the launch from TV studios. Young looked to Crippen for ideas, who, he maintained, *knows more about the Shuttle's computer system than anyone has a right to*. But his Pilot was stumped. Crip had never seen this happen; never even *heard* of it happening. The back-up computer wasn't communicating with the primaries. Another attempt by the back-up computer to synch with the primaries produced only more error messages. Crip reached forward and pushed the reset button. When exactly the same problem reoccurred, Launch Control held the count.

* * *

The crew had been lying on their backs in full pressure suits, helmet visors down, for an hour and a half. It now looked as if they'd be in their ejection seats for a whole lot longer while Houston tried to get to the bottom of what was wrong. There was nothing to do but wait and try to pass the time.

Crip fell asleep and, as he did, his right arm slipped off his lap on to the right console. On the back panel of the flight deck was a switch controlling a back-up signal to the telemetry. No one had imagined at the time the controls were laid out that it might be required during ascent. When that view was revised as the first mission approached, the only way to operate the switch was to tie a piece of string around it and run the other end to the Pilot's seat. It was a one-time deal though. Once the string was pulled, the switch couldn't be unswitched. As Crip dozed, the string got caught in the fastening on the back of his glove. When he woke and raised his hand, the switch pulled. They were just going to have to fly like that. *Going to sleep*, he decided, *was not necessarily such a good idea*.

In the end it didn't matter either way. After waiting in the cockpit for nearly six hours, with no fault in the back-up software discovered, Launch Control restarted the count.

At T minus sixteen the back-up computer tripped again, forcing a decision to scrub the launch.

'Well,' said Crip, after climbing back out of the Orbiter's flight deck on to the service tower, 'this is liable to take months to get corrected.' But it didn't. The problem was identified and a solution tested successfully by the end of the day. It boiled down to a forty-millisecond time skew that occurred at T minus fifteen hours. Caused by different programming priorities chosen by IBM and Rockwell when writing the primary and back-up software, the problem, analysis suggested, was likely to happen just once every sixty-seven times. There was nothing fundamentally wrong, and so, on 11 April, the decision was taken to go for launch the following morning.

Crip still wasn't so sure they'd get away. *Something else will go wrong*, he thought; *we're going to get lots of exercise at climbing in and out.* Neither he nor Young had expected to go first time. On leaving Houston for the Cape, Young had packed a month's worth of clothes into the baggage holder they had slung like an extra fuel tank beneath the T-38's fuselage. As ever, Young tried to make sure he'd thought of everything.

Like *Columbia*'s primary crew, Dick Truly and Joe Engle had flown down to the Cape for the launch. But with the launch scrubbed on the 10th, Truly argued that either he or Joe should return to Houston. He felt strongly that one of them should be in Mission Control during the Shuttle's ascent. *John's healthy, Crip's healthy*, he thought, *I'm not going to fly*. If something went wrong, however, they were trained to the same level of preparation for the mission as Young and Crippen themselves, and in a unique position to contribute. After discussing it with Joe and George Abbey, he decided to fly a T-38 back to Houston, where he would maintain his quarantine, just in case, until the launch attempt the next morning.

He caught up with Crip and Young before leaving the Cape. 'Have a great flight,' he told them.

If that went ahead, as was now planned, on the 12th, it would be just twenty years to the day since Yuri Gagarin had first been fired into orbit on top of a modified R-7 ICBM from the Kazakh Steppes. That they had come this far, from there, so quickly, seemed almost inconceivable. As if to underline the difference between their bird and the tiny falling capsules that launched the space race, while they waited to climb back aboard *Columbia* for their second launch attempt, Young and Crippen spent a couple of hours shooting approaches into the long Kennedy runway in the Gulfstream training jet.

FORTY-SEVEN

Houston, 1981

TWO DAYS LATER. Young and Crippen were once again strapped into their seats aboard *Columbia*. The first cold blue light of dawn was beginning to show on the horizon. An hour and a half until launch.

In their secure room on the 2nd floor of Building 30, Mission Planners Ed Lineberry and Ken Young were all set. Over the previous six months their preliminary operational trajectory for STS-1's orbital phase had become the operational trajectory, the data frozen as nominal. Sister branches had done the same for launch and re-entry. It was sacrosanct. Had been for months. And, while NASA's press kit explained that each day the 'launch windows opened at local sunrise plus 45 minutes and are more than 6 hours in duration', Lineberry and Young knew that, while in theory that was right enough, in practice there was almost no flexibility at all. 0700 hours EST had been chosen very specifically. It was when Lineberry and Young had established that the Shuttle had to launch to optimize the opportunities for the NRO to capture their images of her on orbit.

After the scrubbed launch on April 10th, the two men were

forced to look again at the scheduled lift-off time and make small adjustments to *Columbia*'s planned orbital altitude.

Over the months prior to STS-1, they'd asked one of their engineers on the Mission Planning and Analysis team, Dave Scheffman, to modify a special graphics program that would calculate and display the Orbiter's planform – her underside – in various different orbits, attitudes and lighting conditions. Mainly, they asked Scheffman to model the view from the various ground-based radar and telescope sites from which NASA planned to look at *Columbia*. But just occasionally, they'd ask him to take a bird's eye view from points around an imaginary sphere outside of the Shuttle's orbit. They never gave a reason and they were never asked for one.

Dave's most probably figured out, Young thought, *that something hush-hush is going on*. But Scheffman didn't probe. Just ran the program.

In the two days following the initial launch attempt, using an offline computer in Building 12 next door to Mission Control, he ran views using Lineberry and Young's revised calculations. All good.

It was anything but inside the Blue Cube. With barely an hour and half to go until the launch of the Shuttle, one of the KH-11 KENNEN birds had experienced a major attitude control problem. In racing to correct it, Air Force controllers realized that the required attitude manoeuvres had become uncoupled, causing a significant change in their satellite's orbit. And that would affect the timing of its long planned orbital conjunctions with the Orbiter. The opportunity to capture sat-squared images of *Columbia* was now in the balance.

* * *

An urgent, encrypted message from the operators in Sunnyvale arrived with Lineberry and Young in Houston: The spy satellite would be travelling along a different state vector to the one around which the Shuttle's launch had been designed. As the countdown to *Columbia*'s launch ticked down, Lineberry input the new numbers into the computer to calculate a new lift-off time. Near two and a half minutes later.

Young called Mission Control on the secure line. Flight Director Neil Hutchinson took it in his stride. 'Well,' he said, 'we can adjust the lift-off for that easily when we come out of the built in hold at T minus nine minutes.' Thinking it through he added, 'But what am I gonna tell the Launch Director at the Cape is the reason for that little delay?'

'Hell, I don't know, Flight,' Young said, 'just make up something.' Before Young had a chance to consider further what that might be, Hutchinson's voice cut across the Mission Control open loop: 'Guys, we may have a small lift-off delay of two or three minutes when we come out of the T minus nine hold. The guys in RTCC [the Real Time Computer Complex on first floor of Building 30] need time to reboot a backup computer.'

Com Sup, the Computer Supervisor running the RTCC, came back quickly: 'Hey, Flight, *we* don't have to reboot *any* back up machine . . .'

Flight Dynamics Officer Jay Greene cut in: 'You. Do. Now. Com Sup.'

'OK, FIDO, we're on it.'

Listening in to the exchange alongside Ken Young in the secure room, Ed Lineberry continued to type numbers into the HP desktop, running through pre-programmed equations.

'Hey,' he turned to Young, his steeltrap mind processing the possibilities. The words tumbled out: 'we don't really need to delay lift-off. We can make that two and half-minute shift by just adjusting the two circularization manoeuvres *up* one or two nautical miles.' Spacecraft weren't launched straight into their final operational orbits. Instead, after reaching a safe orbit, they performed a further series of manoeuvres to reach the orbit required for the mission. By pushing *Columbia* a little higher, into a slightly slower orbit, the effect on the conjunctions between the Orbiter and the spy bird would be similar to that achieved by delaying the launch. Lineberry and Young's new recommendation was accepted by Mission Control. *Columbia* would launch on time, but it wouldn't be without some cost.

By altering the distances between the two spacecraft on orbit, Lineberry and Young had torn up the attitude requirements for capturing the pictures of the Shuttle's heatshield that had been so carefully worked out and confirmed with Dave Scheffman's help. But stick to the flight plan as written and all the imaging opportunities would be, at best, less than optimum. But at worst, useless.

The Plan that had taken weeks and months to refine now needed to be rewritten in hours and days. What *Columbia*'s Commander, John Young, had so happily described as 'easy', was now a good deal more complicated.

FORTY-EIGHT

Kennedy Space Center, 1981

'THAT'S A MIGHTY fine piece of speech' was John Young's succinct, appreciative reaction to words read out on President Ronald Reagan's behalf by Launch Director George Page. Page then added a few more words of his own: 'John, we can't do more from the launch team than say, we sure wish you an awful lot of luck. We are with you 1,000 per cent and we are awful proud to have been a part of it. Good luck, gentlemen.' Two minutes later, after a planned ten-minute hold built in by Page to make sure his team was calm and focused, he restarted the count. Outside, the crowds packing the bleachers erected around Kennedy Space Center cheered.

T minus nine minutes and counting.

With his left hand, Jon McBride advanced the T-38's two throttle levers. In the seat behind him, Pinky Nelson clutched his camera. The jet's twin General Electric J85 engines flared as the afterburners sprayed fuel into the jet-pipes to add a few thousand pounds of extra thrust. Chase One accelerated down the runway at Patrick Air Force Base, on the edge of Cocoa Beach, Florida, followed by a second T-38, Chase Two.

T minus seven minutes and fifty-two seconds.

Lying on his back inside the cockpit of the Orbiter, John Young watched the White Room access arm jerk away awkwardly from *Columbia*'s side. It looked like it might jam completely, but after telescoping back a short distance from the fuselage it then swung smoothly out of harm's way. With seven minutes to go, computers on the ground were cycling through diagnostics and checks on every element of the Shuttle stack. Launch Control advised Young and Crippen to lower the visors of their helmets.

'*Columbia*, Launch Control, initiate APU pre-start.'

'Roger,' acknowledged Crip. With authority to light the auxiliary power units, he reached down to begin the sequence to power up the hydraulics to the flight controls and rocket gimbals.

He glanced over at Young. Even as *Columbia* came to life, Crip still wasn't confident they'd get away. *We're going to find something that's going to cause us to scrub again.* He kept his thoughts to himself.

'Control, *Columbia*, pre-start complete,' reported Crip. 'Powering up APUs.'

T minus five minutes.

The last of the three Chase Air Force T-38s, a flying reserve, climbed away from Patrick towards a holding pattern to the south of the Shuttle's flightpath. There, all three would wait at 38,000 feet, ready to pick up *Columbia* on her return to Kennedy in the event of a return to launch site abort.

* * *

Liquid oxygen boiled off through vents at the top of the external tank, condensing in the sub-tropical humidity and curling in ribbons above the astronauts. Behind them, *Columbia* flexed her flight controls for the first time, moving the elevons, body flap and rudder through their full arcs of travel. In the cockpit, Young and Crippen monitored the strip indicators in the centre of the control panel that recorded their progress – another step along the path to ignition.

T minus three minutes, forty-two seconds, and counting.

At two minutes, the venting valves on the external tank were closed, allowing the liquid hydrogen tank to build up pressure for flight.

Seconds later, Firing Room 1 announced, 'You are go for launch,' following up with a call that hadn't made it on to the cue cards: 'Smooth sailing, baby.'

For the first time, Crippen dared to believe they were going to go. 'I think we really might do it . . .' he said to Young. And then his heart began to race, soaring quickly to around 130 beats per minute. Alongside him, his Commander's pulse barely pushed ninety.

T minus thirty-one seconds.

Columbia's primary computers assumed control from the ground launch sequencer and the solid rocket booster nozzles were powered up. As the Shuttle groaned and sighed on the pad, girding herself for lift-off, Hugh Harris, the Firing Room Commentator, counted down over the public address system: 'T minus fifteen . . . fourteen . . . thirteen . . . twelve . . . eleven . . . ten . . .'

A cat's cradle of sparklers fizzed beneath the bells of the

three main engines, burning off any bleeding hydrogen to ensure a smooth ignition. At T minus 3.3 seconds, the taps were turned on. Liquid hydrogen and oxygen streamed into the main engine turbopumps, which near instantaneously spun up to their full 28,000 rpm, forcing high-pressure fuel and oxidizer into the engine's combustion chambers. In quick succession, over less than a quarter of a second, the three main engines ignited, building rapidly in power over the next three seconds. As the three pale flames flared then settled, orange shock diamonds formed, hovering like angry ghosts at the centre of each plume.

We have main engine start.

On the flight deck, Young and Crip felt the vehicle come alive with a smooth rumble. In the centre of the control panel, the strip indicator bars rose rapidly to 90%. Outside the deep window on the Commander's side of the cockpit, the stack cast a moving shadow on the launch tower as the power of the three main engines pushed it forward out of the vertical – the twang. Young and Crippen felt the trident-shape of the Shuttle stack swing forward then rock back through a 2-foot arc. Then, with the three main engines burning stable, in the split-second the stack swayed back to a completely vertical position the two solid rocket boosters exploded into contention.

From silence to near six million pounds of thrust in less than half a second.

And everything changed.

As *Columbia* leapt from the ground, the astronauts were pushed back into their seats with a force unlike anything John Young and the other Apollo astronauts had experienced at lift-off. With the snapshot, all-or-nothing ignition of the SRBs, what had been, for a few seconds, a powerful advertisement for the sophistication of Rocketdyne's engineering was brutally and instantly overwhelmed, becoming instead a bone-shaking, agricultural experience, like driving with flat tyres along a washboard road. The astronauts' cheeks shook, the sound of their ejection seats fighting at their mounts drowned out by the bigger cacophony beneath them. Even inside their helmets, the noise was like standing close to a chainsaw. The simulator hadn't even come close to capturing it.

So jolted around was he by the sharp, metallic vibration of the SRBs, Young had trouble reading the instruments ahead of him. But the boosters were doing their job; the Shuttle was really moving. By the time he called lift-off they were halfway clear of the tower. He'd joke later that he thought he'd swallowed his Adam's apple.

Rainbirds, a series of sound-suppression water cannons, had been laying liquid sheets across the pad from eleven seconds before ignition. The intention was for the water to prevent the shockwave produced by the ignition of the SRBs from bouncing back off the pad and damaging the rear of the Orbiter. The rainbirds didn't even come close to containing it. With a twenty-millisecond pulse of sound, too brief for the crew even to be aware of, the SRBs fired a pair of massive aural pistons – like bolt guns against the pad – that threatened disaster. A wave of overpressure rebounded to strike *Columbia* before she'd made her escape with a violence she'd not been designed to endure.

Tom Moser had been in Launch Control before lift-off, but he knew he had to see, hear and feel the Shuttle's departure. He'd made his way on to a balcony, watching first as the main engines gave life to a boiling, furious cloud before the SRBs seemed to carry *Columbia* out of a weather system of her own making on two solid pillars of white fire. Then, a second or so later, having covered the 3½ miles from Pad 39, a freight train of noise hit him, a physical, super-amplified sound like tearing canvas that pulled at his clothes and reverberated inside him – a noise capable of killing anyone within 800 feet of it.

John Young called 'Tower clear' as the gantry's lightning conductor disappeared beneath them. A couple of seconds later, sitting outside the centre of rotation, he and Crip were turned on their heads when *Columbia* rolled through 180° on to her back as the computers established her in the right trajectory to reach an orbital inclination of 40.3° to the equator. Sunlight flared through the cockpit glass as the Orbiter rolled out of shadow into the rising sun. Accelerating through 100 mph, the stack settled into a pitch of just under 80°. Inverted like this, the load on the wings was reduced as they approached the period of maximum dynamic pressure – Max Q – during the ascent. Still low, the Shuttle stack's awkward cluster of shapes and struts generated fierce turbulence and vortices as it approached transonic speed. As they had so many times in the simulator, Young and Crippen checked that the three main engines were throttled back to 65% to protect the vehicle from the worst of it. Hanging the Orbiter beneath the tank as she climbed, though, also meant that, if a main engine failed, she was in the right attitude for Young and Crippen to perform a return to launch site abort.

From 38,000 feet, Jon McBride had the best seat in the house. He watched *Columbia* soar past him on a

750-foot-long plume from the solid rocket boosters just a minute after she'd left the pad. *This is the greatest*, he thought, urging her on, listening intently to the calls between CapCom and the Shuttle. But he was going to be a lot happier once Crip and Young had passed the point of no return.

In Houston, Mission Control missed the spectacle of it completely. Video feed of the launch was deemed to be a distraction to the controllers, each monitoring their own particular sub-system. Dick Truly, plugged in behind Dan Brandenstein at the CapCom console, watched trajectory data crank up on a monochrome TV display while he listened to the radio calls between Brandenstein and the crew on his headset.

'*Columbia*, Houston,' Brandenstein called, his voice betraying his own excitement. 'You're go at throttle up.'

'Roger, go at throttle up,' Young replied.

Past Max Q, the three main engines came back up to full power. Now travelling faster than sound, *Columbia*'s rapid acceleration was unhindered by the concerns about dynamic pressure that had afflicted her as she bludgeoned her way through the dense air of the lower atmosphere. Ahead of him, Crip watched as small flakes of white debris that had sloughed off the nose of the external tank skittered and stuck around the cockpit glass, some part of him aware that they shouldn't be there.

In Mission Control, Flight Dynamics boss Jay Greene didn't like the look of the trajectory. Ascending in a long open arc, *Columbia* was supposed to be flattening off, trading gain in altitude for a gain in speed. But she was lofting. *We're going to go off the plot board*, he thought. There was nothing anyone could do except ride out the solids then sort it at second stage.

From his ejection seat, John Young noticed that the pitch needle was off-scale high. *Better high than low*, thought Crippen alongside him, as the pitch deviation pushed as far as 6° beyond nominal.

John Young once said 'what you don't know won't hurt you', before admitting 'but sometimes it will, I guess'. Neither he, Crip nor anyone else yet had the least idea of what had happened at the moment the solid rocket boosters had blasted into life. On the ground, the big solids had scorched the grass a mile in every direction. Closer to the epicentre there was buckled steel, melted and torn-away electrical wires, and a lift door that had had a railing driven through it. But it was the effect of the overpressure from the big boosters on *Columbia* herself that, had they known of it, would have been of greater concern.

That there would be a shockwave was understood and anticipated. One-fifteenth-scale tests were staged using the solid rocket motors from Tomahawk cruise missiles. Once assembled alongside a one-fifteenth-scale Orbiter and external tank on a one-fifteenth-scale launch pad, the engineers fired the motors and scaled up their results. In extrapolating the figures from the test, however, the wrong assumptions were made. Once more it was shown that, as the last line of an old air accident investigation from the fifties had it, 'extrapolation is the fertile parent of error'. What hit *Columbia* was a factor of ten higher than had been predicted. As a result, a strut holding the oxidizer tank for the forward reaction control system in the Orbiter's nose buckled, straining the pipes carrying the propellant. That it didn't fail – as the structure of the nose had during stress testing – was a lucky near miss. In the rear of the Orbiter, the body flap, crucial to controlling the Shuttle at hypersonic speeds, was deflected 5° out of position.

Young would later suggest that if he'd known *that* he'd have ridden *Columbia* to safe altitude and ejected. But at this stage, the only real evidence of any problem came from a handful of outlying accelerations recorded from around the airframe at the point of ignition. No one was asking whether there might be any damage to the Shuttle as a result. Nor would they. After surviving the split-second assault from the shockwave at lift-off, *Columbia* was performing like a dream.

'*Columbia*, you're negative seats,' Brandenstein told the crew as the Shuttle climbed through 120,000 feet, now travelling too fast and too high to use the ejection seats. Seconds later, Brandenstein spoke again; it was time for the first major milestone of their ascent to orbit: '*Columbia*, you are go for SRB separation.'

On the flight deck, the computer displays told the crew that the chamber pressure inside the solid rocket boosters had dropped below 50lb per square inch. Fuel gone, their job of giving *Columbia* a leg-up to 160,000 feet was complete. The solids had fought and battered their way through the lower atmosphere, now it was time to drop their dead weight and let the Shuttle's three main engines, grateful for the brawn supplied by the big boosters, continue alone towards their goal.

Two batteries of eight rockets fired for a second. With an explosive thump, the astronauts' view ahead was engulfed by a white-orange flash rolling back past the side window of the flight deck as the two boosters were punched away. Then, silence; the violent metallic grind of the solids suddenly absent. A beat of concern pulsed through Crippen as, with the dramatic end to the fury of the solids, the thought struck him that they'd lost the power of the main engines. A quick glance down at the temperatures and pressure reassured him. *Still doing their thing*.

Young reported the separation of the SRBs over the radio to CapCom.

'Roger on the sep, *Columbia*,' confirmed Brandenstein from Mission Control.

Sitting behind him to his left, Max Faget could contain himself no longer. He leapt up from behind his console and shouted, 'They're off!'

Chris Kraft spun round in surprise. 'Max, for Chrissake, sit down,' he told the exuberant engineer.

With the boosters gone, *Columbia*'s trajectory began to flatten out as she accelerated under the smooth, refined power of the three main engines, leaving nothing but an exhaust of water vapour in her wake. As the speed built up, so too did the gs on the crew, pushing Young and Crippen back into their ejection seats. A little over two and a half minutes after lift-off, *Columbia* was 40 miles high and 40 miles downrange of Kennedy, chasing down a mile every second. And getting faster.

Behind her, the two used-up SRBs continued upwards on a ballistic arc for over a minute, climbing another 50,000 feet before reaching their peak and starting the fall back to Earth. Two specially built 1,052-ton recovery ships, *Freedom Star* and *Liberty Star*, each carrying a team of divers, awaited their return. So too did a Soviet spy trawler. Instead of nets, winches and freezers, the ship was packed with electronic intelligence-gathering equipment, intent on mining what data she could from the launch of America's first Shuttle. A US Coastguard helicopter was on hand to make sure she didn't come too close to the operation to recover the boosters.

Columbia speared through an altitude of 62 miles – the so-called Karman line that NASA recognized as the boundary of space. Barely four minutes after ignition – and fifteen years after being chosen as an astronaut by the US Air Force – Bob Crippen had earned his astronaut wings. His first words were, 'What a view! What a view!'

FORTY-NINE

Chase One, 1981

'COLUMBIA, STAND BY for negative return. Mark. Negative return.'

In the cockpit of Chase One, this was the call from CapCom that Jon McBride was waiting for. Whatever happened to her now, the Shuttle was too far away to be able to make it back into Kennedy. If they lost an engine, Young and Crippen either had to claw their way into orbit or fly the Orbiter across the Atlantic into Rota Naval Station in Spain. With a return to Kennedy no longer an option, McBride turned the Chase Air Force on its heels back towards Patrick AFB. With their flight plans west already pre-filed, they'd be at Patrick only long enough to refuel before taking off again. All being well, in a few hours' time they'd be on the ground at Edwards and ready – in case of an early end to the planned two-day spaceflight – to meet the returning Shuttle.

Young and Crippen were travelling rather more swiftly, now covering 2 miles every second and accelerating fast. No longer fighting air pressure, the Orbiter was gaining another mile per second's velocity every minute. But it still wasn't enough. Six minutes into their ascent, at a height of 400,000 feet – 76 miles high – *Columbia*'s nose pitched down into a powered dive that would trade height for speed. Upside down, Crip watched as Earth's blue horizon, streaked with cloud, seemingly descended into his field of view. Using gravity's pull, the Shuttle lost 70,000 feet of altitude, and in doing so piled on the speed. At a little over 330,000 feet they were lower now than the maximum altitudes reached by the X-15 rocket plane, but moving at 15,000 miles per hour, three times faster. *Columbia* had become the fastest winged vehicle ever built at around the same moment Crip won his astronaut wings. He was being pressed into his seat with a force of 3gs. A trim 160lb, he now weighed 480lb. Next to him, John Young's 5ft 9in frame was pushing 500. *Columbia* herself was suffering similar weight gain.

To ensure the load never exceeded the airframe's 3.5g limit, the main engines, programmed to maintain a constant 3g limit, throttled back again to 65% of their rated power for a second time. Six seconds later, a cluster of three red lights on the main engine panel blinked on as the three main engines shut down for good. In the immediate silence that followed, the two astronauts were thrown forward gently in the straps, bracing themselves against the cowling of the instrument panel as the Orbiter unloaded, the relentless push from the big Rocketdyne engines at an end.

And then, for the first time, Crippen noticed that any loose items, washers or screws that had been tucked away, out of sight of pre-flight inspections, had begun floating around the flight deck.

'OK, MECO,' Young reported – main engine cut-off. It

had been nearly a decade since his last spaceflight ended when he'd splashed down in the *Apollo XVI* Command Capsule. He was enjoying being back.

'Roger, *Columbia*, MECO,' Brandenstein responded.

Over the exchange between the Shuttle Commander and CapCom, the public announcer took the reins: 'Confirm shutdown. *Columbia*, the gem of this new ocean, now in space, not yet in orbit. Standing by for external tank separation.'

Although travelling at an orbital speed at MECO of 17,500 mph, *Columbia* was not yet in a stable orbit. To get there, two smaller 6,000lb thrust Aerojet rockets mounted in the orbital manoeuvring system pods on either side of the tail had to fire. But with no further need for the main engines or the liquid hydrogen and oxygen fuel contained in the external tank, it was time to ditch it. With that gone, the flight ahead gave Young less cause for apprehension than what had already passed. It was the ascent of such a new, untried combination that had preyed on his mind. Now he was back in zero-g and able to marvel, once more, at how beautiful Earth looked from space.

As he waited for the tank to separate, the movement of the main engine bells as they stowed for orbit suddenly pitched the nose up. In response, the powerful reaction control system jet thrusters in the nose fired, thumping through the flight deck like the report of a field gun as they produced a combined 2,610lb of thrust, quickly forcing the nose down. As the three jets fired white plumes tens of feet into space ahead of the cockpit window, Young's thoughts echoed the warning they'd been given in quarantine about the RCS jets – *not 100% corrected*. Adrenalin shot through him. Fearing they wouldn't now shut down on their own, he reached up with his right hand to an overhead panel to kill the system's drivers. Then the roar stopped as abruptly as it had begun.

Young glanced down at the centre console to see that the external tank separation light had blinked off. Neither he nor Crip had even felt the bolts blow as it flew free. But as the Orbiter pitched away, he got his first look at it. *Looks like that tank's been in a war*, he thought as he and Crip watched it float away through the cockpit glass. Three regions of the big bullet-shaped white tank had been burned black during ascent. It was to be expected, perhaps, around the bottom where it was closest to the fierce heat of the five rocket exhausts, but the nose too was swathed in black. The charring here was purely a result of aerothermodynamic friction. Speed alone had been responsible. On top of that, though, there was another area displaying signs of heat damage, and that was the point where the forward strut connected the Orbiter to the tank – the place that had given John Houbolt such cause for concern.

But for the sound of cooling fans, it was now quiet on *Columbia*'s flight deck as she travelled at over twenty times the speed of sound in a loose formation with the abandoned external tank, already beginning its terminal dive across Africa before burning up on re-entry over the Indian Ocean. Of the four-and-a-half-million-pound stack that lifted off Pad 39A just nine minutes earlier, only 214,000lb was left. That was the kind of equation it took to get into orbit.

'*Columbia*, Houston,' called CapCom through the static. 'You are go for nominal OMS 1.'

'Roger that.'

The crew manoeuvred their ship into position for the first firing of the orbital manoeuvring system rockets, a one-minute-twenty-seven-second burn that would push *Columbia* into an elliptical orbit with an apogee – the greatest distance from Earth – of 132 nautical miles and a perigee – where she passed closest to Earth – of 55 miles. As John Young watched the chamber pressure of the orbital manoeuvring rockets build on a strip indicator below his computer display, *Columbia* passed out of range of the Bermuda relay station.

'We'll see you in Madrid,' Brandenstein told them.

After a further four and a half minutes in contact with Madrid, Young and Crippen were on their own until, thirteen minutes later, they acquired the signal from the Air Force tracking station in the Seychelles.

'Well, the view hasn't changed any,' Young observed as *Columbia* sped south-east across the Indian Ocean towards Australia.

'I tell you,' responded Crip, 'John has been telling me about it for three years but ain't no way you can describe it. It's hard to get my head in the cockpit here and do my procedures.'

For a moment the two astronauts were transfixed by the sight of Earth, curving deep blue and white beneath them. Then something seemed to cross Young's mind.

'Hey, Crip,' he asked his Pilot, 'you remember to lock that old pick-up of yours?'

The moment the Shuttle cleared the Pad 39A tower at Kennedy Space Center, control of the mission had passed from Launch Control in Firing Room 1 to Houston. After two years in the spotlight while *Columbia* sat inside the hangar at Kennedy having the 30,757 tiles of her heatshield stuck on one by one, Center Director Dick Smith didn't need much prompting to admit that he'd never 'wanted to get rid of anything any worse in my life'. But Kennedy's role in STS-1 wasn't entirely finished. First of all, out at Edwards there was the hundred-strong recovery convoy, a twenty-one-vehicle welcome party that, dressed in atmospheric protection suits, would make the Orbiter safe at wheels stop. Then at the Cape itself, barely was the Shuttle out of sight before working groups began examining film and video of the launch taken by cameras placed around the launch pad.

Fresh from watching *Columbia* go, Tom Moser was part of the effort, reviewing data streams, high-speed film and stills photography of the launch. It only reinforced the good feeling he had about the mission. Outside, post-launch pad inspectors would, as soon as it was safe to do so, scour the launch site, flame trenches and surrounding area for debris.

The soft acceleration from a second, shorter forty-four-second burn from the two orbital manoeuvring rockets circularized *Columbia*'s orbit at a touch over 132 nautical miles above the Earth's surface. While orbital mechanics dictated that they would never meet, they'd now joined a pair of Soviet cosmonauts who'd been in space for a month already, circling the globe at a steeper inclination of 51.6° in their *Salyut 6* space station.

Crippen unlocked his helmet and pulled it off, placing it inside a soft drawstring bag before putting it aside. That didn't mean on a surface necessarily, just floating in the cabin air alongside him would do. After returning from longer flights it could take broken bottles to remind astronauts that, back on Earth, they had to put things on shelves rather than just let them go when they were done. In zero-g for barely half an hour so far, Crippen was still finding it a novelty.

Next to him, only Young's boots remained. Before unstrapping and floating out of his chair to switch on the cabin fans located on the aft crew station, the experienced Commander had unzipped them and slipped his feet out while leaving the boots' spurs still attached to the ejection seat behind. He advised Crip to do the same. You didn't want to be kicking around the cabin in a pair of heavy leather flying boots. Crip, though, wasn't planning on kicking around at all. Still worried about the possibility of bringing on space sickness, he decided it would be less disorienting to treat the floor as the floor. After reconfiguring the computers for orbit, he unlaced his boots and slipped his feet out. *Nice and slow*, he told himself as he carefully pushed himself out of his seat.

So far, *Columbia*'s temperature had been controlled using a series of flash evaporators that sweated her cool by boiling off water, but they only had a lifetime of a few hours. On orbit, she relied on two pairs of 15-by-10-foot aluminium honeycomb radiator panels that unfolded like a beetle's wings from beneath the graphite-epoxy carapace of the payload bay doors. If they failed to deploy the astronauts would have no option but to de-orbit in a hurry. Finding a runway would be a secondary consideration; with the avionics overheating, they'd just have to get down before things began to fry.

Running the whole length of *Columbia*'s spine, from cabin to tail, the 60-foot-long doors were at the time the largest carbon fibre structures ever flown. The original plan had been to use an aluminium honeycomb, until Tom

Moser calculated that they could shave another 600lb off the structure by pushing the state-of-the-art. Too feeble to operate on Earth without deforming, their success was predicated on operating in the zero-g environment of space.

Bob Crippen's first task was to make his way back to the work station at the rear of the flight deck and check the radiator panels were working. He watched through one of the two toughened glass picture windows as an incision of light opened along the top of the dimly lit payload bay. It all unfolded at a stately pace, until the starboard door reached its full extent, lying, as if unpeeled, just above *Columbia*'s wing, covering both Old Glory and the legend 'United States' that ran down the length of the fuselage in a crisp black Helvetica font. And as he examined the view aft he realized that all was not as it should be. He called for John Young to join him.

Columbia had a problem.

FIFTY

Space Shuttle *Columbia*, 1981

THERE WERE TILES missing from the heatshield.

The fronts of the orbital manoeuvring system rocket pods that bulged out of the rear fuselage either side of *Columbia*'s vertical tail should each have presented a solid chalk-white surface. Instead both were blemished with splinters of black where the tiles were gone.

Crippen continued to cycle the payload bay doors while Young inspected the back of the pods, the black-tiled vertical stabilizer, and what he could see of the critical reinforced carbon-carbon leading edges of the wings. They needed to tell Houston what they knew. One hour and fifty-three minutes into the mission, as they began their second orbit, they were reacquired by the Madrid station. Shooting real-time video that could be seen on the ground, Young asked which camera they were looking at in Mission Control.

'Roger,' Dan Brandenstein confirmed, 'we're looking out the forward camera.'

'OK, we're . . .' Young started again. 'We want to tell y'all here we do have a few tiles missing off both of them. Off the starboard pod, it's got basically what appears to be three tiles and some smaller pieces and off the port pod . . . looks like . . . I see one full square and a few little triangular shapes missing. We're trying to put that on TV right now.'

Less than a minute later, Houston again lost the signal to *Columbia* as she passed out of range of the relay stations in Madrid and Dakar leaving nothing but static in the headsets of those listening in on the ground. Sitting behind Brandenstein at CapCom, Dick Truly now heard only silence around him. *Holy Mackerel*, he thought, immediately wondering what the bottom of the vehicle looked like. *If it's been stripped of tiles then these guys are in deep trouble*.

From the aft crew station, Young's inspections with the binoculars suggested that the hardened silicone adhesive below the tile was still there, offering a measure of protection to the airframe below. That was a good sign. 'Unless it gets up to two thousand degrees,' Young said, sounding confident, 'we'll just melt some of the RTV because that's an ablator.' Working the same way as the old Gemini and Apollo capsule's heatshields, he thought that if the glue – the RTV (room temperature vulcanization) – beneath the tiles was to form an insulating layer as it burned away it might protect the aluminium skin beneath. 2,000° was wishful thinking though. The adhesive was only good to 600°F, but the astronauts had little choice but to agree that they weren't worried about it. Young certainly credited his Pilot with being sanguine about the situation, but, despite what he'd said to Crippen, those missing tiles did make him nervous. An aversion to letting go of any technical issue until it was solved and dealt with had underpinned his success in the astronaut corps. Young's reaction was a combination of sangfroid and concern that would be echoed in Mission Control.

As *Columbia* passed through the blind, Max Faget was eager to try to reassure the room. He was unequivocal: the tiles were not critical; nor would their loss represent a hazard to the vehicle or crew. The Public Affairs Office was quick to share Faget's reaction. But Houston's Chief Engineer had called it wrong on the orbital manoeuvring system pods before. What's more, providing an instant response, he could only pass judgement on what he could see on a fuzzy television transmission from space, not what still remained *unknown*. Faget's dismissal of the possibility that *Columbia* might be in trouble was premature; the truth was it was too soon to say so with any certainty.

No one knew, at this stage, whether the tiles they could *see* were missing represented the full extent of the damage, nor, for sure, the extent to which the tiles they could see were gone might themselves jeopardize the Orbiter's safe return. Furthermore, Faget's instincts about the OMS pods had been shaky. Air Force test data had first shown him to be wrong about the need to protect the leeward side of the Shuttle on re-entry. Then, in an effort to disprove his own theory on that, Faget had said to Bob Ried, Dottie Lee's divisional head, 'Bob, I want you to burn at least a small hole someplace on the lee side.' It was beyond the capability of either the computational fluid dynamics or the hypersonic wind tunnels to reproduce the re-entry conditions with any certainty, but in the tests requested by Faget, the one place Ried's team came close to burning through the vehicle's aluminium skin was on the OMS pod.

While Faget's confident, snap assertion helped calm

nerves, thoughts at the same time began to turn to how best to get to grips with a potentially serious situation. For all his conviction, Faget was going to have to wait for answers like everyone else. But as they'd shown during *Apollo XIII* and *Skylab*, tackling a real-time crisis was a game at which NASA excelled.

In Florida, Tom Moser was still in the Launch Control building, but satisfied that the playback of video and 70mm film from the launch had revealed nothing amiss, he was preparing to drive to Orlando to catch a commercial flight back to Houston. At Johnson he could enjoy the rest of the spaceflight with his colleagues. Then he was called back to look at the footage being beamed back from *Columbia*. There was trouble. The feed was grainy and washed out, but the tarnished white curves of the Orbiter's orbital manoeuvring pods were only too clear. Moser focused on the accusing black cluster of missing tiles and thought: *Oh shit*.

As Dottie Lee watched the first TV images of *Columbia* with other members of the engineering team it should have been cause for celebration. Instead, at the sight of the damaged heatshield, her eyes filled with tears. Feeling sick with anxiety, she phoned her counterparts at Rockwell in California to ensure that there was a parallel effort from the contractor to properly understand the ramifications of the damage to the Orbiter. The drinks brought in for the occasion were immediately forgotten as the team in Building 13 got to work.

John Yardley told Moser he wanted him to get back to Houston to run his team. To get him there, the Shuttle programme boss bumped someone off a jet chartered by NASA to fly senior headquarters personnel back to Johnson. Before leaving the Launch Control building, Moser phoned Houston to make sure that detailed analysis of the missing tiles was already underway. Then he took his seat inside the cramped cabin of the business jet for an uncomfortable, anxious two-hour flight back to Houston, knowing that he was in the dark about the real nature and scale of the problem they faced.

FIFTY-ONE

Houston, 1981

DICK TRULY SCANNED the room, looking at the faces of the people sitting at the tiered rows of flight controller consoles. He recognized Hans Mark. The two men hadn't met before, but Truly already knew more about NASA's next Deputy Administrator than most of those in Mission Control. Because of the security tickets he'd acquired during the MOL programme, Truly was aware that, as Under-Secretary for the Air Force, Mark had also been Director of the National Reconnaissance Office – classified information to which only a handful of people at NASA were privy. Mark was one of the few people, Truly realized, he could talk to about what assets the Air Force might be able to bring to bear on the potential threat facing John Young and Bob Crippen aboard *Columbia*.

Mark was shocked by the revelation that the Shuttle's thermal protection system was compromised. He'd been concerned about its sturdiness from the moment the programme had been approved in 1972. Now, like Truly, his anxiety was about whether they yet knew the full extent of the damage to the tiles. *Were they*, he wondered, *missing from regions of the surface where such a failure could possibly be catastrophic?*

The two men with the TALENT/KEYHOLE security clearances made their way to the edge of the control room, out of earshot of anyone else, to talk about the classified Department of Defense resources that could be made available – which meant KEYHOLE satellites.

While he'd been in Washington, as fears over the robustness of the thermal protection system loomed over the whole Shuttle programme, Hans Mark had told John Yardley that, if it came to it, the Air Force might be able to help. With NASA and the astronauts debating whether or not to fly with a tile repair kit on board, Mark told Yardley that DoD reconnaissance satellites – what Yardley called 'spooky ships' – would be on orbit during the first Shuttle flight. Before ordering any astronaut out of *Columbia*'s cabin with a manoeuvring unit and can of silicone paste, he told Yardley, 'we can get a photograph to see if you've got a problem'. Now, just weeks after leaving the Pentagon and yet to assume his new role at NASA, Mark was officially outside the decision loop; but while no more than what he considered *an educated spectator*, he and Truly discussed the assistance the Air Force could provide. Mark, though, was unaware of the extent to which plans for their involvement had already been developed and that, as they talked, Dick Truly was forced to feign ignorance of just how detailed that preparation had been. Irrespective of how and when it had come about the collaboration between the country's civilian and military space programs, though initiated as something of an academic exercise, had now assumed vital importance.

* * *

In 1962, Captain William 'Ted' Twinting was introduced to the press as 'one of eight seasoned test pilots selected by the Air Force for special astronaut training'. It didn't quite work out that way. In line to fly the Dyna-Soar spaceplane, Twinting, after that programme's cancellation, had to content himself with flying chase for the X-15 before being sent off to war in South-East Asia, flying a hundred combat missions in the Republic F-105 Thunderchief. But Twinting wasn't quite done with space. In 1980, he was given command of the Space and Missile Test Organization at Vandenberg Air Force Base. A year later, as the Department of Defense manager for Space Shuttle Support Operations, Brigadier-General Ted Twinting, still sporting the crewcut he'd worn as a young astronaut student, was sitting at a console in Mission Control, conspicuous in his one-star epaulettes over a crisp white short-sleeved shirt. There was one more thing that marked him out from the others sitting in rows in the control room: at his station, mounted next to his computer monitor, was a red telephone providing a secure, direct link to the Department of Defense. Twinting made the call.

Two hours and twenty-seven minutes into the mission, after a brief pass across Indian Ocean station, *Columbia* was reacquired by the transmitter at Yarragadee in Western Australia.

'And we do have another question concerning that starboard OMS pod,' began CapCom Dan Brandenstein. 'We would like to know the colour of the area where the tile is missing.'

'Roger, it's red.'

'Roger, Crip, we copied red.'

Like John Young, Houston wanted to know the extent to which the layers of thermal protection were gone. That closer inspection confirmed the presence of silicone adhesive simply provided the engineers in both the Structures and Mechanics Division and at Rockwell with more hard information about the damage to the Shuttle. With that, their effort to assess the risk could be more accurate. Barely three-quarters of an hour after first becoming aware of the missing tiles, NASA was preparing a statement about the situation, the public announcer explaining that in addition to post-launch assessment of the launch video and checking surface temperature measurements from the Shuttle for abnormalities, they would also be relying on outside help. 'Department of Defense ground stations, that's Hawaii and Malabar, Florida,' he said, 'will be utilized and obtain ground-based photography for any other potential damage.' He was careful not to share any more information than that.

Codenamed TEAL BLUE and TEAL AMBER, the two DoD telescopes charged with carrying out the 'ground-based photography' performed a role the Air Force described as 'Space Object Identification' and represented the zenith of an effort initiated by the Air Force soon after the launch of *Sputnik* in 1957 to track and photograph spacecraft. The former was mounted 10,000 feet high on Mount Haleakalā in Maui, the latter, 30 miles south of Cape Canaveral. Both Air Force telescopes were used to take high-resolution pictures of orbiting Soviet and Chinese spacecraft. By combining spotting telescopes slaved to a more powerful reflecting telescope, the TEAL BLUE and TEAL AMBER first found their orbiting targets then followed them using computer-controlled tracking to keep the larger telescopic camera smoothly focused on its quarry. The capability of the system was classified, but an earlier generation of ground-based Air Force tracking telescope had taken pictures of the Manned Orbiting Laboratory test article, launched along with the MOL programme's Gemini-B capsule from the Cape in 1966, that were sufficiently clear to reveal how UV light had faded the paintwork on the skin of the space station facsimile. In the fifteen years since, it was safe to assume that the quality of the images attainable using TEAL BLUE and TEAL AMBER was substantially greater.

Alongside the classified facilities in Hawaii and Florida, there were also plans to try to capture pictures of *Columbia* using another Air Force telescope: a deployment mapping instrument located at the end of a 13½-mile fire trail, 4,098 feet above sea level on top of Anderson Peak, Big Sur, California, considered by the Air Force to be one of the few remaining clear-sky areas of the United States. Used to track ballistic missile tests out of Vandenberg Air Force Base, the Anderson Peak DMI was a reflecting telescope with a 24-foot focal length which, like the Hawaii and Florida telescopes, was slaved to computer-controlled hydraulic servos. There had been existing plans to use the Anderson Peak tracking station to follow *Columbia* on her glide across the Pacific towards Edwards at the end of the Space Shuttle's first mission. Now the possibility of pressing the Western Test Range facility into service as early as *Columbia*'s sixth revolution was being explored.

At no point was there any public acknowledgement that NASA might be looking for help from Department of Defense assets in space.

* * *

Above: 'When thoughts turn inward ...'
A watercolour by artist Henry Casselli
captures *Columbia*'s contemplative
Commander prior to the flight.

Above: A very composed looking STS-1
crew walk out to the 'Astrovan' for the
drive to the pad. Walking behind Crip at
the left of the picture is Flight Operations
Director George Abbey.

Left: Young and Crippen as they would
have appeared before the launch –
although the smaller US flags on their
shoulders show this was a picture taken
during an earlier simulation.

The T-38s of Jon McBride's
Chase team had the best view
in the house of the Shuttle's
first launch.

Above: Lift-off. Nine years after President Nixon announced the programme to the nation, at three tenths of a second past seven EST on 12 April 1981 – the twentieth anniversary of Yuri Gagarin's first spaceflight – *Columbia* took to the air.

Right: Into the Black. Riding nearly seven million pounds of thrust, *Columbia* breaks the surly bonds.

Left: SRB Sep. Fuel spent and their job done, the two reusable solid rocket boosters are ejected, leaving *Columbia* under the three SSME engines in her tail. She's ascending past 150,000 feet and accelerating past Mach 4.5.

Below: ET Sep. The empty, discarded external tank, blackened by heat from *Columbia*'s engines and friction, looks like it's been through a war.

Above: After reaching orbit, Crippen and Young unstrapped, floated back from their ejection seats to the aft crew station and opened the payload bay doors.

Below left: Flight Director Gene Kranz, JSC Director Chris Kraft and Engineering Director Max Faget study the STS-1 telemetry inside Mission Control.

Below right: Dr Hans Mark, ex-Director of the National Reconnaissance Office, in Mission Control during STS-1. Mark, aware of the capabilities of NRO satellites, recommended that NASA ask the Department of Defense for help.

The view aft. Obvious on the starboard pod – left of the picture – are the tiles missing from the heatshield. The reasons for their loss were unclear. Whether it signalled tile loss from more critical areas was unknown.

Above: Brigadier General William 'Ted' Twinting, USAF, sits at the DoD console in Mission Control studying TV footage downlinked from *Columbia*. Note the red phone to his left.

Left: The Blue Cube. The Air Force Satellite Control Facility in Sunnyvale, California, from where, inside building 1003, Air Force personnel manoeuvred a KH-11 KENNAN spy satellite to take pictures of *Columbia* in orbit.

Below left: The Hubble Telescope. The space telescope is essentially a digital reconnaissance satellite pointing, instead of at earth, towards the heavens. Hubble provided an indication of the appearance of the KH-11 KENNAN spy satellite.

Below right: Cape Cod pictured from *Columbia* during STS-1. The first minute-and-a-half-long window in which the KH-11 might capture the Shuttle took place over the Atlantic about 150 miles to the south.

Above: Joe Engle and Dick Truly – at the right of the picture – sit behind the Capcom console. When he first saw the missing tiles Truly knew that if there were further tiles missing the STS-1 crew could be in real trouble.

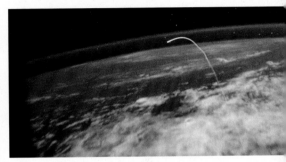

Above: Shooting Star. This 2011 picture of *Columbia*'s sistership *Atlantis* re-entering captures the blistering heat generated by plunging into the earth's atmosphere at 17,500 mph.

Above: As hot as the surface of the sun. The temperature of the pressure wave created by *Columbia*'s re-entry could reach over six thousand degrees fahrenheit.

Right: The Kuiper Airborne Observatory used the infrared telescope behind the open door on the forward fuselage to observe the heavens. In 1981 it launched from Hawaii to capture images of the re-entering Shuttle.

Tally-ho! Chase team leader Jon McBride keeps station on *Columbia*'s starboard wing as she glides towards Edwards. In the back seat, fellow astronaut 'Pinky' Nelson took photographs.

Above: Final approach. *Columbia* over Rogers Dry Lake Bed, pictured from Chase One.

Right: Three greens. With *Columbia*'s gear locked down, John Young flares for landing.

Using the air brakes on the vertical stabilizer to slow her, *Columbia* rolls to a stop on Runway 23. Moments later her ecstatic Commander, John Young, claimed, 'This is the world's greatest all-electric flying machine, I'll tell ya that!'

A spacecraft has landed on earth. The view from inside *Columbia* on landing could hardly have looked more like a scene from a science-fiction movie.

John Young and Bob Crippen leave *Columbia* to grab a shower and a beer before addressing the waiting crowds.

Post-flight. *Columbia* in repose. Half spacecraft, half aircraft. The yellow tractor wouldn't look out of place at any airport. The umbilicals attached like IV drips speak of something more special.

Above: Mission Accomplished. With her return to Edwards *Columbia* wrote the last chapter of a story that began with Chuck Yeager's first supersonic flight in 1947.

Below: After the success of STS-1 and subsequent flights, Bob Crippen, Dick Truly and the rest of the MOL veterans looked forward to a return to Vandenberg AFB in California and launching the Shuttle into polar orbits on DoD missions. It wasn't to be.

On board the Shuttle, John Young and Bob Crippen pushed on with their flight plan. The missing tiles had no bearing on the mission they had to perform on orbit. *There's nothing you can do about it*, Crip told himself, *so don't worry about it*. To help distract him, there was an issue with one of the development flight instrumentation (DFI) recorders. Stuck running, unless it was switched off manually, it would reel through its limited supply of magnetic tape and deprive engineers of valuable re-entry data. While Mission Control explored possible solutions, a workable temporary fix was simply cutting the power. But apart from the need for Crip to keep pulling the DFI circuit breaker in and out, *Columbia*, it seemed, was performing beautifully. As she took eighteen minutes to fly the width of the continental USA, Crippen purged the three hydrogen-oxygen fuel cells of the water they produced as a by-product of generating electricity.

'*Columbia*, Houston, you guys did so good we're going to let you stay up there for a couple of days. You're go for orbit.'

'Let's go for orbit,' Young confirmed. 'This thing is performing just outstanding.'

With confirmation that their mission would continue, the astronauts no longer needed to wear the Air Force pressure suits they had on to protect them in the event of a high-altitude ejection. While Crippen reorganized the checklists, Young floated down to the mid-deck to change into sky-blue slacks and a dark polo shirt sporting the obligatory NASA logo.

The Shuttle's cabin was arranged on two levels. On top was the flight deck which, once operational flights began, would include two seats for mission specialists behind the Commander and Pilot. All the cabin's windows, flight controls and, at the aft crew station, the payload control panels were on the top deck, a close-walled den of screens, dials, buttons and switches. Access to the mid-deck was through a companionway behind the Commander's seat. Here, after seats for four more crew members for ascent and re-entry were stowed on orbit, was a lavatory – or rather a waste management compartment – galley, work station, sleep stations and, biting a large chunk out of the limited space, the airlock that opened into the payload bay behind the crew compartment.

As *Columbia* streaked across southern Africa and out over the Indian Ocean, Crippen made his way down through the hatch to the mid-deck to doff his pressure suit. For two and a half hours he'd been moving around the cabin with the slow, exaggerated care of a man doing an impression of weightlessness. Then, as he pulled and stretched, focusing on the awkward task of extricating himself from the pressure suit, he began to tumble head over heels. Any pretence at keeping his head level was gone, but as he spun every which way he realized with relief and no small amount of joy that, apart from a slight pressure in his head, he wasn't feeling sick at all. There seemed to be no way of predicting which astronauts would suffer from space sickness, but Crip was one of the lucky ones; he was going to be able to enjoy the flight. Like John Young, it turned out that, when it came to space, Bob Crippen was a natural.

In Texas, the Learjet carrying Tom Moser back from the Cape landed at Ellington Field at around 10.30am local time. The engineer caught a lift back to Johnson Space Center and headed straight to Building 13, eager to bring himself up to speed with his team's progress.

FIFTY-TWO

Houston, 1981

THERE WERE THREE separate Flight Control teams assigned to the Shuttle flight, each with responsibility for a different phase of the mission: silver, bronze and crimson; ascent, on orbit and descent. With *Columbia* safely established on orbit, Dan Brandenstein passed his seat at the CapCom console to Bronze Team's Hank Hartsfield, while Brandenstein's Silver Team Flight Director, Neil Hutchinson, joined Gene Kranz, the pugnacious Deputy Director of Flight Operations who would one day be credited with the line 'Failure is not an option', for the change-of-shift press conference. An excited press corps was waiting for them in the space centre's Building 2 auditorium. And there was one subject, above all others, that they wanted to know about.

'Can we,' asked the first questioner, 'get a few words of wisdom on the tile situation? Why did they come off? How many are missing? Are they critical? Are there any other tiles that might be missing? How will you know?'

'That's a lot of questions . . .' Kranz began wearily, sounding as if he had anticipated just what was coming.

Of the forty-five or so questions that followed, over

thirty were about the missing tiles. And of particular interest was just how NASA was going to get hold of the high-resolution pictures that might put everyone's mind at rest. The journalists teased and probed at the subject, looking for different ways in which they might persuade Kranz to reveal more about the help being received from the Air Force. Patient and relaxed, Kranz, projecting a lack of concern, was as open and straightforward as he could be, suggesting that revolution – orbit – nine might provide an opportunity for the camera in Hawaii to capture an image of *Columbia* when the lighting was favourable. But if a question provided him with no wriggle room, he was blunt. Asked about the purpose and resolution of the Air Force telescopes in Hawaii and Florida, he first neglected to answer; then, when the question was repeated, he provided a terse 'that's classified information'.

Throughout the press conference he seemed almost at pains to keep stressing that the help from the Air Force was from 'ground-based optics'. And certainly that help was forthcoming. But, put on the spot about a possible alternative, his answer was qualified.

'You've already mentioned high-resolution photography from Earth,' the questioner began. 'Do you have any plans to take pictures of the Orbiter from satellites already in orbit?'

'We don't have any specific plans for that type of work . . .'

Not yet they didn't.

The specifics may not have been nailed down, but Ed Lineberry and Ken Young were working on it. Using the revised KH-11 and Shuttle orbits, Lineberry and Young calculated new sets of coordinates to be sent up to *Columbia*'s crew. Each new instruction – time of ignition, duration of burn and the Orbiter's pre- and post-burn orientation described in three axes – was verified by Dave Scheffman by running it through his software then sharing the results with Lineberry and Young. Initially, Scheffman ran round with the printed computer graphics in his hand, took them up to the 2nd floor secure room and, not cleared to actually enter, just slid them under the door in a plain envelope. He left a bemused security guard in his wake. Later, Scheffman transmitted images of the printouts from Building 12 via closed circuit TV.

Hans Mark may have supported the idea of turning to the DoD for help, but he was outside the small cadre inside Mission Control who knew the detail of what was being done. He knew well, though, that to capture images of one spacecraft from another, while it could be done, required not just skill but also a good deal of luck. In taking normal reconnaissance pictures of the Soviet Union from space, months could go by before a KEYHOLE satellite was able to deliver photographs of any given location. Frequent, thick, persistent cloud cover over Russian territory often bedevilled the National Reconnaissance Office's efforts. So too did night and day. If a satellite's orbit took it over downtown Moscow at midnight any images would be of limited value. In that situation there was no option but to wait twelve hours until the satellite's orbit coincided with daytime. A further complication was simply that the Soviets knew when the American spy birds were passing over; orbits were tracked and predictable, meaning there was time to hide anything that they wanted to remain unseen. That lack of cooperation, at least, wouldn't be an issue with any effort to take pictures of the Shuttle.

At the time of *Columbia*'s launch, the NRO had four birds on orbit. Two of them, a CIA KH-9 HEXAGON and a KH-8 GAMBIT launched by the Air Force, still used the bucket drop system to return their images for analysis. That alone removed them from contention. The planned fifty-four-hour first flight didn't allow enough time for the two earlier-generation designs to snap the Shuttle, drop the film for the Air Force to snatch from mid-air and transport it back to Federal Building 213 in Washington DC – the unprepossessing home of the National Photographic Interpretation Center – to be developed and analysed, then for any conclusions to be shared with NASA. And that's assuming that the big KH-9, designed to image great swathes of enemy territory, was even capable of an operation as tricky as sat-squared photography. With HEXAGON and GAMBIT out of the equation, the NRO's efforts rested with the digital KH-11 KENNEN, the only member of the KEYHOLE family capable of beaming images back from space for near instant analysis.

In July 1977, Lockheed and Perkin-Elmer had been chosen to build what NASA called a large space telescope (LST) that was to be launched using the Shuttle. Hans Mark, in his capacity as Director of the NRO, was kept closely informed of the programme's development. It was little wonder. In the broadest terms, the new telescope – which, ultimately, would be named after astronomer Edwin Hubble – was a digital spy satellite designed, instead of pointing at targets on Earth, to look at the wonders of the universe. Hubble's 94-inch mirror, already under construction at the Perkin-Elmer plant in Connecticut, would be at the heart of a telescope capable of focusing on a dime in Boston from

Washington, nearly 400 miles away. The diameter of the mirror used in the first KH-11 KENNEN was only 2 inches smaller. So closely related was the technology shared between the two satellites that Mark was unable to discuss the detail of his involvement in Hubble's development with anyone who didn't possess the necessary security clearances.

An imaging satellite like KENNEN or Hubble is not a mass-produced machine. Instead, it has the feel of an instrument, built in a laboratory rather than a factory. There was no concession to style; form followed function. Individually cut and milled panels and components were stitched together with rivets and cross-head screws to lend it a custom-made feel – assembled rather than manufactured. And both space telescope and spy satellite were huge, bare-metal cylinders the size of a Greyhound bus. But another thing they had in common was that neither was designed to take pictures of a fast-moving target on an intersecting orbit. And the challenges of doing that were formidable.

While the cloud cover that blighted the NRO's efforts to take photographs of northern Europe wouldn't complicate any effort to capture *Columbia*, there was a different set of challenges to contend with, unique to the task in hand. Chief among them were those springing from the different orbits flown by *Columbia* and the two KEYHOLE satellites. While the Shuttle circled the world around its middle, never exceeding a latitude of 40° North or South, the two reconnaissance satellites were in a polar orbit. There was no guarantee that those orbits would intersect in such a way that the Orbiter and the satellite were overhead the same point on Earth at the same time. While on paper there was a point in every revolution when their orbital tracks crossed, it was no good if, for instance, *Columbia* passed overhead the Azores twenty minutes before the satellite did. And if NASA and the Air Force were fortunate enough to identify a latitude and longitude at which the two spacecraft crossed at the same time, they also then had to be lucky enough for that to happen in sufficient daylight. In completing an orbit every hour and a half, *Columbia*'s darkness was never more than three-quarters of an hour away. What's more, location and light were just the essential preconditions for being able to take a sat-squared photograph.

Columbia was in a circular orbit. The two KH-11s were in elliptical orbits, swinging in and out as much as 40 nautical miles between the apogee and perigee. Establishing the distance between the Shuttle and the satellite at the point of any encounter was crucial, but it was also subject to rapid change. If occasions when the spacecraft's orbits intersected were found, the moment when one was directly above the other would be a split-second. Immediately before and after that point in time, the distance between them would either be closing fast or lengthening; and this highlighted perhaps the greatest technical difficulty involved in capturing a clear, high-resolution shot of the Shuttle from a spy satellite: the angular velocity between the KEYHOLE bird taking the picture and its target; that is, the speed of *Columbia* from the point of view of the KH-11. It can be well illustrated by the idea of trying to read a station name from inside an express train as it speeds past the platform without stopping. But imagine if that station was itself moving, travelling vertically up or down at the same speed your train is racing horizontally along the track. Now try reading that station name as it flashes past the window of your carriage.

It's been suggested that the KH-11s were capable of handling an angular velocity of just 1.6° per second. The smaller the distance between the satellite and the Orbiter, if and when they crossed paths, the greater the likelihood of the angular velocity exceeding the design limits of the satellite's camera system. Closer wasn't necessarily better.

While over in Houston, Ed Lineberry and Ken Young worked on NASA's side of the equation, inside Sunnyvale's Blue Cube, operators from the Air Force Satellite Control Facility's Special Projects Office did the same. The control problems suffered by the KH-11 KENNEN that had nearly delayed the Shuttle's launch meant their work had to take into account the late, forced changes made to the orbits of both spacecraft. With tiles confirmed to be missing from part of *Columbia*'s heatshield, their work had an urgency to it they had not anticipated. Professionalism had always demanded that the bluesuiters would do everything possible to successfully capture imagery of the Shuttle. Now, though, their work concerned life and death.

As Gene Kranz and Neil Hutchinson continued to field questions about how they planned to capture pictures of the Orbiter, Bob Crippen and John Young returned to the flight deck to prepare the orbital manoeuvring system engines for two further burns to push *Columbia* into her definitive 147-mile circular orbit.

In Mission Control, the white-haired Hank Hartsfield, one of the six MOL refugees who'd joined NASA alongside Crip, was at the CapCom console. Together they'd shared the bitter disappointment of the MOL programme's cancellation, then years in NASA's astronaut corps knowing that there were others ahead of them in the line for spaceflights.

'Hey, Hank,' Crip enthused, 'I guarantee you it's worth all the time you and I have been waiting for it.'

After his frustration at missing out on the *Enterprise* test flights, Hartsfield too was now in training with T. K. Mattingly to fly *Columbia*. For the time being, though, he was relishing working CapCom for the first time since he'd played the role for John Young, nine years earlier, during the *Apollo XVI* lunar mission.

'While we got a few minutes here,' he said, 'I might tell you that we had a pretty big flare pop on the sun here back at 1700 ZULU on the tenth . . . no concern at all for radiation effects, however we do want to caution you that during the next four revs as you pass through the South Atlantic anomaly, there may be enough radiation there's a possibility of kicking off your smoke detector.'

'Good grief, Henry!' said Crip with mock concern. 'That sounds like fun . . .'

'Also, there's a good chance there's going to be a spectacular aurora later this evening.'

'Okey-doke,' answered John Young, before getting back to business. 'And could somebody look at these coordinates and see if they look OK to you?'

While the STS-1 Commander worked the digital autopilot, Crip, anchored in the right-hand ejection seat, reached up to the control panel in the roof of the cockpit to arm the two orbital manoeuvring rockets.

From their own, carefully cultivated sources inside the military, some of the journalists covering the Shuttle mission got word that, as *Columbia* crossed the Pacific towards the California coast to begin her sixth orbit, the telescope on top of Anderson Peak had captured pictures of her.

Unfortunately the results did little but confirm that snaring a photograph of the Shuttle on orbit was a job of a different order of magnitude to tracking the sub-orbital flight of a ballistic missile. While the Western Test Range operators had successfully acquired *Columbia* on orbit, their primary telescope, even at its maximum magnification, increasing its basic 208-inch focal length to 120 feet, just didn't have the power to provide useful images of the Orbiter. As far as NASA was concerned the pictures revealed nothing more than a blurred track across the sky, detailing nothing of her condition.

While Air Force cameras struggled to take pictures of *Columbia* from the ground, inside the spacecraft Bob Crippen was again preparing to transmit TV pictures in the other direction. After a status report on their fuel and water from CapCom, Crippen turned the camera on John Young, sitting in his ejection seat in the cockpit. The Commander was effusive in his praise of his ship: 'the vehicle has just been performing beautifully, much better than anyone ever expected it to do on a first flight . . . just performing like a champ, real beautiful'. Young made a point of thanking the armies of people who'd 'helped get this thing airborne', before handing the baton to his pilot.

'I'd like to echo John's words,' Crip began, talking with similar enthusiasm, before his thoughts turned to the men who'd tragically lost their lives while preparing *Columbia* to fly. 'I think it's only right that we mention a couple of guys that gave their lives a few weeks ago in our countdown demonstration test: John Bjornstad and Forrest Cole. They believed in the space programme and it meant a lot to them. I'm sure they would be thrilled to see where we have the vehicle now. We, I, had to talk to those guys . . .' Crippen's heartfelt message, heard on the ground against a background of static and the intermittent tick of an electronic beep, far away yet close, felt like an appropriate tribute. And it also served as a counterpoint to the copybook enthusiasm that usually characterized scheduled broadcasts from space.

Mission elapsed time was nine hours and twenty-two minutes.

FIFTY-THREE

Houston, 1981

'OK,' CAME THE message from *Columbia*. 'We're all healthy so if you need that Hawaii pass for something other than medical you got it.' All communication between the crew and Mission Control was conducted over an open loop with CapCom, with one exception. As they passed over the Hawaii ground station there was a ten-minute private medical communication scheduled with the Flight Surgeon. Three seconds later they lost the signal from Indian Ocean station.

In Mission Control, after the discovery of the missing tiles and the approach made to the Pentagon for assistance, there had been an animated discussion over whether or not to tell the crew. And how much. Hans Mark listened as people aired different positions, but was in no doubt whatsoever that they owed it to the astronauts to be straight with them. 'We've got to tell them,' he said. Anything else was unthinkable. He wasn't alone; the decision was taken to bring Young and Crippen fully into the picture.

At Young's suggestion, the private conversation between *Columbia* and Houston through the Kokee Park tracking station as she crossed 147 miles above Hawaii was scrubbed. Young thought it had been a waste of time to schedule it in the first place – they hadn't bothered with them during the Apollo missions when they were flying to the moon and back. But without the opportunity for any radio communication over a secure channel, whatever needed to be shared with the crew offline would have to be conveyed via the teleprinter that chattered through reams of folded, stacked paper on the mid-deck. And it was clear, when the ground relay was next restored as *Columbia* acquired the UHF radio signal from Santiago, Chile, that the missing tiles had been at the forefront of people's minds.

On CapCom, Hank Hartsfield conveyed some of the detail of what had so far been learned from the analysis into the OMS pods. He painted a reassuring picture. 'We don't think it will be a concern during the entry,' he told Crip, 'and the structural integrity will be maintained.' In a still thick Alabama accent, Hartsfield explained the work being done by Tom Moser's team studying heat transfer and skin surface temperature data. 'Anyhow,' he wrapped up, 'the bottom line is that we think there is no real problem at all with the pieces of tile or tiles that there seem to be missing off the pod and we don't plan to make any changes to the entry flight plan because of the tile loss.'

'OK, Henry, you could have saved all that for your end-of-mission press conference,' Crippen responded, sounding pointedly unconcerned. 'Sounds good enough,' he added, 'we don't think it's a problem either . . .'

They moved on. However, the situation with the tiles wasn't yet quite as final as the brief exchange between the two MOL veterans had made it sound.

* * *

Inside the Structures and Mechanics Division, Tom Moser's engineers working the problem may have come up with some preliminary conclusions, but they were far from finished. Much of the work was done by the aerothermo-dynamics team, of which Dottie Lee was a member, running computer models in conjunction with arc-jet test data. Of particular concern was trying to establish the extent to which the disruption to the smooth air flow caused by the missing tiles might generate greater than expected local temperatures in exactly the area where the heatshield had been compromised.

But alongside what needed to be done to analyse the impact of the tiles they *knew* were missing, NASA was still uncertain about whether or not that loss represented the full extent of the damage to *Columbia*'s heatshield. To establish that still required help from the military.

Unable to discuss publicly the details of the DoD effort to illuminate the situation, NASA relied on the teleprinter to inform the crew of anything that couldn't be conveyed over open radio channels, and what they might need to do in response. Back in Houston, the requirements of the Blue Cube operators were conveyed to Mission Control via JSC's own Mission Planning and Analysis team in the form of Ed Lineberry and Ken Young. And with that, the carefully worked-out and well-rehearsed flight plan for day two of *Columbia*'s mission was pulled apart and rewritten.

Before loss of signal from Santiago, Hank Hartsfield drew Bob Crippen's attention to a written message sent up from Houston. 'We have a teleprinter message coming up to you,' he said, 'that we would like for you to take a look at, at your earliest convenience.' Through messages like this, the

Shuttle's crew was kept informed of what was going on. And while *Columbia*'s Commander didn't share the same top-level security clearances enjoyed by his Pilot, all that mattered was that *John*, as Crippen put it, *had enough*. The Air Force, though, did guard this sort of information jealously.

While *Columbia* sped east in the direction of the Botswana ground station, Mission Control anticipated contact through Indian Ocean station, which was to follow. 'Indy comm check, Houston comm check,' a female voice requested, 'air to ground UHF.' Greeted with nothing but static, she tried again: 'Sunnyvale comm check, Houston comm check, air to ground one.'

'Houston comm check,' one of the Air Force controllers replied, speaking low and fast, 'Sunnyvale comm check on one.' In contrast to the brightness on air from NASA, the Blue Cube controller was terse, speeding through messages and acknowledging, military style, with a click of the transmit button rather than speaking.

'Houston,' reported Sunnyvale, 'I'm not copying you on air to ground two.'

'OK, stand by . . . you're not copying me on air to ground two at this time?'

'Affirmative.'

Even through this brief exchange, NASA, in trying to establish the line to Indian Ocean station, relayed via satellite through the Blue Cube, was forcing the Air Force space facility into greater public prominence than it was entirely comfortable with; and at a time when it actually seemed to be moving in the opposite direction. Just a month earlier, Sunnyvale had had its security status upgraded from 'controlled' to 'restricted'. That had followed the earlier removal of 3-foot-high chrome letters announcing 'US Air Force Satellite Test Center' from Sunnyvale's entrance in an effort 'to reduce visibility'. How effective this was, given the Blue Cube's unusual and conspicuous appearance, was debatable. Initiatives in other areas were more successful. In setting up what was known as control-mode operation for the Department of Defense missions, command and control was ring-fenced from NASA's unclassified activities using discrete computer networks, secure rooms within Mission Control, codebooks allowing open communication between CapCom and the Shuttle, and encrypted on-orbit video feeds. There were even attempts to avoid any open utterance of the name Sunnyvale. In years to come, astronauts preparing for Shuttle missions carrying DoD payloads would file false flight plans from Ellington to an airfield they had no intention of flying to, only to find some reason to divert en route from their stated destination in to Moffett Field, a couple of miles north of the Air Force space facility in Sunnyvale.

Inside the Blue Cube, pre-flight planning with Houston had shown that only one of the two operational KH-11 satellites was going to be of any use to them. The path of the newest KENNEN, designated 5503 and launched in February 1980, wouldn't, during the remaining hours of Columbia's mission, provide an orbital encounter giving any opportunity to try to capture an image of the Shuttle. Hopes were pinned on a single bird: 5502. And while the necessary conjunctions were there on paper, luck, as Hans Mark had noted, was still a component of securing any sat-squared photography.

While the bluesuiters in Sunnyvale focussed on imaging *Columbia* using the KEYHOLE satellite, operators at another

Air Force facility 2,300 miles to the south-west were already trying to do the same thing from the ground – albeit from over 10,000 feet above sea level on the summit of Mount Haleakalā. As Young and Crippen began their ninth orbit, twelve hours into the mission, it was hoped that, as their spacecraft passed within range of the Maui Space Surveillance Complex TEAL BLUE telescope camera, the Air Force might capture the first usable images of her heatshield. Any success, however, was doomed to failure. The Shuttle, as she passed Hawaii, was still on the flight plan as originally scheduled; there'd been no effort to manoeuvre her into a different orbital attitude. Travelling upside down and tail first, there was simply no way that her underside could be seen from Earth. It was no more visible to the Air Force telescope operators than the far side of the moon.

Sitting in the Pilot's seat of the Shuttle Main Simulator in Building 5, Dick Truly loosened his straps and leaned round behind him, stretching for a circuit breaker with a swizzle stick – a telescopic aid designed to provide access to out-of-reach panels. As had happened throughout the Apollo programme, NASA was trying to work through a problem aboard an operational spacecraft by working it on the ground in the simulator. While the risk posed to the already damaged heatshield from any further missing tiles may have been the main source of anxiety, it wasn't the only concern over the Shuttle's re-entry. Because of the limits of the data provided by the hypersonic wind tunnels and arc-jets used to test the Shuttle, *Columbia*'s first orbital test flight was itself going to be a crucial source of information. Only through her first re-entry could data confirming the validity of all the pre-flight testing be gleaned in real-world conditions. And one of the machines set up to record that data was not turning on and off as planned. On orbit it had been possible to stop the jammed recorder running out of tape prematurely by turning it on and off using the circuit breaker to cut the power, but once Crip was strapped into his seat for the journey home, the back panel was going to be out of reach.

Twisting round in his seat, Truly found he was able to just stretch to the circuit breaker to operate it, but he was in shirtsleeves. Trussed up in the pressure suits Young and Crippen would be wearing for re-entry, he wasn't sure it was doable. For now, though, the faulty tape recorder remained a frustration rather than a worry.

'Perhaps,' joked the public announcer, the crew 'should have carried the Pete Conrad *Skylab* hammer with them.' In referring to the *Skylab 2* Commander's on-orbit repairs, he was unaware that he was also talking about the last occasion the National Reconnaissance Office had used its KEYHOLE spy satellites to help NASA assess the damage to one of its spacecraft.

At twelve hours forty-eight minutes mission elapsed time, *Columbia* picked up the signal from Santiago, Chile – the last scheduled communication with the ground before Young and Crippen turned down the volume of the comms link in order to sleep. At the beginning of the five-and-a-half-minute pass, Mission Control requested a brief private medical consultation. The Flight Surgeon reported that the conversation with the crew over a discrete radio channel revealed no medical concerns of any sort – just as Young had assured Mission Control had been the case two hours earlier. Back on an open channel, Hank Hartsfield flagged the imminent arrival of further written communication – 'a teleprinter message,' he said, 'that'll tell you everything we've done' – and wished them a good night's sleep. There was twenty seconds until loss of signal.

'See you mañana,' Crippen signed off from on board *Columbia*, while in Houston, Hartsfield's Bronze Team leader, Flight Director Chuck Lewis, on finishing his shift, made his way to Building 2 and installed himself in Room 135 for the change-of-shift press conference – and another barrage of questions from reporters about the tiles. They were becoming more forensic, scratching harder at some of the suggestions Gene Kranz had made about possible orbits for taking pictures from ground-based systems.

'I'll tell Mr Kranz when I see him tomorrow,' Lewis joked, 'that everyone's anxious for him to come back over and talk to you.'

Kranz, not this time leading one of the Flight Control teams himself, was responsible for the offline effort to establish the integrity of the heatshield. For now, though, Lewis answered questions about Air Force efforts with the same careful mixture of transparency and evasion that Kranz himself had employed. But asked when NASA expected to draw their final conclusions about the risk from the tiles missing from the OMS pods, Lewis told them 'my guess is around three or four am'. Seven hours away. In fact, the teams working the problem in both Downey and Houston were closer than that to delivering their verdict.

FIFTY-FOUR

Flight Simulation Laboratory, Downey, California, 1981

SINCE COMPLETING HIS nine-month astronaut training, TFNG James 'Ox' van Hoften had spent more time at the Rockwell plant in Downey than he had in Houston. Flying out to the El Toro Marine Corps Air Station on Monday morning before climbing back into the T-38 on Friday night, he spent the week in Mrs Chen's Empire Motel on Lakewood Boulevard. By day he could be found in Rockwell's Flight Simulation Lab (FSL) working on each new iteration of *Columbia*'s flight control software; ironing out flaws, verifying fixes. Working in parallel with the Shuttle Avionics Integration Lab at Houston, FSL tended to focus on the orbital entry software while in Houston their counterparts concentrated on the ascent. Ox van Hoften led the team at Downey with responsibility for re-entry. And after the discovery of the loss of tiles from *Columbia*'s orbital manoeuvring pods, van Hoften was put to work.

Over the hours that followed, van Hoften's team in Downey flew repeated simulated re-entries, configuring the software to represent the effect of different levels of damage on *Columbia*'s ability to come home. If Young and Crippen were to lose some of the reaction control thrusters housed in the pod, then their survival, the FSL simulations suggested, might still be a possibility. If burn-through led to the loss of the whole pod, however, the outlook was bleak. In reporting their findings back to Houston, there was no point in sugarcoating that particular message. If the damage to the OMS pods was too severe, van Hoften said, 'you're going to die'.

Bob Crippen woke up to see his arms floating out ahead of him in the manner of a B-movie zombie. In the background, above his head, Earth appeared to be revolving slowly in the opposite direction. There was something about it he found disconcerting, and he resolved to fold his arms before trying to sleep again. If he could get warm, that was. Crip reckoned he could sleep just about anywhere, and strapped loosely into his ejection seat it should have been a breeze; *weightlessness*, he thought, *the softest bed you could find*. But the cabin temperature had dropped right off. He looked over his shoulder to see John Young lying horizontal in mid-air, just floating around the back of the flight deck behind him. He looked comfortable enough, but, like Crippen, Young couldn't sleep for cold.

The Commander hadn't felt ready for sleep at the time appointed by the flight plan. After Crippen had eased himself into position in the cockpit, Young had drifted around the cabin, checking systems, reassuring himself that his spacecraft was working as advertised. *That's what they paid us for*, he thought. Before trying to sleep he returned to the cockpit and unclipped his Brogan flying boots from the ejection seat and pulled them on. It wasn't ideal, especially when, from time to time, he kicked the control panels, but at least they'd keep his feet a little warmer. Despite the impression from where Crip was sitting, Young managed to do little more than catnap.

While the astronauts tried to sleep, Tom Moser's engineering team at Houston worked late into the night on their analysis of the missing tiles. No one believed that they faced the kind of worst-case scenario from the damage to the pods that Ox van Hoften's simulations had explored. But shortly before 11pm in Houston – mission elapsed time seventeen hours – Moser gathered the teams of aerothermodynamists, materials scientists and structural engineers to review their data. After twelve hours of running thermo-models on computers and reviewing arc-jet data, they were of one mind: *we're OK*. Dottie Lee, managing the aerothermal team's relationship with Rockwell, conveyed the same verdict from their counterparts at Downey: 'you will not burn up the spacecraft'.

Around midnight, Lee finally left the office, climbed into her car and made the short drive home. Subsisting on water and anxiety, she'd not eaten all day. Exhausted, she pulled open the refrigerator door and stared at the contents, almost past caring what she ate. She opted for a leftover patty sausage, then threw ice into a tumbler and poured herself a well-deserved Tanqueray gin. After finishing what counted for a meal, she showered and went to bed, already anticipating the sound of the alarm clock that would wake her in three hours' time.

As Lee and the astronauts slept, in Houston and Sunnyvale NASA and the Air Force continued to work the problem.

Analysis of the orbits flown by *Columbia* and 5502, the sole KH-11 KENNEN satellite capable of capturing the image NASA required, suggested that there would be just three opportunities to do so before the end of Young and Crippen's mission. The first, during *Columbia*'s nineteenth orbit, was out over the Atlantic, east of Philadelphia and south of Portland, Maine. The second, an hour and a half later, saw the orbital encounter between the two spacecraft move 1,500 miles west as *Columbia* tracked across Missouri and Illinois, while the KEYHOLE bird swept down towards Arkansas from the north. The last chance was barely an option at all, offering a marginal window for the satellite as it bisected the mid-west through Idaho and south towards Nevada, after *Columbia* had already crossed its path on her way through Utah and Colorado. Three crossroads in the sky. Three chances. Using the encryption machine hidden away in the Mission Control Center secure room, the Air Force passed details of what it required to NASA, where, once processed by Ed Lineberry and Ken Young, it was passed on over a discrete line to Mission Control for Flight Director Neil Hutchinson and Flight Dynamics Officer Jay Greene to disseminate more widely as laundered, unclassified information.

Irrespective of what reassurance came out of Building 13, the one thing Tom Moser's Structures and Mechanics team couldn't do was say whether or not the tiles missing from the manoeuvring pods represented the full extent of the damage to the heatshield. Having failed in any attempt to capture a usable picture of *Columbia* on day one of the mission, Mission Control needed to make sure that if there was an opportunity for Department of Defense cameras, whether ground-based or in space, to photograph the Shuttle, she was presenting her best side.

In Building 30, Neil Hutchinson's Flight Control team incorporated the revisions to the next day's flight plan. During the day's first change-of-shift press conference, Hutchinson and Gene Kranz had said they were hoping to avoid making any changes to the flight plan, but as Hutchinson's Silver Team considered the new orbital attitudes required of *Columbia* by Lineberry and Young in Mission Planning and Analysis alongside tests of the reaction control thrusters already planned for day two of the mission, it was clear that they were going to have to make a few changes. It looked as if the astronauts were going to be busy. A long teleprinter message detailing the changes to the flight plan was prepared, ready to be uplinked to the crew once they were awake.

As *Columbia* passed across Australia on her fourteenth orbit, day once again blinking into night, Young and Crippen activated their computer displays and prepared for the second day of their mission, grateful to be moving around the cabin again – warming up. As the Shuttle passed out of range of Australia's Orroral Valley ground station on a track north-east towards Quito in Ecuador, there were thirty-two minutes of their scheduled sleep period remaining.

Mission elapsed time was twenty hours and eighteen minutes.

FIFTY-FIVE

Space Shuttle *Columbia*, 1981

ON BOARD *COLUMBIA*, the sound of rocket engines erupted in Young and Crippen's ears – the roar of a launch broadcast over the RT from Mission Control. Then, over the rumble, came the familiar sound of the public address system: 'She's beautiful ... she's cleared the pad ... do you copy, Houston?'

'This is Houston,' came the reply at the moment drums and bass guitar struck up a bouncing country and western rhythm over the background. 'We copy,' continued the commentary as a guitar, pedal steel and vocal joined the song. Performed by twenty-two-year-old country and western singer Roy McCall and written by a Shuttle technician, 'Blast Off *Columbia*' was chosen by Mission Control as the day two wake-up music. Some of the technical detail in the lyrics was a bit iffy, but the song captured the mood, urging the crew on with repeated hollers of 'Alright!' and claiming that, thanks to *Columbia*, Young, Crippen and God's grace, America was once again number one in space.

'Alright!' Crip repeated, making a mental note to give a copy to his mom to put on the jukebox in Crippen's.

'Morning, gents, how's the Silver Team this morning?'

'Well, we're just fine,' replied Dan Brandenstein, back at the CapCom console on Flight Director Neil Hutchinson's watch. After enquiring about the cabin temperature and the need to capture some missed data from the previous day's tests of the reaction control thrusters, Brandenstein got to the meat of the situation. 'There are going to be some time-line changes today in the CAP [crew activity plan]. They don't start until twenty-four hours MET [mission elapsed time] so we have a message that's being ginned up that will request all those changes.' Sandwiched in among other more mundane procedures, there was no suggestion from Brandenstein that it was anything out of the ordinary. But it marked the beginning of a series of manoeuvres that it was hoped might provide critical evidence of the condition of *Columbia*'s heatshield.

'OK,' Crip acknowledged at the end of the transmission. Nothing out of the ordinary.

From Quito, the Orbiter swept north-east within range of the Bermuda tracking station before crossing the Atlantic on her way to Madrid. On board, Young and Crippen reconfigured the flight deck after their sleep period, adjusting the window blinds and preparing, once again, to align the inertial measurement units, the sensitive collection of gyro-scopes and accelerometers that recorded the spacecraft's changes in attitude to feed into the flight control computers. Measured against snapshots of space taken by the on-board star-trackers, the three highly tuned instru-ments had proved accurate so far. With the manoeuvres being planned for day two it was vital that they should continue to perform.

'*Columbia*, Houston, talking to you through Madrid. We have you for about seven minutes.'

'OK, loud and clear.'

'Roger, and we have a state vector coming your way. And there's also a teleprinter message on its way with some photo information . . .' A combination of three-dimensional attitude and velocity, the state vector recorded the Shuttle's position on orbit at a given moment in time.

While Mission Control prepared the numbers for the first of a revised series of reaction control system burns that would reposition *Columbia* in orbit, John Young looked down through the rear windows at the aft crew station as she flew 147 miles above the Straits of Gibraltar. 'Looks to me as if we're right over Rota, Dan,' he said, identifying the US naval base in Spain that had been earmarked as their destination in the event of a transatlantic abort during ascent.

'That's affirmative.'

Young never tired of the view from space. Throughout the mission, as he looked down from *Columbia*, he pulled out the world map they carried on board to check on anything of interest, only to find that by the time he'd retrieved the map and opened it out, whatever had caught his eye was long gone, vanished behind him as his ship fell around the globe, velocity and gravity in perfect balance.

Twenty-five minutes later they were over Australia. And as *Columbia* came into range of the Yarragadee ground station near Perth, Dan Brandenstein passed the crew the details of the first reaction control system burn. At present, while the crew knew changes to the flight plan were coming, the scheduled timeline remained in place. From the CapCom console, Brandenstein ran through the numbers for the required burn attitude in three axes – roll, pitch and yaw – the Delta V, that is the change in velocity, the time of ignition and the duration of the burn. Crippen took them down and repeated them back.

'Roger,' Brandenstein confirmed, requesting another check on the inertial measurement units before *Columbia* went over the hill from Yarragadee.

On this orbit, Young was to manoeuvre the Orbiter into a position known as the gravity gradient attitude. By positioning the Shuttle in the vertical relative to the surface of the Earth, her nose pointing down, the pull of gravity itself helped keep her stable, which reduced the need to use the thrusters to keep her in the desired orbit; as a result, the demand on their precious supply of propellant was reduced. But it also, as far as Young was concerned, made *Columbia* a bigger target. *Space*, he thought, *was a debris meteoroid shooting gallery; an environment very similar to combat.* He was a lot happier flying tail-first, presenting the Orbiter's rocket bells in the direction of travel. But with her nose pointing towards the centre of the Earth and circling like the hand of a clock, *Columbia* was able to maintain her attitude and orbit without a single correction for over three hours. And with her wings canted 120° to port, she would also be pointing her belly directly at the rising sun as she flew over-head Pad 39A to begin orbit seventeen, some twenty-four hours after launch.

Inside control rooms at the Malabar Test Facility, 45 miles south of Kennedy Space Center, Air Force personnel operating the TEAL AMBER telescopic camera were preparing to track and image *Columbia* as she crossed Florida. At nearby Patrick Air Force Base, on the edge of Cocoa Beach, a jet was kept on standby to carry the results of their efforts back across the Gulf of Mexico to Ellington Field.

Waiting in Houston, Gene Kranz thought *it's going to probably take longer to get the film here than it is to analyse it . . .*

At twenty-three hours mission elapsed time, as *Columbia* settled into orbit seventeen, the Flight Control team shift changed again. At CapCom, Astronaut Joe Allen of the Crimson Team replaced Dan Brandenstein.

'Good morning, Joe,' Crippen greeted him. 'It's about time you all came to work . . .'

In return, Allen was able finally to confirm that their flight plan had been changed. 'Curious to know if you have Message Eleven on board,' he added as the Shuttle flew across Australia again. 'It's a pretty major change to the timeline and . . . prepared to answer questions, when and if.'

'OK, I'll tell you what, John is down on the mid-deck now. Meanwhile, I got a little piece of Slim Dusty and "Waltzing Matilda" for our friends down under . . .'

'Let her rip!'

And as the familiar sound of Australia's alternative national anthem, sung by their most popular country and western star, struggled through the static on the S-band radio, backed by the additional electronic beeps that accompanied transmissions from space, *Columbia* disappeared out of range over the Pacific, on her way back to Cape Canaveral for the first time since blasting off the previous day.

On the mid-deck, John Young pulled Message Eleven off the teleprinter and began reading. *Mission Control Center*, he thought, *had really redone the whole flight plan*. Normally he'd have questioned the wisdom of changing so much mid-mission, but this time, of course, he knew that the Flight Directors were trying to solve a real-time problem. It was entirely clear what they were trying to pull off, and why. Potentially, his life and that of his Pilot depended on it.

By the time *Columbia* arrived overhead Merritt Island, Young was back in his ejection seat, performing checks on the flight controls. Alongside him, his flight data file floated freely near his head.

'Bob, *Columbia*,' Joe Allen called up. 'If you look down now you'll see Cape Kennedy perhaps. There was a tremendous launch from there yesterday, which you may not have seen . . .'

Looking down towards Florida from the cockpit, Bob Crippen reached for the camera as he picked out the Vehicle Assembly Building and long runway built for the Shuttle from orbit.

While Crip snapped away at familiar landmarks on Earth, next to him John Young performed checks on the rotational hand controller – the joystick. And, at the Malabar Test Facility below them, Air Force technicians operating the TEAL AMBER camera tried and failed to capture an image of *Columbia* as she passed overhead. For all that it was hoped that the attitude of the Shuttle would help facilitate NORAD's efforts to take the picture, the report received by Gene Kranz in Houston claimed that, on this pass, the elevation level was too low. For now, the aircraft on standby at Patrick to spirit any pictures back to Houston remained firmly on the ground. There was just one more chance to capture an image from Malabar, on orbit twenty-one. Otherwise, hopes of getting an image using ground-based assets now rested with the TEAL BLUE telescope in Maui.

There, too, arrangements had been made to courier photographs back to Houston as fast as was possible. The problem was capturing the Shuttle in the first place.

'The weather isn't cooperating,' Kranz told reporters at the end of Neil Hutchinson's shift, 'and between the time we'd like to get set up and make sure we don't have any obscuration from the ground sites, then we look at the sun angle, look angle and range to the spacecraft, we've got a tough geometry problem.' The veteran Flight Director's tortuous syntax seemed to reflect the difficulty he was experiencing in trying to answer the journalists' queries without straying into areas he knew were off-limits. He remained calm and collected through further persistent questioning about the tiles, but in the end his frustration at being unable to deal more conclusively with the issue showed itself. 'We're working the daylights out of this thing trying to understand it,' he admitted, 'but for me it's unfamiliar territory.'

The Department of Defense either would or would not succeed in being able to tell NASA what they had no other way of knowing. Until they did, though, the sword of Damocles would hang over *Columbia*. Whatever happened, the next five orbits would be key.

As the ground link to Madrid began to crackle and fade, Bob Crippen thumbed the transmit button one last time. 'Joe,' he said, 'one additional request. If you could zap me up another copy of that timeline change, I would appreciate it. So John and I each have a copy.'

Clearly it was going to need their full attention.

'Roger that.'

FIFTY-SIX

Space Shuttle *Columbia*, 1981

As *COLUMBIA* SWEPT over the continental United States on orbit eighteen, Young and Crippen's conversations with Mission Control became increasingly frequent. Following a successful first test of the reaction control thrusters, the scheduled timeline called for two further burns, RCS 2 and RCS 3, at 14.22 and 15.22 GMT respectively. Both the times and the final post-burn attitudes into which the spacecraft would settle following these manoeuvres were changed. A fourth RCS burn was also added to the flight plan and scheduled for 17.12 GMT – or a mission elapsed time of one day, five hours and twelve minutes. In each case, Young and Crippen had first to accurately take down the coordinates, ignition time and velocities from CapCom, then input them into the digital autopilot. With barely an hour to go before TIG, the time of ignition, on RCS 2 burn, Crippen pushed Houston for clarity. 'We do not have a RCS 2 PAD [pre-advisory data] on board,' he told them. Despite an earlier suggestion to John Young over Orroral Valley that the PAD with the details the crew needed to perform the manoeuvre was already on board, they were still waiting for the numbers from Mission Control.

As the Orbiter completed her run across the country, passing out over the Atlantic towards Africa, Crip looked down from the flight deck to see the familiar contours of the north-eastern seaboard. 'I think we just passed over Long Island, there,' he reported.

'Roger.'

Ahead of *Columbia*, just over 1,200 miles to the east, the NRO's KH-11 KENNEN, number 5502, shot across her bows, streaking south towards Brazil on an orbital inclination of 96.82° and at an altitude a little under 50 miles higher than the Space Shuttle. In Sunnyvale, California, Air Force satellite operators were preparing their bird for its first shot at capturing an image of the Shuttle. Before the two space-craft met again, in an hour and a half's time, after both had completed another orbit, Mission Control and the astronauts had to make sure *Columbia* was in position for the second attempt.

Passing in range of the Bermuda ground station, Crippen pushed CapCom for details. 'About when can we expect an RCS 2 PAD from you?'

'Well, how about one second from now,' Joe Allen came back. 'We got it ready for ya.'

'Alright, let's see if we're ready for it. OK. If you got time, we got the time.'

'OK, here it comes . . . burn attitude two three nine-ah. Post-burn attitude is four eight decimal six, one six nine decimal zero, three zero four decimal two. And attitude time is three hours and fourteen minutes. Read back.'

The three numbers making up the post-burn attitude represented roll, pitch and yaw. All were measured clock-wise from a nominal zero in which *Columbia* was travelling forward in the direction of travel, upright and wings level. The instructions from Houston would position her on her back, flying tail first, her nose tilted down, her underside pointing up towards a point in space to the south-east of her position. Equally important was the attitude time given – three hours and fourteen minutes. This was the moment at which the Air Force operators from the Blue Cube Special Projects Office had determined that the window during which their bird could capture its image of *Columbia* closed. The point at which the two spacecraft were closest – just 158 miles from each other – came a little over two minutes earlier, at 15.11:52.

So by 15.14 GMT – the 'three hours and fourteen minutes' given by Joe Allen from Houston – the orbits of the KEYHOLE satellite and the Shuttle would have already crossed; the KENNEN would be looking from behind the Orbiter as she raced away. But while, ultimately, the Shuttle would quickly travel out of range of the KH-11 camera, it was not necessarily the moment when the orbits actually crossed that the Air Force was gunning for. Instead of proximity alone, Sunnyvale prioritized reducing the angular velocity between the spy satellite and its target and the position of the sun. Taking a picture from a latitude below *Columbia* – and from behind her – meant that to face towards the satellite's telescope mirror, the Orbiter's belly would also be facing in the direction of the afternoon sun from the south as it arced from east to west, lit up like an artist's model.

John Young repeated the vital details of the message from CapCom, confirming 'end time is three fourteen'.

'Read back is good and we'll see you in six minutes over Madrid.'

* * *

In Mission Control, as *Columbia* tracked down across Africa and on across the Indian Ocean towards Western Australia, Dick Truly and Hans Mark followed the radio exchanges between the astronauts and CapCom. Later, for public consumption, Mark, his hands tied by the levels of classification attached to the KEYHOLE programme, would say only that NASA, ultimately, had no choice but to trust in God and hope for the best; but he and Truly both had the necessary security clearances to appreciate the significance of *Columbia*'s upcoming change in attitude. With ringside seats, they clustered round the consoles to follow more closely the efforts of Young and Crippen to position the Orbiter, able only to imagine the work being done in parallel by the Air Force controllers in the Blue Cube. And now, even as Young inputted the numbers for the second burn into the digital autopilot, at CapCom, alongside Truly, Joe Allen prepared the PAD for the third reaction control burn, needed to manoeuvre *Columbia* into position for the encounter with the spy satellite that would follow on orbit twenty.

On the flight deck of the Shuttle, as she picked up the signal from Yarragadee, the tasks were beginning to stack up. While, twenty minutes off the planned timeline, Young prepared to manoeuvre the Orbiter, Crippen reconfigured the flight control computers and stowed the radiators.

'If you got time on your hands,' began Allen, 'I can read up to you the RCS test sequence for number three.'

'Stand by one, Joe.'

'Roger, no hurry.'

While Allen waited to pass his message, Bob Crippen organized paper and a pen.

'Both John and I are standing by to copy,' Crip confirmed.

As Allen read carefully through the complex sequence of instructions, Crip took them down before repeating the details back to CapCom. 'The postburn att is three one one decimal two, one eight two decimal seven, zero one three decimal zero. End time is four four four, over.'

The time of ignition was 15.42 GMT. As before, the roll, pitch and yaw numbers would position the Shuttle with its underside facing the satellite as it passed overhead the Shuttle. Once again, the end time given – 16.44 GMT – was two minutes later than the point at which *Columbia*'s orbit took her closest to the NRO bird. This time there would be just 67 miles between them as their trajectories crossed.

Joe Allen listened to Crip read back the numbers, checking them against those on the sheet in front of him. This time the numbers didn't tally.

'OK, Crip, three corrections,' Allen began, but as he tried to fix the errors, he lost the signal to the Shuttle. '*Columbia*, this is Houston,' he called up in the blind, 'we're having a little trouble copying you . . .' Each time he made a transmission, the static was punctuated by a shrill electronic bleep.

When Allen next made contact with the spacecraft over the Orroral Valley tracking station there was a further correction to make. Instead of configuring *Columbia* for a manoeuvre using the batteries of small reaction control thrusters, the crew were set up to use the two powerful orbital manoeuvring system rockets in the tail. '*Columbia*, we are showing you about seven degrees out in yaw,' he pointed out, 'and it's an RCS burn coming up and we're showing OMS selected at the moment.'

'OK,' Crip replied, caught by surprise. 'We ... yeah, you're right, you're right. Thank you.' He was annoyed with himself for screwing up, but conscious that there was no damage done.

In losing the signal at Yarragadee, Allen had known he'd pick up *Columbia* again at Orroral Valley, just thirty seconds' flight time away. But Orroral was the last opportunity to communicate with the crew until they were picked up by Hawaii forty minutes later – and that was just twenty minutes before the first encounter with the NRO satellite. If *Columbia* had to be in the right attitude at eleven minutes past three, it needed to be sorted out now. And as she sped north-east towards the limits of the Australian ground station's transmitters, there were discrepancies between Mission Control and the crew. Allen passed up the correct roll, pitch and yaw values.

'Stand by one,' responded Crippen, sounding wrong-footed, 'you are implying we're not in the correct attitude?'

'That's affirm,' Allen said, before once again reading out the post-burn coordinates.

'OK, Joe, I guess we're confused, I guess we did not understand that you wanted us to go to the post-burn attitude. You want us in the post-burn attitude right now?'

'Multi-axis burn always use post-burn attitude,' Allen confirmed as Young and Crippen began to lose the signal from Orroral Valley. As he tried to help them orient the Shuttle for their encounter with KENNEN 5502, Allen eventually had to accept they were out of range. '*Columbia*,' he told them, 'we are in the blind . . .'

FIFTY-SEVEN

Space Shuttle *Columbia*, 1981

COLUMBIA LANCED NORTH-EAST across the equator, 144 miles above the Pacific. With the autopilot programmed, at mission elapsed time of one day, two hours and forty-two minutes, John Young moved his hand controller out of the detent to initiate the burn. The ignition boomed through the hull of the spacecraft, shaking the cabin as clusters of jets in the nose sent angry white plumes of exhaust streaming into the blackness ahead. *A hundred times worse*, thought Young, *than it ever shook the Lunar Module*. Ahead, his view of Earth curving away beneath him in blues and whites began to swing around the picture through the layers of cockpit glass.

In the vacuum outside there was only silence as the Shuttle began to rotate around three axes, all the time continuing arrow straight at 17,500 mph along her orbital trajectory. To be heard inside, the thrusters needed to resonate through the substance of the airframe and cabin air. In pitch and roll the movement was smooth. Around the 'z' axis, though – yaw – there seemed to be minor oscillations, almost as if she was walking. *It just didn't feel clean*, Young thought after the manoeuvre, his long experience in spaceflight giving him a seat-of-the-pants appreciation of his ship's movement. It was something the simulator hadn't predicted – it would need looking at. As the Shuttle continued her measured tumble head over heels, she was reacquired by Mission Control.

'Hello, *Columbia*, Houston, back with you through Hawaii, over.'

'OK, we read you loud and clear and we are manoeuvring to the post-burn attitude.'

'OK, copy that.'

Between Hawaii and the coast of California, opposing sets of reaction control jets fired against the direction of the Orbiter's rotation to stabilize her attitude. In space, coming off the flight controls simply left a vehicle to continue its movement until a positive force was applied to stop it. But as the spacecraft was reacquired from the ground by the Buckhorn ground station, and after the confusion that had crept in earlier during the pass between CapCom and the astronauts, the news was that all was as it should be.

'OK,' Young confirmed, 'we are in attitude.'

'OK, John, it looks good.'

Now it was up to the National Reconnaissance Office.

To see the Northern Lights from Thule Air Base, military personnel stationed there had to look south. Positioned in northern Greenland, less than 14° south of the North Pole and 750 miles north of the Arctic Circle, the American military's most northerly facility was a lot closer to Moscow than it was to Washington. Icebound and dark for much of the year it was not an easy environment in which to work. In March 1972, rocks the size of baseballs were whipped up and thrown against buildings like cannonballs as Thule was battered by the fastest low-altitude winds ever recorded on Earth. The reading on the an-emometer was 207 mph and climbing before the instrument was destroyed by the force of the storm. But Thule's remote, unforgiving location also made it one of the country's most vital military bases. Described in one newsreel as 'this nation's greatest military secret since the mounting of the Normandy invasion', construction of the base was a massive engineering feat – sometimes compared to the building of the Panama Canal – complicated by extreme conditions, frozen ocean and the absence of a port. The Strategic Air Command Base was home throughout its life to squadrons of fighters, bombers, missileers, and elements of the Ballistic Missile Early Warning System in the shape of SAC's 12th Missile Warning Squadron. It had also, since 1961, been next-door neighbour to the Thule tracking station, part of the global network of facilities serving the Air Force Satellite Control Facility at Sunnyvale.

Three and a half miles to the south-east of the air base, the tracking station was a small mushroom farm of radio and radar antennas housed in white spheres of different sizes, augmented by masts, dotted around a collection of low-rise buildings, their shallow-pitched roofs as often as not blanketed in thick snow. Until its existence was declassified in 1969, it was referred to simply as Operating Location Five.

Because of its unique position near the North Pole, Thule was able to provide telemetry on nearly every orbit flown by NRO satellites launched out of Vandenberg.

Now, as the Shuttle crossed the continental United States on a track that took her from south of Portland, Oregon, across the mid-west and southern shores of the

Great Lakes towards Philadelphia, KENNEN number 5502 was passing directly overhead Thule towards Newfoundland. At their consoles, beneath the NRO bird's orbit, the Thule operators – callsign POGO – tracked its progress, their data relayed back to the Blue Cube in Sunnyvale via the Defense Satellite Communication System network.

In the exchanges between Mission Control and the Shuttle, there was no suggestion that anything significant was in train. In fact, there seemed to be a studied indifference to it.

'John, we've got a call from Building 4,' began Joe Allen, as *Columbia* approached her first orbital encounter with the KH-11. 'It's Monday morning here and they want to know if they should go ahead and start the pilots' meeting without you.'

'Pilots' meeting?' replied John Young. 'I never knew whether it made any difference whether I was there or not! Is it Monday morning already?'

'Would you believe that?' Allen played along, before pointing out that the line was bad. 'Tell John,' he said to Crip, 'his communication was a little garbled.'

'He's been accused of that before,' Crip teased as he looked down towards the eastern seaboard from the flight deck. 'We're just coming across Cape Cod once more.'

While Young and Crippen bantered with Joe Allen on CapCom, they knew they were barely thirty seconds from the window in which Sunnyvale controllers would first try to capture a sat-squared image of *Columbia* on orbit.

It was rubbing your tummy and patting your head on a grand scale. Performing complex, computer-controlled orbital ballet – against the clock – in a $10 billion spacecraft while never dropping the attaboy, A-OK, no-sweat demeanour with ground controllers that was expected by the public. However they felt, whatever they were doing, the astronauts had to stay unruffled, unhurried and *cheery*. It was as if a combat pilot were also required to exchange pleasantries on prime-time radio while flying a dangerous, demanding mission over enemy territory.

And if that weren't enough, the two astronauts had to keep their counsel on perhaps the most remarkable aspect of their fleeting encounter with the NRO spy bird: they'd seen it. As 5502 plunged south over a hundred miles distant, Young and Crippen, their view unhindered by atmospheric moisture, caught sight of it glinting in the sunlight through the windows of *Columbia*'s flight deck. And they couldn't tell a soul.

There was no signal or acknowledgement of whether the Air Force had been successful in their effort to capture a photograph of *Columbia*. The paths of the two spacecraft simply crossed as swiftly and silently as they had been destined to from the moment the Shuttle was successfully established in an orbit determined only after the crew had already been strapped into their seats. That, during their brief encounter, the Air Force team in Sunnyvale had commanded their big KH-11 reconnaissance satellite to try to steal an image of the Orbiter as she passed was a moment which, in itself, was incapable of providing any reassurance. But whatever the combination of light and shade gathered by the KENNEN's powerful reflecting telescope and collected by its arrays of hundreds of thousands of electronic receptors, it was now on its way via SDS relay to Area 58, the NRO's Aerospace Data Facility East in Fort Belvoir, Virginia.

* * *

A minute before loss of signal through Botswana, Bob Crippen reported 'we're back in OPS 2 and getting ready to go to the post-burn attitude'. Crip reached forward to the centre console, pushed the OPS button on the keyboard, keyed in the three digits to bring up a new computer operating mode and pressed the PRO button at the bottom right of the panel to confirm it. Because of the limited memory capacity of the computers, different flight modes demanded the scrubbing and reloading of different software throughout the flight. The reaction control burns were performed using OPS 3, and Crippen was already on the case.

'OK, *Columbia*,' Joe Allen responded, 'we were hoping you would tell us that.'

As the Shuttle crossed the Indian Ocean, John Young pushed his hand controller through the detent to fire the third reaction control system burn. Once again, the thrusters boomed through the cabin like howitzers as *Columbia* began to manoeuvre into the attitude that she'd maintain until her next encounter with the NRO's spy satellite, itself now sweeping round over Antarctica.

Mission elapsed time was one day, three hours and forty-two minutes.

The little city of Unionville was the county seat of Butler County, northern Missouri. It labelled itself 'The Hunting Capital of Missouri'. Today that seemed appropriate enough, as Unionville's 2,000-strong population were located directly below a celestial pursuit that, in Mission Control in Houston, Flight Director Gene Kranz compared to trying to find a needle in a haystack. As dawn broke on 13 April, *Columbia* approached the Pacific North West to begin orbit twenty, and her next pass across the continental United States. At the same time, coming down across the Hudson Bay towards Ontario was the NRO's KENNEN 5502 spy satellite. Less than ten minutes later, at 16.42 GMT, it was mid-morning in Unionville when, unbeknownst to the residents going about their business below, *Columbia* and the KH-11 crossed paths in the skies overhead.

As the spacecrafts' orbits bisected, Joe Allen was already relaying the numbers for the next manoeuvre to the astronauts. Mission Control wanted them to maintain the same post-burn attitude for the fourth reaction control burn. 'End attitude time is one day, six hours and fourteen minutes.' 18.14 GMT marked the last time there would be any opportunity to use the NRO's satellite to capture a picture of *Columbia*'s heatshield. John Young repeated the numbers.

'OK, John, read back is correct. And, once again, do the burn in the current attitude.'

'Alrighty.'

Everything looked good. The crew had the details of their reaction control burn – one that had not been required in the original mission timeline. But, over three orbits, Houston had rewritten the details of the flight plan in real-time to ensure that *Columbia* presented herself as the best possible target for the NRO's cameras. Still, it all seemed in hand. Summing up, the Houston Public Affairs Officer explained: 'Information for the fourth reaction control system test-firing was passed up, the delta velocity [change in speed] is two feet per second, duration of the burn is four seconds, ignition time is one day, five hours, twelve minutes.' It sounded easy.

'OK,' confirmed Young as Houston reacquired *Columbia*

through Dakar, 'we're at three zero two, the burn is all fit up and the targets look OK.'

'OK, very good,' replied CapCom. Then, after a brief conversation with Crippen, they pressed to transmit again. 'OK, John and Crip, we want to modify your RCS burn PAD . . .'

Mission Control wanted to shake things up again. And just to complicate things further, on the next orbit Young and Crippen were scheduled to be talking to the Vice-President, George H. W. Bush.

FIFTY-EIGHT

Houston, 1981

FOR THOSE IN Houston trying to ensure that the orbital attitude of the Shuttle would place her in the gaze of both celestial and terrestrial Department of Defense cameras, the situation was not straightforward. Orbits twenty and twenty-one offered a cluster of possibilities that would have been more easily accommodated had they been more spread out. As well as the only three KH-11 passes offered up during STS-1, there was also another chance to capture *Columbia* using the Air Force TEAL AMBER telescope on Malabar. But in order to be in position for both, the Orbiter would have to flick-flack quickly to present her belly in two entirely different directions. With the Shuttle already over Australia and less than fifteen minutes' flight time from Hawaii, there was now some urgency about relaying the news to the crew.

'*Columbia*,' said Joe Allen, 'we're going to ask you for another RCS manoeuvre using the current attitude.' As the Orbiter arced north-east up over the Pacific and in and out of radio range, Allen worked hard to convey the necessary information. 'We're also sending up a one-line teleprinter message to you there we'd like you to glance at before you get to Hawaii.'

Then, just as Allen was on the verge of transmitting the numbers for the new manoeuvre, the comms link began to fail.

'*Columbia*, Houston . . . *Columbia*, this is Houston, over,' he persisted, getting nothing but static in return. 'Hello, *Columbia*, this is Houston, do you read?'

Eventually, barely decipherable through the crackle, John Young's voice made it back to CapCom: 'We're here.'

'You're weak, but we're basically in the blind,' Allen told the spacecraft's Commander. He pushed on: 'We'd like a body axis burn at the time five hours and forty-two minutes and thirty-five seconds.' After reading off the solution for the new manoeuvre, he finished by telling Young, 'Do the burn in OPS 2 please, over.'

'Houston, you're cutting out,' Young said after a long pause.

Allen repeated the details, and again finished by telling Young he wanted the burn done without first reconfiguring the Orbiter's computers with the appropriate software. It remained academic.

'That whole thing was completely broken,' Young told him, his voice faint against the background hiss. 'We didn't hear a word you said there . . .'

The link may have been poor, but, as *Columbia* approached the limits of the Yarragadee ground station's range, they were about to lose it completely. With one eye on the clock, Allen tried again. 'And that's in OPS 2,' he emphasized as time ran out, uncertain whether the crew had captured the message, 'and we'll see you shortly through Guam.'

Bob Crippen hadn't plugged into the comms link as Young scribbled down the particulars of the hastily added manoeuvre from Allen's last transmission.

In order to try to lighten the load on the astronauts, Mission Control had decided not to ask the crew to reconfigure the software out of OPS 2. But it was a mistake.

When he returned to his seat, Crip, who'd been so central to developing the software in the first place, urged his Commander not to go ahead with the burn in that set-up. Young was struck by his Pilot's conviction, but at the same time he felt Mission Control had been clear: 'do the burn in OPS 2'. Houston, he figured, *knew something I didn't know*.

'Well, I'll do it,' he told Crippen.

With the time of ignition looming, Young brought up the MNVR EXEC – 'Manoeuvre Execute' – display and selected the reaction control jets from four different engine options. He loaded the data and time of ignition into the digital autopilot, chose 'Manual', then reached up to press the 'Execute' button in the centre console.

At 17.43:45 the Commander used his hand controller to trigger the burn, expecting to feel the acceleration as 1,740lb of thrust produced by two downwards-pointing jets mounted in the right-hand orbital manoeuvring system pod pushed *Columbia* into her new attitude. But when the fire command was sent, nothing happened. Young watched the delta-V monitor on the instrument panel ahead of him hold stubbornly at zero before a 'jet fail' message flashed up on the warning display between him and Crip.

Oh, we've really messed up now, he thought.

* * *

– 433 –

Two minutes later, as the Shuttle flew into range of Guam, John Young thumbed the transmit button to tell CapCom what had happened, or rather what had failed to. Because of the attempt to perform the burn in OPS 2, the reaction jet drivers, necessary to fire the thrusters, were not switched on. As a result no chamber pressure was sensed, and that triggered the fail light. Bob Crippen did his best to mask his frustration.

'I thought we all knew that it was not satisfactory for doing burns,' he chided.

'OK, we did know that it was only semi-accurate,' Joe Allen confessed.

'Semi-totally-*un*accurate,' Crippen responded, a hint of impatience creeping into his voice.

There was still time to correct the mistake, however.

'I'm going to reselect the jets,' Crip said.

'OK, very good.'

And shortly after Mission Control lost the signal from Guam, after powering up the jet drivers, the astronauts completed the manoeuvre into the required post-burn attitude, barely a minute before being picked up as she passed north of Hawaii.

'I've discovered,' Crippen had told CapCom earlier in the mission, 'that there's lots of clouds around the world.' And now, as Air Force personnel at the Maui Optical Station and in Malabar tried to track and focus on *Columbia* as she passed within range of their telescopes, they were proving to be an issue. Both sites had been chosen because of clear air and good weather. Today, though, during one of the few orbits in which the Shuttle's trajectory combined with favourable lighting conditions to make an image a possibility, the Air Force was obliged to contend with scattered cloud around 4,000 feet and visibility of only about 7 miles. It was making a difficult task even harder.

'White House, Houston . . . White House, Houston.'

'Houston, White House, testing one, two, three, four, five, five, four, three, two, one.'

'White House, you're loud and clear.'

'Houston, you're loud and clear also.'

A month before the launch, Bob Crippen had just returned to the Astronaut Office after a 5-mile run when he got a message that John Young wanted to see him. 'Say, Crip,' Young told him with a smile, 'the Vice-President wants to go run with you!' After Crip had pulled on his running shoes again, the picture taken of him and Young, dressed in regulation NASA T-shirt and shorts, running with George Bush, became one of his favourites. The Vice-President, like both astronauts, was a naval aviator whose interest in the upcoming spaceflight was evident. As well as running round Kennedy, Young and Crippen were able to pull the strings necessary to get him into the cockpit of *Columbia*, already standing vertical on the pad. Now, with President Reagan recovering from the attempt on his life at the end of March, the job of conveying the nation's appreciation of the astronaut's efforts fell to the Vice-President.

'John and Crip, we have a telephone call coming into the space network from the White House for the crew members of the spaceship *Columbia*. We would like to patch them through. If you would, please, Mr Vice-President, go ahead.'

There was an easy familiarity about the brief conversation that followed, with Bush teasing Crippen about his racing pulse at lift-off. 'I couldn't understand that,' he said, 'I thought he was a calm guy out there, you know, and now look at him . . .'

'Right, right,' agreed Young.

As Navy fliers, all three men had been schooled in a tradition in which excitability was frowned on. Crippen's heart-rate was an easy target. But, when it came to the way the mission was unfolding, the Vice-President's full résumé provided him with an unusual appreciation of things. Alongside his time flying Grumman TBM Avenger torpedo bombers for the Navy, he could also point to a year in which he'd served as Director of the CIA.

Jimmy Carter's election had brought George H.W.'s time at the agency to a premature end, but Bush had still been at the helm for the launch of the first KH-11 KENNEN satellite, number 5501, on 19 December 1976. During the transition period following Gerald Ford's election defeat, he'd also briefed the incoming Democratic President on the capabilities of the CIA's revolutionary new reconnaissance platform prior to it capturing its first imagery – the pictures shown to Carter on his first full day in office. Now, at precisely the moment he was talking to the astronauts from the White House, the spacecraft he'd helped usher into service with the National Reconnaissance Office was passing within range of *Columbia* for the last time. And as the conversation with the Vice-President came to an end, thoughts on board the Shuttle turned to coming home.

'The only bad part about it, Joe,' Crippen said after the link to CapCom was restored, 'is we're going to have to come down.'

'Well don't come down in that attitude,' Allen joked.

Crippen laughed. It wouldn't matter what state the

heatshield was in if they tried to re-enter upside down.

After Tom Moser's Structures and Mechanics team established, late the previous night, that there would be no risk to the Shuttle during re-entry from the sixteen tiles known to be missing or damaged from the two orbital manoeuvring pods, he'd had to share their conclusions with the project management team. Carrying the analysis and calculations with him, he presented his case to John Yardley, Chris Kraft, Aaron Cohen and Bob Thompson in Building 1. They challenged him on every point but, ultimately, were reassured. Moser's reward for being so convincing was to be asked by Kraft to join the next change-of-shift briefing alongside Don Puddy, boss of the Crimson Team, and Gene Kranz in the Public Affairs Office auditorium inside Building 2. Moser had never been asked to sit in on a briefing of that magnitude and knew well the focus there had been on the integrity of the heatshield in all the preceding press calls. But confident of his ground, he felt comfortable about the prospect of sitting in front of the room.

As he prepared himself, Mission Control turned their focus to another problem that, although attracting less attention, had been nagging both them and the astronauts throughout the flight.

Every thirty-three minutes, Mission Control had re-minded Bob Crippen to pull the circuit breaker on the faulty flight recorder required to capture the vital data from the Shuttle's re-entry. It was hoped, following Dick Truly's efforts in the sim using the swizzle stick to pull it in and out, that this process might continue after Crip had strapped himself back into the ejection seat for the return home. But Truly wasn't optimistic that, once his friend was constrained by his bulky pressure suit, he'd be able to make the stretch.

While John Young sat in his ejection seat, running a comms test with the new Flight Control shift, Hank Hartsfield, on CapCom, reminded him that Crip needed to try reaching the circuit breaker.

'I'll tell him, Henry,' Young said. 'He's still downstairs off comm.'

After pulling himself into the tan pressure suit on the mid-deck, Crippen drifted up through the companionway to the flight deck above, then floated across the cabin, using his arms to stabilize himself as he pushed himself down into his seat. After a quick comms check, Crippen tried to reach the circuit breaker, but without success. 'I might accidentally do it,' he told Hartsfield. 'I can just barely lay the top of the swizzle stick on top of it, but I can't get any pressure on it. I got my neck rotated through 180° to do *that*.'

Hartsfield kept pushing, suggesting that maybe Crip should try loosening his straps.

'Yeah, I can do that,' Crippen replied, with an exasperated laugh. He could get up out of his seat if they wanted him to, he added. 'I don't understand what it is you're all asking. When is it you were thinking you might ask me to do that?'

About four minutes before they fired the rockets to begin the Shuttle's de-orbit, Hartsfield told him.

Crippen's reaction was immediate and firm. 'Uh huh,' he said, dismissing the possibility. 'I got APUs to start and a burn to get off.'

Every second of the astronauts' time was accounted for. 'Not one place in the flight plan,' Young observed, 'where it says "go to the bathroom" even.' On day one he'd had to christen the waste management system between the two rocket burns that established *Columbia* in her final orbit – an effort that had impressed Crippen with its efficiency. If Mission Control wanted them to add anything to the flight plan, they had to find ways of making time to do it.

'OK, Crip,' Hartsfield said, backing down, 'don't bother trying to work at it any more now,' before explaining more carefully why they were spending so much time trying to work the problem.

The worry back in Houston – the worst-case scenario – was that if they failed to capture the re-entry data they needed from the first flight they might have to consider adding another mission to the orbital test flight programme. And with Plan A – Crip and the swizzle stick – now out of contention, the management's thoughts turned to the possibility of asking the crew to start removing panels in order to try to disconnect the offending flight recorder and swap it with one of the other two carried on board *Columbia*.

FIFTY-NINE

Houston, 1981

After a brief introduction from Don Puddy, who'd just come off shift, Tom Moser was given the floor to deal with the issue that was top of the media's agenda. Sitting alongside the two veteran Flight Directors, in front of a packed auditorium, he provided a full account of the work done by his team and their conclusions, always conscious that he was talking to a particularly well-informed group, many of whom had covered the space programme since its inception.

The tiles missing from the manoeuvring pods would not, Moser said, 'have any effect on the safety of the vehicle on entry'. But while he could be unequivocal about what was known, what was still unknown about more critical areas of the heatshield dominated the rest of the briefing.

Gene Kranz was asked what measures might be taken if damage to these more critical areas of the thermal protection system was discovered. His answer revealed the limits of the options available. The astronauts could, he told them, purge and depressurize any of *Columbia*'s tanks and systems that might be at risk from burn-through; there was room to slightly alter the re-entry profile except that, he admitted, it was already planned to be about as benign as possible. Re-entering with even a fraction of a degree more sideslip in an effort to reduce the thermal stress to a damaged area of the heatshield would only increase the danger elsewhere by shifting the point of peak heating from the reinforced carbon-carbon nosecone to a less resilient part of the airframe.

At first glance, the presence of *Salyut 6* in orbit might have looked like a potential lifeboat, but the harsh realities of celestial mechanics meant it was no such thing. Circling in an orbital inclination of 51.6°, the Soviet space station would have needed around 20 tons of propellant to move her into the same 40.3° orbit as *Columbia* – as much as the total mass of the space station itself. The change in velocity – the delta-V – necessary to make the manoeuvre would have been around eight times the speed at which the average airliner flies. A shift in its orbital plane was about as thirsty a manoeuvre as any spacecraft could be required to perform. And in this situation it was impossible.

That left the possibility of keeping Young and Crippen on orbit until some plan for bringing them home could be improvised. And while, through careful conservation of the resources on board, *Columbia*'s time in space could be extended, it was anything but indefinite; days not months.

The only realistic option was confirmation that the critical areas of *Columbia*'s heatshield were undamaged. As Kranz stressed throughout, 'we have no hard evidence at this time that leads us to suspect that we have any problems with the underside of the Orbiter'. The trouble was, neither did he, Puddy and Moser have any hard evidence to the contrary.

Asked whether the Air Force had yet successfully managed to take pictures that determined there was no tile loss from the underside of the Shuttle, he could only admit they had not. The most recent effort to capture the Orbiter using the TEAL AMBER telescope at Malabar on orbit twenty-one had failed to yield any useful data. 'Officially, and organizationally, operationally we have no usable photos obtained from the ground stations,' Kranz said, but that statement didn't shut the door completely. And it was clear that some of the reporters in the room had sources suggesting that, while ground-based telescopes may have drawn blanks, the Department of Defense might still be in a position to provide imagery of *Columbia*.

'We're working with the Air Force in using available resources,' Kranz explained, declining to expand on what those DoD assets might be. 'Any further discussion of this subject,' he said, 'is classified.'

As he spoke, there were still hopes that the classified efforts of the Air Force controllers at Sunnyvale might yet provide the photographic reassurance NASA wanted, even if the details of how they might do so had to remain off-limits to the public.

'*Columbia*, Houston. Can you listen now?'

'OK, Hank, I guess I can listen up,' replied Crippen.

'OK. Our management has met and talked about this DFI problem and we have concluded that the data that we could get on that is very important to have.' Hartsfield emphasized the significance of the development flight instrumentation recorder that was proving so profligate with its supply of tape. It was supposed to be recording the thermal data from the point at which *Columbia* began her re-entry from space through to when she emerged, after a communications blackout, in atmospheric flight. Impossible to test properly in ground facilities, it was the most elusive and uncertain part of the Shuttle's envelope.

As the date for the first flight launch had approached, Bob Crippen had lobbied hard for an in-flight maintenance capability to be carried aboard the Orbiter. Now, providing Mission Control could reorganize the flight plan to give him the time to do it, he was enthusiastic about his chance to road-test the tool kit. John Young was more sceptical. It didn't make sense to go messing around changing out big components on the vehicle unless, he thought, they were flight critical. And the flight recorder, as much as the information it captured was valuable, had no bearing on whether or not *Columbia* and her crew were going to make it home in one piece. It was not a flight safety issue.

While Young stayed up on the flight deck, Crippen floated down through the hatch on to the mid-deck to get to work.

From CapCom, Hank Hartsfield read through the details of a task that hadn't been practised on the ground. 'On step eight,' he said, 'what we want to do is change the order of removing those connectors so that we don't . . . so that we reduce the possibility of pulling the pins and the wires, and the order we want is as follows: eight . . . seven . . . six . . . one . . . three . . . four . . . five.'

It sounded as if he was transmitting the step-by-step instructions for defusing a bomb.

'Henry,' John Young cut across them, 'you want me to give torquing angles now?'

While leading Crip through trying to change out the flight recorder, Hartsfield still had a job to do helping the Commander fly the Shuttle. 'Yes, sir,' he replied without breaking his stride, 'we've got fifty-five seconds. Shoot 'em to me.'

Young just had time to pass his message before *Columbia* lost the signal from Botswana.

On the mid-deck, Crip was having a hard enough time of it to stop whistling the tunes with which he'd set about the job. For all the discussion of precision disconnection, the problem facing him first demanded brute force. Most of the twelve fasteners holding each panel to the bulkhead just wouldn't budge, and without the help of gravity to root him to the spot, he was struggling to gain the purchase he needed to make inroads on the stiff Allen-head screws. With the temperature in the cabin now restored to a comfortable level, he was beginning to work up a sweat.

As *Columbia* tracked north-east across the Himalayas en route to Hawaii, Crip called up to his Commander to see if he might have any more success in moving the screws and shifting the panel. Young drifted down head-first through the hatch, completely at home manoeuvring around in weightlessness, but after trying to plant himself in front of the bulkhead, his efforts to loosen the screws were no more successful than his crewmate's. Young imagined they'd been tightened up and painted in by some 300lb guy. No chance, he thought, that they could *expect a pair of astronauts that only weigh nothing in zero gravity to get the thing off*.

After wrestling with it for nearly forty minutes, when they reacquired a radio link to CapCom, they broke the bad news. 'OK, Hank,' Crip said, 'we've got a small problem, or a big problem, depending on how bad you want DFI.'

If it had been mission critical, if they'd said 'change out the computer', Young maintained, he and Crip would have worked all night to get the job done. *If that panel had to come off, we'd have sawed it off with a bone saw.*

But even that might not have been enough. In deciding that the mission wouldn't fail on account of his work, the technician responsible for installing the panels had used Loc-tite glue to seal it shut. There was no way *his* panel was going to be the one that shook itself loose. The astronauts would have been lucky to be able to drill them out.

'We'd better give up on it,' Hartsfield agreed, 'unless somebody here comes up with a Eureka in the next few minutes.'

The failure to change out the recorder was a frustrating end to a day that had borne almost no resemblance to the flight plan. And yet it had also been encouraging. *Columbia* had shown herself to be abundantly capable and flexible, responding with alacrity to all that had been expected of her and more.

Bob Crippen returned the panel hiding the flight recorder to the condition in which he'd found it, stowed the tool kit and flight procedures file which he'd had to hand as he worked, then pushed himself back up through the companionway to the flight deck. After three-quarters of an hour without a view outside, he was drawn to the sight of the stars through the cockpit windows. Previous astronaut groups had used a planetarium to study the heavens – knowledge of which was necessary as a back-up for the spacecraft's inertial navigation and guidance systems. Crippen and the rest of the MOL guys hadn't been so lucky, relying instead on the simulator. John Young, though, as with lunar geology, had pursued astronomy with enough fervour to become expert. As he and Crip had trained for STS-1, the veteran astronaut had shared his understanding of the stars with his Pilot. Throughout the mission, during the time they'd spent beyond the range of the ground-station network, Young had been pointing out stars to his rookie crewmate. Now, gazing out from *Columbia*'s cockpit at a constellation he recognized as the Southern Cross, bright and unfiltered by Earth's atmosphere, Crip was moved by how dramatic it looked. Fifteen years after the Air Force had selected him to be an astronaut, it was a small moment of wonder that would from here on colour his memories of his first spaceflight.

With the timeline compressed because of the attempt to change out the recorder, there was little more to do than eat and prepare for their second night's sleep on orbit.

'We all look forward to seeing you tomorrow,' Hank Hartsfield told them, 'we're excited about it. We understand you're buying.'

'Well,' Crippen laughed, 'you might be right . . .'

Seconds later, the crew lost the signal through Santiago. Because of the ground track of orbit twenty-five, it would be nearly an hour and a half before, returning to Chile, they could have any contact at all.

The astronauts' sleep period began at one day, thirteen hours MET. As Young and Crippen settled in their ejection seats, blinds drawn over the deep cockpit windows, in Houston, Flight Director Neil Hutchinson's Silver Team replaced the Bronze Team at the Mission Control consoles. After plugging in and shadowing the last minutes of Hank Hartsfield's shift, Dan Brandenstein sat down at CapCom for what he anticipated would be a quiet shift. There were ongoing efforts to come up with a solution to the flight recorder problem. One possibility was simply to trip the master switch and shut down all three recorders until just prior to re-entry. Aaron Cohen and the programme management were weighing up the pros and cons of that. At the same time, a team of astronauts were expecting to work through the night in the sim to see if they could jury-rig an appendage for the swizzle stick that might give Crip the reach and control he needed to operate the circuit breaker from his seat.

On board *Columbia*, Young and Crippen turned down the ground-to-air comms link. For the next seven hours they just had each other, and the silence of space, for company.

SIXTY

Houston, 1981

AS SILVER TEAM settled into their nightshift with the big screens and banks of consoles inside Building 30, the outgoing Flight Director, Chuck Lewis, left Mission Control and walked to the small briefing room in Building 2. There, once again, Gene Kranz was sitting in on the change-of-shift press conference. He was already braced for what was coming and, sure enough, the first question concerned the timing of the last opportunity to take pictures of *Columbia*. While Kranz conceded that, conceivably, it might be possible to get some coverage on revolution thirty-five, he didn't think it would be necessary. 'We've examined all data that's available,' he told the reporter from the *Minneapolis Tribune*, 'and we've concluded we've got no basis for altering our plans.' It was a line he clung to like the Fifth Amendment for the rest of the briefing. While the trouble with the flight recorders added a little variety, it was only a minor distraction from what had dominated proceedings all day. One reporter suggested that perhaps Kranz appeared to be sounding more confident than he had at the morning briefing. 'Mainly because maybe I feel this is the last time I have to do this,'

Kranz dodged, 'because tomorrow we're all going to be interested in the landing phase of the mission. Now if you want me to read that statement once more I will . . .'

'Never seen a dead horse get beat so much,' joked Chuck Lewis. But they had barely started on Kranz. As in previous briefings, he referred only to ground-based optics and to the quality of the images that had been seen at Anderson Peak. What had sounded obliging and informative earlier in the day now seemed to shine a light on what was not said. Reporters had got the message that there were no usable *ground-based* images. And that the unclassified Air Force telescope at Anderson Peak, imagery from which Kranz so readily discussed, wasn't up to the job.

'You give,' as one of the journalists pointed out, 'a very careful answer to that question.'

'You're right,' Kranz agreed.

'Let me hit it nose on: do you have any data indicating an anomalous condition associated with the tiles other than those which we've seen on TV on the OMS pod?'

'I'll read you the statement again if you want it . . .'

'OK, let me go at it this way then: are you under instructions not to go beyond that statement?'

'I've stated I can't go into further detail on this subject.'

'Why can't you?'

'Because I said I can't go into further detail on this subject.'

'Well, I mean, is God compelling you not to go beyond it or is someone else saying "you shalt not go beyond that statement"?'

Kranz paused, thinking how best to answer. 'I think I identified in an earlier activity that we were using DoD resources.'

'Do you, then, have confirmation?'

'I'll read it one more time . . .'

'You've mentioned twice now this blip transversing the sky,' the reporter continued, focusing on Kranz's description of the images from the Western Test Range telescope on Anderson Peak. 'Is that kind of the sum total description of all you've gotten today? From *all* these resources?'

The question, dripping with disbelief, prompted laughter from the rest of the press pack.

'I think that I identified from the optical sites that we didn't get any data that we believed usable. I think Anderson Peak . . . the data is going to be played back and you can form your own opinion on it. Any other data, I just can't discuss.'

SIXTY-ONE

Houston, 1981

AT FIRST, IT was just a ripple around Mission Control, little more than a rumour: *we don't have to worry any more.*

In the second row of consoles, sitting alongside CapCom Dan Brandenstein as his back-up, fellow TFNG recruit Terry Hart asked, 'Why don't we have to worry any more?'

'Can't tell you,' came the answer, as if it were some kind of parlour game, before once more stressing 'you don't have to worry any more'.

But there was only one thing that had been nagging away throughout the mission and that was the state of *Columbia*'s heatshield. The risk from the missing tiles on the manoeuvring pods had been dismissed quickly and completely by Tom Moser's engineers. But, despite Kranz's calm assurances at briefings throughout the flight, the effort to establish beyond doubt whether or not there was a far more serious risk to *Columbia* and her crew had rightly been regarded as an emergency. It just wasn't one which, because of the highly classified nature of the resources being brought to bear, was allowed to play out in full view. In the Mission

Control room, though, the curtains were pulled back a little on the secret world.

The reporter who'd detected a change in Gene Kranz earlier had been right. About an hour after the buzz first began to spread around Building 30, the Deputy Director of Flight Operations walked into the Mission Control room. The blond crewcut and widow's peak still framed his boxer's face; the trademark waistcoat, tailored by his wife, Marta, was present and correct; but Kranz, whose thin-lipped intensity had come to personify Mission Control, seemed unburdened.

After the pad fire that had claimed the lives of John Young's friend Gus Grissom and his *Apollo I* crew, Ed White and Roger Chaffee, Kranz had told flight controllers that, whatever any accident investigation established, 'we screwed up . . . we should have caught it . . . we are the cause'. The words that followed became known as the Kranz Doctrine. 'Flight Control will be known by two words,' he'd said, '"Tough" and "Competent". *Tough* means we are forever accountable for what we do or what we fail to do. We will never again compromise our responsibilities. Every time we walk into Mission Control we will know what we stand for. *Competent* means we will never take anything for granted. We will never be found short in our knowledge and in our skills. Mission Control will be perfect.'

And this was how NASA had responded to the missing tiles. From the moment the damage to *Columbia*'s thermal protection system was discovered, Kranz had led the charge to deal with it. He'd hoped for the best, but had assumed nothing; they'd *worked the problem*.

Now, as the astronauts on board the Shuttle slept, Kranz had returned to his domain carrying the last part of the jigsaw: photographs of the bottom of the Shuttle. And these were very definitely not the kind of amorphous streaks Kranz had been prepared to describe to reporters. The Flight Director had a sheaf of high-resolution pictures taken hours earlier by the NRO's KH-11 KENNEN spy bird. What, during the *Skylab* crisis, had taken days to turn round had this time, relayed through Area 58, the Aerospace Data Facility East at Fort Belvoir, arrived the same afternoon.

The detail was stunning, showing clearly that there were none of the black pockmarks on *Columbia*'s belly that would indicate there were further missing tiles. Crew safety, Kranz had always maintained, was the first priority; in leaving no stone unturned he now had the evidence to illustrate it.

'How did you get those?' Kranz was asked.

'I can't tell you,' he responded, no more forthcoming about their source than he had been with the press about their existence.

Joe Engle and Joe Allen, both in the control room when Kranz brought in the pictures, examined them. But of the two of them, only Engle, like Dick Truly, also with them in Mission Control and who also had the opportunity to look at the images, knew in detail how they'd been captured. 'Special methods', 'national technical assets' or just plain 'DoD' was as much as any one of them clustered next to the rows of consoles, as far as Allen was aware, had been told about their source. Truly and Engle knew otherwise. And Truly, through his own experience with the NRO, was able to attach a programme name to the 'DoD resources' Kranz had alluded to with reporters. He knew the nature of the spy bird's day job, its capabilities, and how good fortune, along with skill, had contributed to Air Force controllers' success in capturing the pictures that confirmed that his close friend Crip and his Commander were safe. Relief flooded through him; the anxiety he'd felt at the prospect of the missing tiles being symptomatic of a more serious problem immediately washed away. Around him, too, the mood was transformed.

The Shuttle's fine, Terry Hart told himself, reflecting the view of the whole room; *it's going to come back fine*.

For Ken Young in the Building 30 vault there was relief. With his division boss Ed Lineberry, the burden of capturing the photos had fallen most heavily on his shoulders. Now confident of *Columbia*'s return he could enjoy the moment and note with satisfaction just how closely they resembled the images modelled on Dave Scheffman's computer.

The pictures did not stay in the Mission Control room for long. And, ironically, Hans Mark, the man who had reassured NASA's John Yardley in Washington that the Department of Defense had the capability using its 'spooky ships' to take pictures of the Shuttle on orbit, did not see the results with his own eyes. But he did learn when he returned to Mission Control that 'it's OK'. He didn't need to be told more than that – the rest of it he could fill in himself. As had been apparent during the Iranian hostage crisis, the NRO's KEYHOLE satellites couldn't actually intervene when things went wrong – they just provided information. Sometimes, though, knowledge alone was cause enough for celebration.

Bob Crippen was jolted awake by the insistent electronic *bing bing bing* of an alarm. For a moment he was disoriented, unsure where he was or what he was doing.

'What's that, Crip?' John Young asked anxiously, also woken by the alert.

Crip tried to force himself awake, eventually shaking off his slumber sufficiently to break out the big malfunction book. For now, *Columbia* was out of range of any of the ground stations. Whatever was happening was something the crew were going to have to sort out themselves. Crip checked to see where they would next pick up CapCom. Ascension Island. And he began rifling through the 'mal' book in search of the source of the problem.

Ahead of him, at the bottom of the matrix of warning lights in the centre of the instrument panel, the APU temperature light glowed. While only fire and a loss of cabin pressure triggered the sirens or klaxons of a Class 1 alarm, the auxiliary power units that moved the Orbiter's flight control surfaces were still crucial. Without them, once she returned to atmospheric flight, *Columbia* would be uncontrollable. There were three APUs. Like virtually every other critical system, redundancy was built in. Shut down after reaching orbit, one and three had both been restarted to make sure the crew were confident of having at least one functioning system. The problem now was with APU 2, the concern being that, if the temperature dropped too far, it would be impossible to start it again when it was needed.

Following the procedures laid out, Crip floated back to the aft crew station to switch on the heaters. It seemed to have no effect.

'*Columbia*, Houston.' As the Shuttle passed into range of the S-band and UHF transmitter on the remote British mid-Atlantic island, Mission Control, their own screens displaying the APU problem, made contact. On CapCom, Dan Brandenstein told them to switch on the heaters.

'OK, well, our heaters are A-Auto,' Young replied. 'Crip did that first thing.'

'Roger, *Columbia*. Our switch scan shows the switch still in Bravo.' In Mission Control, the telemetry was suggesting that Crippen's action had failed to initiate the changeover. 'I recommend you cycle the switch,' Brandenstein suggested.

On board the Orbiter, Crippen tried again.

'OK, that's done now.'

'We show it per nominal,' said Brandenstein. 'Nominal' was the word every astronaut, engineer and official wanted to hear describing every aspect of a spaceflight – everything was just as it should be. 'And,' he continued, 'the temp's rising.'

'Sure is. See you guys tomorrow.'

'Good night, *Columbia*.'

No one likes things that go bump in the night – and in space least of all. But with the scare dealt with, Crippen returned to his ejection seat, while Young, after having his sleep interrupted, decided he'd stay up for another hour or two, drinking in the view and taking photographs – the archetypal extra-terrestrial, just squeezing the juice from his last few hours in space.

There had been so many memorable sights from orbit: the jagged, snow-covered relief of the Himalayas; boiling thunderstorms mushrooming high above the Amazon; smoke drifting into the sky from the caldera of Mount Etna; and sunlight catching the swell of the ocean at dusk, picking out ships hundreds of miles away. Crip had thought Dasht-e-Kavir, the great salt desert in central Iran, was the most striking sight, its magnificent swirls of red, ochre and white mineral deposits looking more like the surface of Jupiter than anywhere on Earth. For Young, though, it was the Bahamas. They glowed like emeralds, he thought, certain that his photographs could not do justice to the beauty he was able to pick out with his own eyes.

They were on orbit twenty-seven, mission elapsed time one day, sixteen hours. When their sleep period ended in five hours' time, they would, almost immediately, begin preparing for their return to Earth. Even with their heat-shield intact, it remained re-entry, peak heating and hypersonic flight that was the corner of the Shuttle's flight envelope about which least was known. *Columbia*'s return would provide the mechanism for learning it. Not for nothing was STS-1 described as 'the boldest test flight in history'.

SIXTY-TWO

Hickam Air Force Base, Hawaii, 1981

As DIRECTOR OF Ames Research Center, Hans Mark identified a handful of programmes that he felt demanded priority: the series of *Pioneer* space probes; the Illiac super computer; and the nascent but promising tilt rotor programme that would one day evolve into the Marine Corps' MV-22 Osprey which, in the wake of the failed attempt to rescue the American hostages in Tehran, Mark once again advocated as a machine with 'extraordinary promise'. And there was the Kuiper Airborne Observatory.

The KAO started life as an effort by Lockheed to sell its C-141 Starlifter military airlifter to civilian operators. After barely 700 hours in the air it became apparent that there was insufficient commercial interest to sustain the project. But when, in 1970, NASA went looking for a large four-engined jet capable of carrying a big infra-red telescope for high-altitude astronomy, the mothballed jet offered the perfect solution. In 1974, after extensive modifications, she emerged painted in NASA's elegant blue and white house style, and carrying a 36-inch reflecting telescope which could observe the heavens from an altitude above most atmospheric interference. Named in honour of infra-red astronomy pioneer Gerard P. Kuiper, the Director of the Lunar and Planetary Laboratory at the University of Arizona, the KAO's telescope had studied Halley's Comet and total eclipses, discovered rings around Uranus, a heat source inside Neptune, Pluto's atmosphere and water vapour in comets. Observers on board the aircraft had explored the formation of elements inside supernovas, mapped the Milky Way's galactic centre, and examined the structure of star-forming clouds. And in 1976, before Mark left Ames for the Pentagon, his researchers began to look at whether the jet might be used to capture infra-red images of the Space Shuttle as the spacecraft re-entered Earth's atmosphere. By August of the following year, a report concluded it was feasible. The experiment was christened IRIS – infra-red imagery of Shuttle.

Three and a half years after the publication of the study, the long, low, high-winged shape of the KAO was parked up on the apron at Hickam Air Force Base in Hawaii in the pre-dawn, deployed west across the Pacific from Moffett Field in readiness for *Columbia*'s return. Through the night, technicians had prepared her for a mission which had been rehearsed using radar controllers on the ground and, in the absence of the Shuttle on which to focus the telescope, stars and planets. Already the sensitive infra-red detectors were being pre-cooled inside a vacuum flask – a dewar – to –452°F using liquid helium, while the detector's cores were chilled further to just four-tenths of a degree above absolute zero using helium-3.

Joining the KAO's aircrew and scientists for the flight were the aircraft's mascots: two cuddly toys – a koala bear and a kiwi – gifts from colleagues down under.

* * *

While the team from Ames prepared for their contribution to *Columbia*'s return, in Houston, Dan Brandenstein pressed play on the day's wake-up music. Thankfully, Young and Crippen were already awake when a mad blast of Texas radio was beamed up as their day three alarm call. After a brief, deceptive snippet of birdsong, the trumpets began, followed by a lot of shouting.

Get out of the rack!

Both of you could use a shower.

C'mon, John, after five missions you ought to have this down by now!

Be nice to him he's fifty years old . . .

Crip, you've waited for this for twelve years. If you don't wake up now you'll miss the whole darned thing!

It still paled in comparison to the kind of things that were used as wake-up calls in the sims. Most of those were definitely not for public consumption.

With the cacophony over, Bob Crippen thanked Brandenstein for his help in the night. 'I was so asleep in the middle of the night I couldn't think,' he admitted.

'Roger.'

'We got all of your teleprinter messages off there. We appreciate that.'

Unable to broadcast via CapCom details of the NRO's spy satellite photograph seen in Mission Control the previous night, if the good news was passed on to the crew about the integrity of *Columbia*'s heatshield, it was contained in the reams of paper Young and Crippen had woken up to from Houston. And if Crippen was going to be more fulsome in his appreciation, then that too would have to wait until conversations could be held in private on the ground. Further reassurance was also required as it appeared that the efforts of both Mission Control and the crew to solve the problem with the APU heater had been unsuccessful. Although *Columbia* could return with just two functioning auxiliary power units, test data suggested that even if the temperature continued to drop on APU 2, they were still, even if it was a little sluggish at first, going to be able to get it started. As Brandenstein looked forward to the end of his shift, it was a relief not to have replaced one problem with another.

'We're twenty seconds from LOS [loss of signal],' he told Crip. 'The Crimson Team will pick you up at Indy at twenty-two plus fifty-four. See you guys later.'

'OK,' Crip replied, 'we're just mighty glad that you did not have to practise any of those things that you know so much about there . . .'

It was a feeling strongly shared by Brandenstein. *It was*, he thought, *a load off everybody's minds*. After he'd seen the photographs brought in by Gene Kranz, he could hand over the CapCom console to Joe Allen secure in the knowledge that he wasn't passing his replacement a world of trouble.

Ames Research Center's Kuiper Airborne Observatory was just one of a fleet of aircraft being readied for *Columbia*'s return. Already parked up on the Dryden flightline at Edwards were the four T-38 jets belonging to Jon McBride's Chase Air Force. A squadron of helicopters carrying soldiers, doctors, astronauts and rescue workers was ready to control crowds and respond to an emergency.

They were also joined by one of the Gulfstream Shuttle Training Aircraft, sent to Edwards to bring Young and Crippen home to Houston after their arrival. It now also had to conduct the weather flight as the pilot scheduled to do the job was going to be absent. Dick Truly's callsign would have been Weather West.

Truly had been assigned to fly a weather flight over Edwards in a T-38 in advance of the Orbiter's return. With good weather forecast over the Californian high desert base, though, it was agreed that his presence was surplus to requirements. Instead the astronaut chose to stay in Mission Control with Joe Engle for the landing. He wasn't sure why, but for some reason he couldn't put his finger on re-entry had always caused Truly more anxiety than ascent. He wanted to watch the landing from the nerve centre of the mission, feeding on the detail of it. First to know of any developments.

'On your next pass over the US,' Dan Brandenstein suggested, 'do your own weather flight.'

'We will take over for Weather West,' Crippen laughed.

In the Mission Control room, Truly smiled and pointed out that it was actually going to be the other way round – on *Columbia*'s next flight, Truly, with Engle sitting alongside him next to the CapCom console, was actually going to be taking over from *him*.

Clocks were started to count down towards the ignition of the orbital manoeuvring system rockets to initiate the spacecraft's re-entry.

As Crippen bantered with Joe Allen at CapCom through the Bermuda ground station, the Shuttle was beginning orbit thirty-three. Mission elapsed time was two days, one hour. On their next revolution, Young and Crippen would climb into their David Clark S1030 pressure suits and verify the positions of the switches on the mid-deck before returning to the flight deck for the last time in readiness for coming home.

At Hickam AFB, the Kuiper Airborne Observatory sat waiting. Beneath long wings, angling down from her shoulders towards the hard-standing below, the launch crew moved purposefully around her. Her fuel tanks were full; the super-cooled dewar carrying the IR detectors had been loaded and attached to the telescope.

With less than half an hour before their scheduled departure time, the flight crew walked out to NASA's unique variation on Lockheed's big airlifter, climbed on board and strapped into their seats. On the flight deck, the aircraft's captain, Dave Bark, commanded 'Before engines start check-list', prompting a scripted call-and-response between him, his co-pilot and their flight engineer as they set up the jet's systems for flight. Outside on the ramp, a second engineer, the scanner, verified the KAO's responses and directed the launch crew to remove the landing-gear pins and main-gear chocks. With the checks complete, Bark was ready to start the four Pratt and Whitney TF-33-P-7 turbofans. He pressed the starter button, then moved the fuel and start ignition switch to 'Run'. Behind him, his flight engineer routed high-pressure air through the engines then checked the temperatures and pressures as each jet spooled up, generating a building whining growl in the dark of pre-dawn.

In what had once been the aircraft's cargo hold, the team of telescope technicians sat in a windowless cabin in front of a mosaic of switches and cathode ray tubes mounted on grey panels running along the length of the fuselage. Around them were exposed structural spars and junction boxes, as much a victory for function over form as a university electrical engineering lab. The bags of flight food brought on board – potato chips, cookies, soft drinks – didn't suggest nutrition was as huge a priority on the long mission ahead.

With a nudge on the throttles to move them forward out of idle, Bark began to taxi. As the big Lockheed jet rolled towards the runway, his co-pilot next to him set the wing flaps at 75% and armed the spoilers. Minutes later, as the KAO accelerated past 120 knots, Bark pulled gently back on the yoke to raise the nose. As they accelerated away into the night, the crew raised the gear, pulled the power back to 92% and, after raising the flaps and settling into the climb, handed control to 'George', the C-141's autopilot. Given the precision flying to come as they attempted to track *Columbia* as she streaked past Hawaii, it made sense to lighten the load while they still had the opportunity to do so. The success or failure of the KAO's mission would be measured, ultimately, in milliseconds.

SIXTY-THREE

Space Shuttle *Columbia*, 1981

ON BOARD *COLUMBIA*, the astronauts continued their preparations for re-entry. After the wake-up call, life on board the Orbiter had continued noisily as John Young test-fired each of the forty-four reaction control jets needed to control the Shuttle before she descended into air thick enough to allow the aerodynamic flight controls to bite. Then, after firing up one of the auxiliary power units, he cycled the engine gimbals and flight control surfaces before checking the spacecraft's guidance and navigation systems. Crossing the Pacific on orbit thirty-five, Bob Crippen stood at the aft crew station. With the radiators stowed and latched over Australia, he initiated the closing of the payload bay doors. As he worked, data was dumped from *Columbia*'s computers and captured at Orroral Valley and Hawaii ground stations for post-flight analysis.

'Those doors are closed up and locked,' Crip reported, 'just as they were supposed to.' He then asked whether Mission Control had the numbers for the de-orbit burn to trigger their return to Earth.

'We sure do,' Allen confirmed, 'in fact when you're ready, we are ready.'

For the first time since main engine ignition at Kennedy, what followed from CapCom began to weave the language of air traffic control – of aviation – in among that of space-flight. Along with the inertial attitudes and burn duration, Allen also passed up the wind speeds and directions waiting for *Columbia* over California and told them to plan for a 'left-hand turn to Eddie 23' – the 22,175-foot-long runway on Rogers Dry Lake Bed at Edwards.

As John Young prepared to manoeuvre into the right attitude to use the star-trackers in *Columbia*'s nose to pin-point his ship's position against the celestial map for the last time, he looked down at the continental United States passing beneath them.

'It looks like we're going right over Chicago,' he said. 'Or are we lost in space again?'

Outside Edwards Air Force Base there was a 6-mile-long bumper-to-bumper line of vehicles trying to get into the base for *Columbia*'s return. Thousands of cars, RVs and pick-ups were already parked up in the lee of dry, pitted hills dotted with sagebrush. An estimated quarter of a million people had decided they wanted to witness America reclaim her lead in space. Many had arrived the previous night, pitching tents on the cracked earth around the western edge of the Rogers Dry Lake Bed behind a mile-long wooden security fence. Fuelled by beer, barbecue and anticipation there was a festival mood, and after dark the campers were entertained by a country and western band laid on by the Air Force. Some stayed up all night under the stars, drinking, smoking and talking, before then sitting out on folding chairs under dry, azure morning skies waiting for the double-barrelled sonic boom that would herald *Columbia*'s arrival. They looked as if they were lining a beachfront. Announcements from the Public Affairs Officer at Houston echoed from a PA system to keep them up to date with the spacecraft's progress.

By eight o'clock Pacific Standard Time on 14 April, the sun was already beginning to heat the surface of the lake bed. Soon the haze would soften and distort the view across it.

On their thirty-sixth revolution, after verifying the position of the switches at the aft crew station and stowing loose items before gravity had the opportunity to throw them around the cabin, Young and Crippen propelled themselves gently forward, towards the cockpit. They pressed themselves down into their ejection seats, strapped in, then pushed their feet into their flying boots. Crip was glad that his earlier decision to leave them connected to the boot spurs meant he'd dodged a job which in zero-g would only have been more awkward. Hundreds of hours flying F-104 Starfighters, he thought, and *I never could put those spurs on*.

Mission elapsed time was two days, four hours and twenty-nine minutes.

As the time of the Shuttle's scheduled arrival approached, the levels of activity increased. Air Force helicopters circled overhead. One of the two Gulfstream training aircraft climbed away from Runway 24 on its way to report back on the weather at altitude. And, following extensive briefings inside the Dryden crew room, Jon McBride and the seven other pilots and backseaters of the Chase Air Force deployed to Edwards walked outside to their line of immaculate white T-38 jets. Wearing blue NASA flightsuits, each sporting the

Chase team patch, they carried out pre-flight checks of the aircraft before climbing up ladders into their cockpits and lowering themselves into their ejection seats. McBride pulled on his red, white and blue 'Captain America' helmet and strapped into his shoulder harness, lapbelt and parachute lanyard. He plugged into the jet's radio and intercom and connected himself to the oxygen supply. Behind him, his backseater and fellow TFNG recruit Pinky Nelson did the same.

Using estimates of the Shuttle's ground track from Mission Control, the Kuiper Airborne Observatory held station in a pattern north of Hawaii. On board the big NASA jet, Dave Bark and his flight crew were ready for any late updates on *Columbia*'s expected re-entry that would require them to reposition prior to the rendezvous. Even without last-minute changes, success required precise navigation based on accurate data from Houston, skilful handling and seamless teamwork between the aircrew on the flight deck and the telescope operators behind. KAO needed to be flying at 475 knots and 41,000 feet on a path parallel to the Shuttle as the spacecraft overtook her. From their own position, cruising 16 miles south of the Orbiter's descent, there would be a window of just sixteen seconds between the moment the IRIS tracker could lock on to the Shuttle at a range of 50 miles and the point at which *Columbia* passed out of the heat-seeking telescope's field of view. After two and a half days in space and a hypersonic descent from orbit, the maximum allowable cross-track error was just 2½ miles, while the difference in speed between the two flying machines would be over 9,000 mph. Bark's last update on *Columbia*'s predicted trajectory would come ten minutes before the entry interface, the point 400,000 feet over the western Pacific where she began meeting resistance from the upper atmosphere. Her encounter with the airborne observatory would take place nearly twenty minutes after that. The margins were fine, the opportunities for error great.

When the *Apollo XIII* capsule re-entered the Earth's atmosphere after a slingshot around the moon that had saved the lives of her crew, there was a period of around six minutes during which communication between the spacecraft and Mission Control was impossible. Because the capsule was forced to re-enter at a shallower angle than originally planned, Gene Kranz and his team of flight controllers in Houston had endured a blackout that persisted for nearly a minute and a half longer than expected. After all that had been done to bring Lovell, Haise and Swigert home, the uncertainty over the fate of the astronauts was agonizing. But while it may have been more persistent than predicted, the blackout itself was no surprise. In fact, it was a direct consequence of the capsule's forceful collision with Earth's atmosphere. And it would be no different for *Columbia*.

When the Orbiter's nose and underbelly collided with the upper atmosphere at a speed of nearly 5 miles per second, a shockwave would form ahead of her. And between the bow of that shock and the thin boundary layer flowing over – and protecting – the surface of *Columbia* herself, atmospheric gases, heated by their violent compression to temperatures of over 6,000°F – comparable to the surface of the sun – would begin to disassociate. Electrons, stripped from the molecules contained in thin air around the Orbiter, would then envelop her in a super-heated sheath of ionized gas, or plasma. And when the density of the plasma cocoon around the Shuttle exceeded the level at which radio communication was possible, *Columbia* and her crew would be cut off.

In the Mission Control room, it was impossible not to be aware of a tightening mood as the entry interface approached. Sitting behind the CapCom console, Dick Truly and Joe Engle were plugged in, listening to what remained to be shared with Young and Crippen during the final hours of their spaceflight. Always more anxious about the return, Truly could feel his own tension mounting, despite the reassurance the satellite pictures from the National Reconnaissance Office had provided. Sitting ahead of Truly and Engle, Joe Allen and Rick Hauck scanned the telemetry coming from *Columbia* on their monitors. Allen pressed the transmit button.

'*Columbia*,' he said, 'your burn attitude looks good to us and everything aboard looks good to us. You are go for de-orbit burn and we'll go LOS in thirty seconds here. We'll talk to you in Botswana in about five minutes.'

'OK. We understand ready for de-orbit burn. Thank you now. That's the best news we've had in two and a half days. And we've had some mighty good news . . .'

Jon McBride checked his fuel settings, reached forward with his left arm and pressed the right-hand engine start button. Behind him, the first J85 engine whined as the needles on the temperature and pressure gauges in the cockpit swung clockwise. He set the throttle at 47% rpm, then repeated the procedure for the port engine. With both engines idling he checked his flight controls, speed brakes and trim switch, before requesting taxi clearance from the tower. He advanced the throttles to 75% and began rolling forward. He dipped the brakes and continued, setting the T-38's flaps at 45% for take-off as he taxied towards the runway, steering the nosewheel with his rudder pedals. Alongside him, Chase Two, Three and Four settled into a brisk procession along the taxiway. En route, the crews of the lithe, athletic-looking little jets closed their canopies as they approached the runway threshold.

Nineteen minutes earlier, Young had manoeuvred out of the top-sun attitude he'd used to try to chill *Columbia*'s belly before re-entry. Now, as she completed a three-minute pass across the Botswana tracking station, she was flying upside down and tail-first in anticipation of the de-orbit burn that would upset the delicate balance of speed and gravity that had kept them in orbit. At ignition minus one minute, Young made a final adjustment to the attitude, checked the manoeuvre targets on the CRT screen ahead of him, and armed the two orbital manoeuvring system engines. Then, at ignition minus fifteen seconds, he reached forward to the computer keyboard in the centre console and pressed the EXEC button – *execute* – to begin the countdown.

SIXTY-FOUR

Space Shuttle *Columbia*, 1981

ONE HUNDRED AND forty-five miles above the Indian Ocean, fuel and oxidizer ignited inside the combustion chambers of *Columbia*'s two Aerojet OMS engines. The combined 12,000lb of thrust generated by the two hypergolic rockets burned against the Shuttle's direction of travel, gently pushing Young and Crippen back into their seats – a subtle suggestion of the gravity that would soon reassert itself. As the engines fired, Young and Crippen monitored the computer displays in front of them. The crucial number was that of their current perigee – the lowest point in their orbit. The two-and-a-half-minute rocket burn barely slowed *Columbia*, shaving just 200 mph off her 17,500 mph orbital velocity. Instead, its effect was to alter the shape of her orbit from a circle to an ellipse with an apogee, or maximum altitude, that remained the same, but a perigee that brought them low enough to ensure their capture by the atmosphere.

At the termination of the burn, *Columbia*'s return from space was inevitable, but not instant. Still travelling at over 17,000 mph, the entry interface with the 'sensible' atmosphere – the 400,000-foot altitude at which friction would begin to slow her progress – remained half an hour away. But this was the point at which the conclusions drawn from both the analysis done by Tom Moser's engineering team on the missing tiles and the study of the NRO's satellite photographs would be put to the test. And while those difficult-to-capture pictures had been able to demonstrate that the heatshield was in place as designed, they were not able to highlight areas where the design of the heatshield itself might be inadequate. That would remain unknown until *Columbia* put theory into practice.

John Young typed in the pitch, roll and yaw data to the autopilot. Following the attitude indicators on the eight-ball display ahead of him he used the rotational hand controller – the joystick – to pitch the spacecraft into the high angle of attack in which she'd re-enter. From her tail-first, belly-up attitude, *Columbia*'s nose arced silently anti-clockwise like the needle on a dial through a semi-circle at a rate of just 0.2° a second. Through the cockpit windows the two astronauts watched their view of Earth disappear over their shoulders until Young halted their progress, his attitude indicator telling him he had the required zero-degree roll, zero-degree yaw, and *Columbia*'s nose pitched up at an angle of 39°, braced for the worst re-entry could inflict on her. That was less than twenty minutes away.

The four Chase team T-38s waited, poised in a diamond formation at the threshold of the runway, their engines turning and burning, generating a billowing heat haze in their wake. In the lead jet, Jon McBride set the horizontal stabilizer to the take-off position, thumbed the transmit button and requested take-off clearance from Edwards tower. Moments later, riding the brakes, he pushed the two throttle levers forward to full military power. With the little Talon straining at the leash, he took his feet off the brakes, pushing the throttles through the detentes to light the afterburners. He and Nelson were pushed back as the aircraft leapt forward in unison with her three sister ships. As they reached 135 knots, all four pilots pulled back on the sticks and the formation lifted off as one, tucking up the gear straight after leaving the ground. McBride kept them low to gather speed before climbing away into clear blue skies. Almost immediately, Chase Three and Four peeled off towards the south where they would wait in a holding pattern as back-up, ready to fill the breach if required. McBride and Nelson, followed by Dick Gray and Pete Stanley in Chase Two, continued their climb towards 40,000 feet where they were to wait, flying circuits, crossing the initial point just south of Edwards lake bed with each northbound leg.

'*Columbia*, this is Houston through Yarragadee. We're standing by.'

'Burn was on time and nominal,' was all the Shuttle's Commander offered Joe Allen in reply. Even by Young's standards it was succinct. But if the spacecraft's ascent had been a concern to the veteran astronaut because of the possibility that something might go wrong requiring him to take manual control of the situation, during the long glide across the Pacific he knew for sure that he'd be flying *Columbia* on her return. As he concentrated on what lay ahead, his clipped report was understandable.

Alongside him, his Pilot remained relaxed.

Joe Allen told them: 'You'll like to know that four chase aircraft just launched from Eddy and are coming up looking for you.'

'Check Six,' Crippen joked, repeating the age-old fighter pilot's instruction to watch your tail for enemy aircraft. 'Yeah, we ought to be there in about forty-five minutes.'

'That's what they're hoping,' Allen replied, before quickly adding 'and we're *sure* of it.' The slip went unnoticed by a buoyant Crippen, apparently unfazed by their imminent return, and trying to capture his last memories of what remained of the first flight.

'Joe, whoever said that space was black was not kidding ya. It's *really* black.'

'Roger, you've convinced us.'

'Yeah, but you are so easy. It is my great scientific observation that did it to you.'

The signal from Yarragadee was lost thirty seconds later.

From inside the cockpit of Chase One, Jon McBride monitored two different radio channels. He kept the volume on the link to Vandenberg's radar controllers turned up, with a line to Mission Control in Houston down lower in the mix. It was the Western Test Range radars at Vandenberg that would first get a fix on the returning Shuttle, before handing over to their counterparts at Edwards who would then control the rendezvous with *Columbia*. As he and his wingman flew circles in the air between Edwards and Rocket Ridge – the nickname given to the engine test facility on Leuhman ridge, a few miles to the east of the Dry Lake – he passed back a real-time report on the conditions at 40,000 feet. Winds were favourable. There wasn't a cloud in the sky.

Approaching Guam, the astronauts worked their way through the entry switch checklist. With a precision hard-won in the simulator, Young and Crippen reached around the banks of switches and buttons that surrounded them, preparing the guidance and navigation system for atmospheric flight, dumping propellants from the forward reaction control system which, unlike the rear reaction control jets, wasn't going to be required after re-entry, and cycling the rudder, speed brakes and elevons for five minutes to prepare the hydraulic system. They pulled the safety pins out of their ejection seats and stowed them. Now armed, the seats were ready to use once they reached a safe speed and height with one pull of the 'lil' handle'. Then they ran through the entry switch checklist again. *One armed paper hanger* was the phrase that went through Crip's mind.

Two separate clocks counted down in the Mission Control room in Houston. One recorded twelve minutes and twenty-five seconds to entry interface; the other, that there were just forty-five minutes until, from *Columbia*'s current position over the South Pacific, John Young was due to bring her home to Edwards. Over the Flight Control loop during the comms blackout period between Yarragadee and Guam, Flight Director Don Puddy told his Crimson Team what he wanted from them when *Columbia* arrived. After the nosewheel touched down, he said, they had exactly fifteen seconds to 'whoopee', then it was back to work until the vehicle was made safe.

Flying at 41,000 feet over the Pacific north-east of Hawaii, the Kuiper Airborne Observatory had already conducted a fifteen-minute practice run along *Columbia*'s expected ground track. Following updates from Dryden via the USAF's high-frequency radio net, the big NASA four-jet was now again tracking east by north at 475 knots along a path 16 miles south of where the Shuttle would dive past at hypersonic speed. Just forward of the port wing root, the 36-inch telescope pointed up through an open hatch in the C-141's fuselage. Isolated from the movement and vibration of the aircraft on an air-cushioned bearing, the telescope seemed to be in constant motion. In reality, its restlessness was relative. The telescope's mounting kept it still while the aircraft responded to the ripples and swirls in the air ahead.

Joe Allen reacquired *Columbia* over Guam just as Bob Crippen spooled up the last of the three auxiliary power units before reconfiguring the software for entry. The Shuttle was 85 miles high and travelling just shy of twenty-four and a half times the speed of sound.

'Hey, *Columbia*,' Allen signed off, 'we are fifty seconds from LOS. Everything looks perfect going over the hill. Nice and easy does it, John. We're all riding with you.'

'Roger that.'

'Ten seconds until LOS. We'll see you at about Mach 12.'

'Bye-bye. Looking forward to it.'

And with a crackle, static overwhelmed the radio link.

Columbia and her crew were now on their own. Soon Young and Crippen would be the only two people alive who knew whether or not they were alive or dead.

His movement restricted by the layers of his Nomex pressure suit, Young reached forward to the coaming above the instrument panel and punched a button to switch the Orbiter's body flap from manual to automatic.

SIXTY-FIVE

Houston, 1981

INSIDE MISSION CONTROL there was tension you could chew. The Mercator projection world map that had recorded *Columbia*'s orbital track around the globe on big screens in front of the consoles was taken down. In its place went a graphic showing the Shuttle's velocity and ground track across the Pacific towards the West Coast. The former was also included in the flight dynamics data displayed in front of Joe Allen and Rick Hauck at CapCom. But for sixteen minutes, from the point when hot plasma gas engulfed the Shuttle, there was nothing.

Behind the CapCom team, Dick Truly was, by his own admission, sweating bullets, conscious that there had never been a Shuttle model in a hypersonic wind tunnel that was bigger than a toy. *And now*, he thought, *here's the Shuttle itself, half the size of a house. Enormous.* He looked around the room. *Tense City.*

Sitting alongside Hans Mark at the programme office console, his friend Deputy Programme Manager Milton Silvera had similar concerns. A veteran NASA engineer,

Silvera had once walked beneath the vast flat-iron canopy of *Columbia*'s underside and, struck by the difference in scale with the 12½-foot diameter of the Apollo capsule, had had to ask himself: *Do you really know what you're doing?* He wasn't sure if the answer was yes.

Rick Hauck welcomed the distraction of a radio call from Jon McBride in Chase One.

'Chase, this is Houston,' he responded, 'go ahead.'

'We're set up in orbit pattern for a nominal two three with a left turn,' reported McBride, confirming that he was expecting the Orbiter to cross Edwards from the west then double-back to land on Runway 23, which ran towards the south-west.

Hauck told his colleague that *Columbia*'s de-orbit burn and entry trajectory were nominal. He confirmed that he'd call again, whether or not they'd yet heard from the crew, on the assumption that the astronauts' return was going as planned. McBride acknowledged. Then there was silence, except for the background static through his headset. It felt like his heart was beating at triple speed.

As Young tweaked *Columbia*'s attitude with the reaction control jets, Crippen caught sight of small flashes of light in his peripheral vision through the right-hand window. He figured it had to be light from the little reaction control jets reflecting off the sparse fringes of the atmosphere. It was the first indication that the vacuum of space was beginning to fill with molecules. At around 350,000 feet on a track passing north of Hawaii, *Columbia* began to experience microdeceleration produced by an almost insignificant dynamic pressure of just a tenth of a pound per square foot;

but with her plunge into thicker air, that figure was starting to rise quickly. All being well, they were barely fifteen minutes away from wheels stop at Edwards.

It was still dark outside, and Crip was first to notice the pale pink glow through the sides of the cockpit glass as they fell through 330,000 feet. Both astronauts pulled down and locked the visors on their helmets, sealing their pressure suits, ready to inflate and protect them if burn-through led to a loss of cabin pressure. A deeper orange glow began to build around *Columbia*'s nose.

It was nothing like the inferno produced by the ablative heatshields that protected the capsules that had previously returned Young from space, but it was mesmerizing. *A bunch of happy ions*, thought the Commander.

Next to him, Crip felt as though they were flying down the length of a giant neon tube. But the apparent gentleness of the light show did not properly reflect the ferocity of the conditions outside *Columbia*'s cabin. If they'd turned to look through the windows in the roof of the flight deck, they might have seen the pink-orange glow of the long plasma plume their ship was leaving in her wake.

Instead, Young kept his eyes on the instruments ahead, looking for anything in the position of the flight controls that might signal trouble. For now, the readings were nominal.

'I don't see it any more,' Crip said, as the rising sun ahead of them overpowered the pale colours flowing off *Columbia*'s heatshield. At the same time, the sensors recorded that the dynamic pressure from the atmosphere had risen to 10lb per square foot. The computers deactivated the RCS thrusters used to control the Shuttle in roll. Responsibility for that now fell to the big elevons hinged across the trailing edges of the delta wings. *Columbia* was beginning her transition from spacecraft to aircraft. She was also approaching peak heating. And, unknown to the crew, super-heated plasma had begun to snake its way through gaps between the tiles in the starboard main landing gear door.

The trajectory of a spacecraft's re-entry is a trade-off. By re-entering more steeply greater peak heating is experienced, but for a shorter time. At the same time, the greater rate of deceleration of a steep re-entry imposes higher g-loads on the vehicle and its crew. On their return from the moon on *Apollo XVI*, John Young and his crew had been subjected to over 7g. The Shuttle, by contrast, wasn't designed to endure that kind of load and so, returning along a shallower re-entry corridor, was forced to stay hot for longer. But there were further consequences of coming in shallow in a winged vehicle like *Columbia*, the first being that as the density of the air increased, the lift generated by those big delta wings meant she might skip back out of the atmosphere like a stone bouncing across a smooth pond. The other was simply that she had to find a way to bleed off her excess speed. Both issues were dealt with through a series of long, sweeping S-turns known as roll reversals. By banking on to her side, the lift generated by *Columbia*'s wings no longer worked to push her higher, but instead to change her direction, pushing her in the opposite direction to where she was pointing her belly. The angle of attack she was using to slow her down remained the same, but there was no longer a danger of skipping and further increasing load on the heatshield.

At 255,000 feet and Mach 24, *Columbia* stood on her starboard wing as she rolled through 70° into the first of those big sweeping turns. Through the flight deck window, the crew watched the Shuttle's nose begin to track right across the curve of a horizon silhouetted against the blackness of space by the rising sun. But almost immediately Young had something other than the view from the cockpit to occupy him. The slip indicator beneath the eight-ball display ahead of him pegged all the way over to stops and stayed there. At its limit, the instrument was only able to show a sideslip of 2.5°, but it looked like it was trying to record something more substantial. In fact, *Columbia*'s tail had fishtailed out by 4°. In the simulator, Young had seen high yaw angles at hypersonic speed develop into rolls that had torn the wings off. What was happening now was unexpected and unnerving.

Travelling three times faster than any winged flying machine had ever flown, *Columbia* was feeling her way. In the thin high-altitude air her attitude was controlled using a combination of the aerodynamic surfaces – the elevons and rudder – and the remaining active thrusters. But she was also travelling over twice as fast as the best available wind tunnel at Ames had been able to replicate. The effect of yaw thrusters firing into the heated, disassociated airflows over the wing and OMS pods had been mispredicted. Now, every time the RCS jets fired in an effort to try to stabilize the oscillation, they were also, at the same time, contributing to it.

Nor was it just the risk of losing control that was of concern as a result of the sideslip. Any more than plus or minus a degree of sideslip had the effect of moving the stagnation point – the area where the heat of re-entry is at its greatest – from the Orbiter's reinforced carbon-carbon nose to a part of the heatshield less capable of resisting the near 3,000°F surface temperatures.

It took over forty seconds for *Columbia*'s flight control system to bring her swinging tail to heel after the trim integrator had intervened to prevent the first skid from exceeding limits beyond which it would have been impossible to recover.

Young and Crippen knew they'd just experienced a serious encounter with a characteristic of the Orbiter's high-altitude, high-Mach flight that had, until now, been unknown. 'It certainly got our attention,' Young would say later. That *Columbia* was able to cope with what was, in the circumstances, a wild oscillation was testament to the work done by Crip's MOL comrade-in-arms, Hank Hartsfield, with teams of Rockwell and NASA engineers, to build a flight control system that was capable of accommodating uncertainty.

And there was still more of that to come.

After a briefing earlier in the morning, the recovery convoy had driven out to a point 2½ miles south of where *Columbia* was expected to roll to a halt. Twenty-one vehicles strong and manned by a hundred personnel from Kennedy Space Center, they were the only people allowed within 100 feet of the Shuttle until they could establish she was safe to approach. As well as detectors capable of sniffing out toxic or flammable fumes escaping from the Orbiter, they also carried powerful wind machines to disperse any dangerous gases. On orders from Flight Director Don Puddy in Mission Control, the convoy crew began pulling on their self-contained atmospheric protection ensemble – SCAPE – suits, which gave them the appearance of a biohazard emergency response team. Above them, the sky throbbed with the sound of helicopter blades thumping the air. On board the helos were more men in SCAPE suits, their motto 'That Others May Live'.

They were the USAF's own Special Operations team; their day job, combat rescue – recovering downed aircrew from ocean, desert and jungle. Many of them had earned their spurs in Vietnam. For months, though, Air Force para-rescue jumpers – PJs – had been training with NASA to perform the same role for the Shuttle in the event of a Mode 8 contingency. If *Columbia* landed hard or overshot the runway, the PJs were on hand to go in and pull Young and Crippen out. Wearing protective clothing in case of leaking hydrazine from the broken bird, teams of PJs had also deployed to Kennedy and the STS-1 diversion strip at White Sands. Assigned to each group were astronauts. In the time they'd worked with the pararescue men they'd developed procedures and checklists peculiar to the demands of the spacecraft. Together they'd rehearsed lifting people strapped to backboards out of the overhead windows of a mock-up Orbiter. Riding with the PJs in the four helicopters on standby at Edwards was TFNG John Creighton, on hand to provide any specialist technical advice needed in the event of an emergency.

As *Columbia* had battled the sideslip, the elevons had worked in tandem with the yaw jets. On a display in the centre of the instrument panel below the warning light matrix, needles indicating their movement rose and fell. Young and Crippen watched the gain fall as the flight control system eventually got the upper hand. But in the seconds that followed the yaw oscillation another reading caught their eye. The strip indicator recording the angle of the body flap began displaying an unexpectedly high reading.

Following the first roll reversal, the possibility of skipping back out of the atmosphere was past. The wings once again rolled level, *Columbia* remaining locked tight in a 40° angle of attack throughout, using the expanse of her underside as an aerodynamic brake to bleed off over 6,000 mph, dissipating heat as she went, to bring her speed down from Mach 24 to nearer Mach 14. So that she could exert sufficient control in the thin upper atmosphere the elevons controlling pitch were assisted by the large body flap that, like a hinged barn door sitting horizontally beneath the three main engine rocket bells, provided the extra force required.

The aerodata book predicted that the body flap would deflect to 7° during *Columbia*'s deceleration. Young and Crippen watched the needle on the instrument panel rise through 10°. Then 12°. The limit of the body flap's arc of travel was 21° but the needle kept rising through 15°. If its maximum deflection proved not to be enough they wouldn't be able to hold their 40° attitude and that, like everything else during re-entry, had repercussions for the heat *Columbia* would have to endure. At 16° the body flap found the angle it needed to hold the Orbiter in her attitude, then stayed there for the next twelve minutes. While a deflection of over twice what was predicted would expose the flap itself to excessive heating, it would remain local. Unless it moved further it was anomalous but safe.

But through the gap in the tiles on the starboard main landing gear door, the hot plasma gas was getting to work on the structure beneath, scorching and charring the filler bar below. Underneath that, the aluminium skin of the door structure began to soften and buckle.

* * *

Standing behind the CapCom console, Joe Engle and Dick Truly looked on anxiously over the shoulders of Allen and Hauck. Engle, in a dark waistcoat over his shirt and tie, stood with his hands on his hips. He looked more relaxed than Truly to his left, whose hands moved constantly to and from his face.

In the Shuttle's cockpit, with their speed bleeding off, Young and Crippen could feel themselves being gently pressed into their ejection seats. On the instrument panel, the small dial recording their acceleration in gs began to move clockwise off zero. As they plunged across the Pacific the astronauts monitored the vehicle's systems, reporting their observations to each other in clipped exchanges that prompted no more than 'Roger' or 'Verified' in response.

Flying beneath them was the Kuiper Airborne Observatory. The seven-strong crew aboard the KAO had done everything right. As *Columbia* overhauled them, the IRIS tracker locked on then passed through the field of view of the acquisition camera. A digital feedback loop should have then allowed the tracker to follow the Orbiter while the main telescope was brought to bear, but it didn't happen. In the brief seconds available to the telescope operator sitting at the main console in the C-141's hold, a systems failure prevented the 36-inch telescope moving to the necessary elevation to capture the speeding Shuttle in its focal plane. It was a bitter disappointment. They knew they were exercising the capabilities of the KAO near its limits, but that was of no consolation to Bark and his crew as they hauled back to Hickam empty-handed.

In Mission Control, the room watched anxiously as the

clock counted down to the moment when they expected to make contact with *Columbia* again. Rick Hauck pressed transmit to speak to Jon McBride. 'Chase, stand by for the Mach 9, 9,700 feet per second, call.'

'Chase One.'

'Stand by . . . *Mark.*'

'Chase,' Hauck came back a short while later, 'disregard the Mach 9 call, we'll do it again shortly.'

'Wilco, wilco.'

'Stand by for mark on 9,700 feet per second.'

As *Columbia* descended through 200,000 feet, her speed coming down, the heat generated by her passage dropped to a point where air molecules, allowed time to get out of the way, were no longer torn apart by her bludgeoning progress. Instead of shedding electrons, the air began to flow smoothly around her. No longer in need of the protection offered by the paving of black tiles on her underside, the Orbiter unloaded, her nose coming down from the high 40° angle of attack to a point where the leading edges of her wings began to slice into the air ahead. She was now gliding.

On the flight deck, no longer veiled by the plasma layer, Young and Crippen caught the last part of the exchange between Hauck and McBride: *Stand by for mark on 9,700 feet per second.*

'Hello, Houston,' John Young interrupted, '*Columbia*'s here!'

At CapCom, Joe Allen sat up with a jolt, a wide smile triggered on his face. 'Hello, *Columbia*, Houston's here, how do you read?'

'Loud and clear,' confirmed Young.

He and Crip were diving in towards Edwards at over ten times the speed of sound at an altitude of 188,000 feet.

'*Columbia*, you've got perfect energy and perfect ground track,' Allen told them.

'Roger that.'

The Shuttle was shedding around 1,000 mph of velocity and nearly 20,000 feet of altitude every minute. The deceleration pressed the two astronauts deep into their ejection seats.

'We got the coastline in sight,' Crip reported as he accepted the navigation data from military TACAN – tactical air navigation – beacons on the ground. He'd been measured so far in his reaction to the view from the cockpit, but as they flew down the San Joaquin Valley, north of a pillowy covering of cloud extending south towards LA, he was more aware of their speed than he had been previously. They were really travelling. And he let his exhilaration make itself known. 'What a way to come to California!' he exclaimed.

At CapCom, Joe Allen just whooped and grinned in reply.

SIXTY-SIX

Space Shuttle *Columbia*, 1981

PRIOR TO THE flight, John Young was concerned about whether the computers would have the necessary finesse to fly the S-turns in the lower atmosphere. As *Columbia* dived to the south-east past Bakersfield, he took manual control and, at nearly five times the speed of sound and 112,000 feet, he eased her into a 40° bank to the left. Holding his left hand up in front of him to keep the sun out of his eyes, he applied back pressure to the hand controller and felt the mounting gs push him back into his seat. He never pulled more than 1½g, but after two days of weightlessness it felt like more. And, for the first time, after years of 'flying' the simulator, he discovered what it was *really* like to fly the Shuttle. Now it all felt fluid; real not academic. He could feel it through the seat of his pants as *Columbia* responded smoothly to his control inputs, cutting gracefully across her ground track. He watched the clouds going by through the side window and thought to himself *by golly, we really are turning*. He was enjoying himself as, coming through 89,000 feet at Mach 2.8, he levelled the wings briefly then rolled smoothly into a new S-turn to starboard. Young's manually flown roll reversals may have been included in the flight plan both to manage energy and prove the fly-by-wire flight control system, but there was still something exuberant about the way *Columbia*, on her return from space, was carving through the sky like a swooping fighter jet.

Ahead, Jon McBride was already committed to the final three-minute circle required to deliver him to the point above Edwards where he would intercept the spacecraft. As he flew the inbound leg to the rendezvous point he tried to pick out *Columbia* as she descended through 50,000 feet from the west. He knew not to be distracted by the sight of her. When he first made visual contact, the sun picking her out against the clear blue sky, she was still barrelling in at near twice the speed of sound. He had to trust the procedure and the controllers on the ground, not instinct, to make sure he rolled out on time and on target. But there she was, looking beautiful.

He thumbed the transmit button on the T-38's control column. 'Shuttle's in sight,' he told Edwards radar.

When *Columbia* sliced across the Rogers Dry Lake Bed, she was still travelling faster than the speed of sound, but as her speed dropped towards Mach 1, Young began to experience an unexpected buffeting – sometimes an early indicator of an impending stall. He put his hand up again to try to see the angle of attack indicator against the glare, but it was no good. With the sun dazzling him through the cockpit windows there were too many panes of glass, he thought, between him and what he was trying to look at. He pulled up the visor on his pressure suit helmet to get a better look. Alongside him, Crip did the same. But, as the speed dropped towards Mach 0.85, the buffeting was already receding. He asked Crip if he thought it could be something to do with the speed brakes which, along with the hand controller, was the only tool he had to control the unpowered Orbiter's speed and angle of attack.

On the lake bed below, the crowds cheered as the double crack of a sonic boom echoed across the lake bed like gun-fire. Dragged along the boom carpet left in the Orbiter's wake, the shockwave signalled that they were about to get what they came for. In unison, heads turned up to catch a first glimpse, binoculars clamped firmly to faces. On the tops of RVs and campers, amateur photographers were perched precariously with their tripods, ready to capture the arrival from space.

Jon McBride rolled out of his three-minute circle alongside *Columbia*'s starboard wing and dropped the T-38's landing gear so that his slippery little jet could keep station on the big blunt-nosed Shuttle. He pressed the transmit button: 'Chase One coming aboard at 30,000.' Sitting behind him, Pinky Nelson was already shooting off film on the Hasselblad, while in the back seat of Chase Two, flanking the Shuttle on the port side, Pete Stanley shot TV footage.

From the flight deck of the Orbiter, Crip caught a glimpse of the white dart-shaped jet at their side, before it dropped back out of view. *Tally-ho!* he thought.

Once he was stable in formation off the starboard wing, McBride manoeuvred his aircraft back and below the Orbiter, careful to avoid vicious, spiralling vortices streaming off the tips of *Columbia*'s wings that had the power to flip the little Talon upside down. While McBride concentrated on keeping out of harm's way, Nelson pointed his camera up at the Shuttle's underside and snapped away, making a record of any damage to the tiles prior to what yet might be added from stones kicked up off the lake bed by the landing gear.

'Looks real good underneath,' he reported.

With soft hands, John Young pulled *Columbia* into a sweeping left turn around the heading alignment circle, an imaginary 20,000-foot-diameter vertical cylinder of air that, if he followed it round an arc of 225° then levelled the wings, would place him on long finals, 10 miles out from the threshold of the runway. Deprived, in the simulator, of much of the feel of a real aircraft, Young felt he'd always had trouble flying the HAC. This was different, though. *The Orbiter*, he thought, *is a joy to fly*. She just went wherever he pointed her then stayed rooted there without complaint or demur until he moved the control stick or rudder pedals.

Next to him, as Young pulled *Columbia* through the turn, Crip caught a glimpse of the people and cars waiting for them at Edwards. And as they got closer to the ground he suddenly had a strong impression of how quickly they were bleeding off their remaining speed. *Man*, he thought, *we're slowing fast*. They'd lost over 17,000 mph in half an hour, but only now was it beginning to feel as if their ride was slowing down.

Young rolled out on to the glidepath for Runway 23, the nose of the Orbiter pitched down in a steep 20° dive, correcting with a touch of left elevon. Ahead of them, the expanse of the lake bed, smooth, flat and pale against the surrounding scrubland, invited them home.

'Right on the glide slope, approaching centreline,

looking great,' reported Joe Allen, the telemetry giving him a complete picture of *Columbia*'s approach, despite being hunkered down in Mission Control, 1,400 miles away in Houston.

In much closer proximity, Jon McBride called 'Mark fifteen' as he passed through 15,000 feet on the Shuttle's wing. His specially calibrated instruments gave Crip a chance to check the numbers against the Orbiter's own. As Young kept *Columbia*'s nose aimed at a point just short of the end of the runway, Crippen reeled off the airspeed and altitude so that his Commander didn't have to take his eyes off the picture ahead. Young's pulse during lift-off had led Flight Director Neil Hutchinson to joke that he must have been asleep. Now, though, as he concentrated on the single chance he had to bring in his 85-ton glider, Young's heart began to race, hitting 135 beats per minute, something it hadn't managed in any of his four previous spaceflights.

'Nine thousand, 280 knots,' CapCom reported.

'Everything looks real good,' confirmed Chase One. 'Five thousand, 290.'

At 2,800 feet, Crippen called '282 knots'. It was 3 knots slower than Young had planned. He was beginning to get the impression that *Columbia* was a better glider than anyone had expected her to be. *We'll tuck in the boards*, he thought. By closing the speed brakes that butterflied out on either side of the rudder to see how she flew without the drag they generated, they'd collect the data they needed to test that theory.

Descending through 1,750 feet, Young applied back pressure to the control stick, pulling the Orbiter's nose up into the pre-flare for landing. They were still flying 100 knots faster than they needed to be at the point of touchdown, but by trading that excess speed for height, Young could slow her descent to bring *Columbia* in on a flatter approach for a smoother, gentler landing. Young held her on a 1.5° glide as the runway, outlined in black on the lake floor, raced towards them.

Crippen reached forward towards the switch to arm the landing gear, flipped the cover, and pushed it. Then he did the same with the button to the right to lower the gear. It snapped down, faster than either astronaut had anticipated. A trio of green lights on the instrument panel blinked on to confirm the wheels were locked.

'Gear coming,' reported Jon McBride from the cockpit of Chase One. 'Gear down.'

Young held her stable, nose high, losing speed, while she seemed to float above the length of the long runway, reluctant to meet her racing shadow skimming along the lake bed beneath her. *Columbia* sailed 1,300 feet past the designated touchdown point, escorted by McBride's T-38. From the flight deck, high above the main gear beneath the wings, Young was grateful for the commentary from Chase One as he felt for the runway below.

'Fifty feet . . . forty . . . thirty . . . twenty . . . ten . . .' McBride reported, his voice crackling with radio static; 'five, four, three, two, one, touchdown.'

In Mission Control, as *Columbia*'s main wheels kicked up rooster tails of dust behind them, Flight Director Don Puddy told his team, 'Prepare for exhilaration . . .' On the big screen in front of them, the Shuttle continued to speed across the lake bed on her main wheels, her nose held high.

* * *

Crip reckoned that, as they kissed the lake surface sinking at just 1½ feet per second, it was about the smoothest landing he'd ever experienced. *John really greased it in*, he thought. But his Commander's job wasn't quite done. As they raced down the length of the runway, Crippen marked off the speeds while Young maintained a careful balance. Keeping the nose high allowed him to use *Columbia*'s underside as a big aerodynamic brake, slowing her down without straining the brakes. If he left it too long, though, he'd lose the lift he needed to bring the nose down gently and the Orbiter would simply slam down hard on to her nosegear. For ten seconds, Young maintained his back pressure on the stick, before relaxing, carefully guiding his ship's nose down towards the desert floor.

Careful not to ignore his own flying as he kept station alongside, McBride looked on. 'Nosegear's ten feet,' he reported. 'Five, four, three . . . *touchdown*.' The hydraulic leg on the nosegear compressed and bounced gently, the twin tyres streaming soft pale dust behind them. 'Welcome home, skipper!' McBride finished. And with *Columbia* safely on the ground, he poured on the coals, pushing both throttle levers forward to accelerate away. As he pulled the little T-38 into a climb, the world's first winged spacecraft rolled to a halt beneath him, shimmering like a mirage in the heat of the high desert.

'Do I have to take it to the hangar, Joe?' an exuberant Young joked as *Columbia* slowed to wheels stop, her nose low, at the intersection with Runway 15/33.

'We're going to dust it off first,' Allen replied with a smile in his voice.

'This is the world's greatest all-electric flying machine,

I'll tell ya that,' Young drawled, in a voice that somehow evoked the rich history of flight-testing over the Mojave Desert. 'It was super.'

SIXTY-SEVEN

Houston, 1981

'WHEELS STOP ON the *Columbia*,' confirmed Convoy One, and Don Puddy made good his promise.

'All controllers, you have fifteen seconds for unmitigated jubilation, and then let's get this flight vehicle safe,' he said.

And on his command, the tension bottled up and wrung out in Mission Control over the two and a half days since the Shuttle's launch was released in an eruption of cheering and clapping.

An elated and relieved Dick Truly raised his right arm and punched the air.

In the days of Gemini and Apollo it would have been the cue for cigars to be fired up. But while big smokes were pulled out and sat next to the ashtrays built into the consoles, they remained unlit.

'Time's up,' said Puddy, shutting it down as quickly as he'd triggered it. But though he put a lid on the most ecstatic celebrations, it was impossible to keep it sealed tight while, for another three-quarters of an hour, responsibility for the spacecraft remained with Houston.

As *Columbia*'s long speed brakes slowly hinged shut down the height of her vertical fin, Convoy One, leading the twenty-one-strong recovery convoy, reported its progress across the lake bed towards her. Rolling in line astern across the haze of the lake bed towards the waiting Shuttle, thick piping suspended from the front of the largest trucks in the convoy lent them the appearance of bull elephants leading their herd across an ethereal African plain, kicking up powder as they went.

On board *Columbia*, the joy was no less in evidence than in the Mission Control room.

'I can highly recommend it,' Bob Crippen reported back to CapCom on the virtues of spaceflight. The way he was feeling, he said, 'if we can bottle it we can make a million!' For all his excitement, though, the spacecraft's Pilot stayed in his ejection seat, throwing switches and powering down systems. He shut down the APUs, reached up to the over-head panel to deactivate rocket and thruster circuits, then reconfigured the computer software for the last time.

His measured progress through the post-flight checklist was in stark contrast to his crewmate. With adrenalin still coursing through him following the landing, Young was simply unable to stay still. He unstrapped, safed his ejection seat and climbed up and down the ladder that ran between flight deck and the mid-deck below. But, even after removing his helmet and gloves, he had no choice but to remain sealed inside his ship. Until Convoy One declared the Orbiter safe, there was no place to discharge the euphoria he felt. As Young ran back and forth whooping, Crippen sat in his ejection seat watching in amazement. He'd never seen his Commander like this before. No one had.

'I'm looking right out the hatch window and I don't see anybody out there,' Young said, his face pressed against the thick layers of glass. 'I can make it down there if you'd like me to open the hatch and jump out . . .'

Listening to him from the cockpit, Crip thought his Commander might just do that. Usually so deliberate in everything he said and did, it was all now just bubbling out of him.

'You just can't believe what kind of a flying machine this is,' Young continued, 'it is really something special.'

If the Shuttle's Commander was finding it hard to contain his exhilaration, the crowds lining the western edge of the lake bed were unable to. People charged forward; the make-shift fence, an acknowledged boundary until now, was swept aside. On foot, on motorbikes and in cars, people streamed out on to the lake bed to try to get a closer look at *Columbia*. Overwhelmed, the small security detail called the PJs. The Air Force helicopters, already airborne and ready to go as the Shuttle landed, skimmed across the airfield, dropping down low, using rotor wash to churn up a dust storm that might stem the automotive tide. Strapped into the back of one of the helos along with the Air Force para-rescue men, John Creighton could see the chopper pilots were having a ball. Like airborne ranchers herding cattle, the helicopters swooped and shimmied to head off and round up the exuberant motor derby, finally gaining the upper hand before any of the vehicles got closer than a mile to the waiting spaceplane.

White and black against the pale earth and tan of her

surroundings, *Columbia* looked almost like a living thing, albeit helpless and dependent; a patient in intensive care – monitored, connected up, tested, purged, pumped and swarmed over by the white SCAPE-suited technicians. None of it was happening fast enough for Young though, standing down on the mid-deck still scanning the view from the porthole for any sign of the ramp stairs being rolled up against the side.

'We're still here, you know,' he said. 'And if we're going to get this thing operational, this is one of the parts we're going to have to work on a lil' more. You know what I mean?'

'John, I'm pushing as hard as I can,' Rick Hauck told him.

'Well, what are you pushing?!' Paying passengers, Crip laughed, wouldn't be kept waiting this long.

'You're not going to have to wait for your luggage when you get off,' Hauck encouraged – unless, of course, he pointed out, it had been left in Florida. Or they'd forgotten their baggage tags.

From the confines of *Columbia*'s cabin, the astronauts imagined Hauck as the last man in Mission Control, left behind after everyone else had packed up and gone home, just to entertain them. The picture was far from accurate, although Hauck might have been one of the last in the control room having to resist joining in the celebrations.

On the central main screen that during the mission had lent the room the feel of a military command bunker, controllers projected a slide linking NASA's latest triumph to her past. Alongside both the Mission Control and STS-1 flight patches a swoosh swept up through the words Mercury,

Gemini, Apollo, Skylab and Apollo-Soyuz towards Shuttle. 'THE FUTURE IS NOW' it declared. Its appearance prompted another round of clapping and cheering.

Through their own celebrations, those at the back of the room heard a faint tapping coming from behind them. Turning round, they saw Houston Director Chris Kraft standing behind the windows of the adjoining VIP room. Pressed up against the glass was a sheet of foolscap paper on which he'd written, in fine-lined ink, *We just got infinitely smarter.*

Still up on the flight deck, Bob Crippen felt his ears pop at the moment the hatch was finally pulled open. Below him, John Young, obliged to wait for a flight surgeon to come on board to check him before de-planing, wasn't prepared to hold out for the medic's OK. He walked straight out and, shaking off attempts by the ground crew to help him down the stairs, came bounding down, as eager to get away as a champagne cork from a bottle. Grinning broadly, and carrying his helmet, he seemed more interested in *Columbia* herself than in the welcoming party at the bottom of the stairs. Stepping down, he immediately cut left and underneath, dwarfed by the size of the spacecraft above him. He looked up at the heatshield, wanting to inspect it for himself. Pacing backwards, his gait made awkward by the pear-shaped bulk of his pressure suit, he pointed, exuberantly jabbed his fists, then stopped at the main landing gear, pausing only to kick the tyres as if he were a fighter pilot out on the squadron flightline. Only later would the heat damage to the buckled starboard undercarriage door be discovered. For now, there was nothing to spoil his party. Satisfied, he carried on back, turning and looking up at the cluster of rocket bells that towered above him. They'd served him and Crip well.

'I think John's out there doing a post-flight,' Hauck told Crip, who was still waiting in the cockpit to hand over *Columbia* to another pair of astronauts before getting out.

We did it! thought Young. *We pulled it off!* Punching the air with both hands, he returned to the nose of the vehicle, handing his helmet to white-overalled support crew before crouching down in front of the nosegear next to a yellow airfield tow truck and looking back down his ship's centreline. The smile was still fixed to his face as he came back round to the ramp stairs to talk to Director of Flight Operations George Abbey, there to greet him wearing blue NASA overalls and sporting his trademark crewcut. As had generations of pilots before them, they both used their hands to illustrate the conversation about *Columbia*'s performance. There was still no sign of Young's Pilot though, so the Commander, leaving Abbey on the lake bed, got back up into the Orbiter's cabin to check on his progress. Crip was going to be out in a minute.

As he finally pushed himself out of his seat, Crip felt the full weight of his pressure suit for the first time. He climbed down the companionway to the mid-deck and emerged into the mid-morning sun. Unsure whether or not to steady himself with the handrails, he climbed down to meet Abbey who wrapped an arm round his shoulders, shook his hand firmly, and told him 'good job'.

The astronauts stepped into a van and were driven to a specially erected reception centre where a brief post-flight medical was followed by three good things: they met their wives, they took their first shower since leaving Kennedy, and they were both handed a cold beer. It would fortify them for the speeches to come.

Crip's emergence from *Columbia* had elicited further cheers and clapping from those standing in Mission Control watching the scenes from Runway 23 on big screens. But the Shuttle was now in the care of the Kennedy Space Center team, who would begin the job of preparing her for a return to the Cape. And her next flight. It was new ground for NASA – *the future*.

As the ceremonies in Edwards drew to a close, Young and Crippen – as passengers this time – boarded the Gulfstream with their wives for a flight back to Houston. And after twenty years as a pilot, and fifteen as an astronaut waiting for his first spaceflight, Crip's logbook collected a unique new entry. It read:

April 12, Orbiter Veh 102 54.4, night time 20, Young KSC 39A-RW 23 EDW Lakebed

With the TV coverage drying up, a jubilant Max Faget made his way over to where Hans Mark was sitting with Aaron Cohen and Milt Silvera. In 1969, he'd told a small gathering of hand-picked engineers 'we are going to build the next-generation spacecraft'. Now, something like his balsawood and paper model had been made real. Smiling irrepressibly, he told the three men sitting at the programme management console: 'Let's do it again!'

Although he had not been in the Mission Control room, Tom Moser had watched *Columbia*'s landing from inside Building 30 which housed it, sharing in the moment with his colleagues. As people began to move around the building again following the astronauts' safe return, able at last to leave the edges of their seats, Moser was stopped by Bob Thompson. The Programme Manager in Houston with overall responsibility for the whole Space Shuttle system – Orbiter, external tank and solid rocket boosters – it was Thompson who, after the tiles were discovered to be missing from *Columbia*'s OMS pods, had ordered Moser back to Johnson to lead the team analysing the risk posed by their loss.

For over two years Moser had lived and breathed those troublesome silica tiles. Through the difficulties of bonding them to the Shuttle and the discovery of the densification process that had solved it, to the last-minute claim from Langley's John Houbolt that to launch was courting disaster, Moser and his team had kept on working the problem, overcoming each fresh challenge thrown at them. The behind-the-scenes effort from NASA, Rockwell and the Air Force prompted by the discovery that tiles were missing from *Columbia*'s heatshield had merely been the climax of a long-running drama.

Thompson wanted to know the answer to just one question: 'Did you really think they were going to work? The tiles?'

And for Ken Young and Ed Lineberry there was no official word of thanks. Nothing from beyond the ranks of those at JSC who'd been cleared to know about The Plan until, a year later, Young's secretary brought him a small package wrapped in brown paper. Inside was a white coffee mug. On it was printed a light blue rectangle, framing the silhouette of a swooping bird in the same colour. Top left was a dime-sized yellow moon. Beneath the graphic were the words:

A DECADE OF EXCELLENCE

23 SEPT. 1971 – 23 SEPT. 1981

A much appreciated token of gratitude from the black world, Young kept it on his desk. But the source of the unfamiliar-looking mug remained classified. If asked about it he could only shrug and say 'I don't know. Somebody just left it here after a meeting, I guess.'

SIXTY-EIGHT

Edwards Air Force Base, 1981

IN THE EUPHORIA that followed *Columbia*'s arrival at Edwards, John Young made a brief speech to the crowds before returning to Houston. 'We're really not too far,' he said, 'the human race isn't far from going to the stars.' That's how it felt. America was top of the pile again with a machine that might make possible the vision first laid out by Wernher von Braun in *Collier's* magazine in the fifties. The Chief Astronaut's bold claim was no more or less than a recognition that the Shuttle had the potential to be the 'enabling technology' espoused by von Braun as he'd talked to Hans Mark and the crew of a yacht sailing in San Francisco Bay over a decade earlier; the Shuttle could provide the means through which an orbiting space station might become a waypoint to the moon and beyond.

The first step towards that, though, began with the Orbiter's historic second mission. And, ten days after landing at Edwards, in front of reporters at Johnson, Young and Crippen introduced Joe Engle and Dick Truly before handing them a big golden ignition key, purporting to be *Columbia*'s.

'Would you buy a used spacecraft from this gentleman?'

Truly joked as he and Engle accepted the key from Young.

Six months later, on 12 November, *Columbia* carried Truly and Engle into orbit from the Cape after a launch delayed for ten minutes by the requirements of the NRO. Two days later, after Engle had taken manual control for much of the re-entry, they touched down at Edwards. In completing the mission, Engle, following his experience flying the X-15 over Edwards decades earlier, became the first and only man to have flown into space aboard two different winged spacecraft – something no Flying Tiger ever managed.

The STS-2 crew were forced to return three days early because of a fuel cell failure, but for those who'd always maintained that, as a stick-and-rudder pilot, Engle was in a class of his own, the mission was a triumph – especially after Dick Truly announced, as they prepared for their first hypersonic manoeuvre on the edge of space, 'Joe Henry, I can't see a damn thing . . .'

The flight plan had called for Truly to put on a fresh scopolamine patch to ensure he wouldn't suffer from motion sickness during the sequence of twenty-nine planned high-Mach manoeuvres. He'd not suffered any space sickness so far and was about to say 'screw it', before he put thoughts of rebellion aside. But the drug had got in his eyes and soon he was struggling to make out the numbers on the cue cards. He was squinting and blinking back the pain and tears to try to clear his vision.

This is going to be a pretty interesting entry, thought Engle, as he drew on an intuitive memory of the series of manoeuvres he'd been tasked to fly; *we got a fuel cell down, we got a broke bird, we got winds coming up at Edwards, we got no sleep, we're thirsty and dehydrated, and now my pilot's gone blind . . .*

But as he set up the Orbiter to land on Rogers Dry Lake,

Engle thought it was one of the greatest feelings he'd ever experienced. Edwards was in his blood and he was unable to resist a nod to past glories by putting in an unscheduled call to base air traffic controllers. 'Eddy Tower,' he announced over the radio, causing confusion back at Mission Control in Houston, 'it's *Columbia* rolling out on high final. I'll call the gear on the flare.'

'Roger, *Columbia*,' came the deadpan reply, 'you're cleared number one. Call your gear.'

Just like the good old days.

Columbia's third mission, flown by Commander Jack Lousma and Pilot Gordo Fullerton, followed just four months later, at the end of March 1982. This time, an encrypted message from the NRO arrived just before Christmas. Containing details of a major reshuffling of KEYHOLE satellite usage, it forced Ken Young to pull colleagues back from their holidays to rewrite a flight plan they'd thought was settled. Three months after STS-3's successful conclusion, T. K. Mattingly and Hank Hartsfield blasted off from Kennedy on the Space Shuttle's fourth mission and the last of the series of two-man orbital test flights.

On 4 July, after their seven-day mission, Mattingly brought *Columbia* in to land at Edwards AFB. On the ground the President was waiting to welcome them. In his speech, delivered in front of *Enterprise*, the leading edge of her wing draped in blue cloth, Ronald Reagan declared that the Shuttle was now 'operational'. Nor were *Columbia* and *Enterprise* the only Orbiters at Edwards for the Independence Day celebrations. The dramatic climax to the President's speech was provided by the departure for Kennedy of the recently completed third Orbiter.

'*Challenger,*' the President declared, 'you're cleared for take-off.' And on cue, the 747 carrier aircraft, with *Challenger* on her back, began her take-off roll along the dry lake bed, kicking up a thick wake of dust behind her, before climbing into the sky and past the ceremony. 'You know, this has got to beat firecrackers,' Reagan grinned, prompting more cheering from the happy, flag-waving crowd.

In his speech, the President drew attention to a payload aboard *Columbia* that had been sponsored by the Air Force, highlighting the importance of the Shuttle to national security and to developing space as a means of 'maintaining the peace'. The Shuttle programme had been born and kept alive because of the qualified support of the military. Ultimately, it was decided, despite much Air Force opposition, that the Shuttle was to be the sole launch vehicle for the Department of Defense. And with the introduction of Hans Mark's cadre of manned spaceflight engineers, a projected first launch from Vandenberg in 1985, the creation of Air Force Space Command and the construction of a new Consolidated Space Operations Center in Colorado, and the prospect of an improved KEYHOLE satellite, known variously as advanced KENNEN, CRYSTAL, IKON or KH-12, designed specifically for the Orbiter's payload bay, it appeared that enough had been done to bind the Pentagon to the Shuttle programme. But the real contribution of the military to the Shuttle programme was in the people it provided, none more so than a nearly lost generation of astronauts who had seen the Air Force fail in its ambition to pursue its own manned space programme, but who ended up being at the heart of NASA's programme.

As well as being central to the development of the vehicle itself, the small group of refugees from the Air Force

Manned Orbiting Laboratory programme, so reluctantly taken on by NASA in 1969, formed the core of all the early crew assignments. A MOL veteran was part of the crew of the first eight Shuttle missions. Bob Crippen and Dick Truly were given command of STS-7 and STS-8 respectively. Of the first twenty-four flights, two-thirds included astronauts from the USAF's stillborn manned space programme. They flew aboard the first flights of *Columbia*, *Challenger*, *Discovery* and *Atlantis*, conducted the first night launch and landing, the first landing at Kennedy Space Center, and the only landing at White Sands. MOL guys commanded crews including the first female and African-American astronauts.

In the exchange between NASA and the military, however, it didn't all flow NASA's way. Throughout the Shuttle's development it appeared that it would go on to become as much of a military asset as a civilian one. Things didn't quite work out as had been envisaged. And while the Shuttle was certainly to perform a substantial number of useful military missions, it did not end up providing the Pentagon's only access to space. In fact, it's possible that the Shuttle's main contribution to national security came rather more indirectly, but no less significantly for that.

In the autumn of 1985, two Soviet cosmonauts, Igor Volk and Rimantas Stankyavichus, sat in the cockpit of a large black and white delta-winged aeroplane at the threshold of the 17,700-foot-long runway at Zhukovsky Flight Test Centre near Moscow. Their aircraft, labelled BTS-002 – an acronym for 'big transport aircraft' – looked *a lot* like NASA's Space Shuttle, but clustered around the tail were four jet engine nacelles, each containing a 17,000lb thrust Lyulka AL-31 turbofan more usually found powering Soviet Air

Force Sukhoi Su-27 Flanker jet fighters. After warming up the engines, Volk and Stankyavichus pushed forward four throttle levers and began their take-off roll. Six thousand feet along the runway, Volk, the new machine's captain, pulled back gently on the stick and raised her nose. As soon as 'Number Two' was airborne he reduced the back pressure on the stick a touch as his co-pilot raised the landing gear. The Soviet Union's Space Shuttle programme was off the ground.

Unlike NASA's prototype Orbiter *Enterprise*, 'Number Two', although similarly unable to reach orbit, took off and climbed under her own power. Without an aircraft that, like the Boeing 747, was capable of carrying the Shuttle on its back, Soviet designers had no choice but to conduct their series of atmospheric flight tests this way. Over the next two and a half years, twenty-four test flights, flown without serious incident, provided sufficient confidence to look forward to *Buran*'s first spaceflight. But the effort came at an enormous cost.

As a launch vehicle for the NRO's spy satellites, the American Space Shuttle was always expected to make a major contribution to national security. More potentially belligerent missions resulting from the President's 'Star Wars' Strategic Defense Initiative remained theoretical, but even without that prospect, the Shuttle programme had a crippling effect on the country's Cold War opposition: the Soviet Union. By frightening them into launching their own Space Shuttle programme, which they could neither afford nor divine any practical purpose for, it committed them to a project estimated to have cost between fifteen and twenty billion roubles. At late eighties exchange rates that was a figure marching north of $30 billion. It was a sum of money the beleaguered Soviet economy, suffering from years of what was dubbed the Age of Stagnation, could ill afford. In trying to meet a perceived threat from the American Shuttle, the Kremlin was prepared to pay over five times what it had cost NASA to develop their own vehicle, while drawing on the resources of a gross national product that was barely half that of their Western rival. And while the development of the *Buran* Space Shuttle did not, in itself, bankrupt the USSR, it was symptomatic of the unsustainable pressure created by Ronald Reagan's massive investment in the US military that certainly did hasten the creaking political system's collapse. Ultimately, *Buran*, flying unmanned, would complete just two orbits before returning to the runway at Baikonur Cosmodrome in 1988. She never flew again and, following the break-up of the Soviet Union in 1991, the whole programme was consigned to the scrapheap.

SIXTY-NINE

THE PROMISE OF the reusable Shuttle was that in providing regular access to space it would also reduce the cost of it. That never happened. The idea that it might be possible to recover the Shuttle from the runway, dust her down, bolt her to another external tank and set of solid rocket boosters then relaunch her was never realistic. While those who instigated the programme had imagined the redundancy and reliability built into the Shuttle might allow launches when, for instance, one of the four computers failed to synch with the others, it never happened. Between each flight, the Orbiter was pared back to the bone, refurbished and rebuilt. Even systems and components that had worked perfectly were subjected to further testing that, ironically, could hasten the moment when those parts *did* need replacement. This of course cost further time and money.

Even accepting that the Shuttle would not be providing 'cheap' access to space, NASA had to endure intense pressure on the programme throughout its life. It was largely of the agency's own making – a consequence of the promises necessary to get the programme approved in the first place. In explicitly setting out to build a workhorse capable of regular launches and to meet the demand from government, commercial and military customers, the Shuttle programme managers now had to show that they had done so. The idea that the Orbiter was 'operational', however, didn't sit comfortably with many at NASA, none more so than their Chief Astronaut. As far as John Young was concerned, the Shuttle was and remained an experimental vehicle. 'Space machinery,' he would later be explicit in saying, 'is not airline machinery.' The point was well illustrated when Young himself returned to *Columbia*'s flight deck as Commander of STS-9, the first mission to carry the European Spacelab inside the Shuttle's payload bay.

Four hours before re-entry, following a ten-day mission – the longest yet flown by the Shuttle – Young fired the thrusters to reorient the *Columbia* for re-entry. At the moment the jets boomed through the vehicle, one of the primary flight computers failed, its crash signalled by the white cross displayed on the monitor ahead of him. Six minutes later, the next firing of the thrusters triggered the failure of a second computer. This was enough to get Young rattled. His legs trembling, he could only let *Columbia* drift in space while he and his Pilot, Brewster Shaw, ran through checklists, reconfiguring and rebooting the computers in an effort to recover them. Their return to Earth seemed impossible until the astronauts, after eight hours of trouble-shooting, were eventually able to load the re-entry software on to two of the computers and fire the manoeuvring rockets for a delayed re-orbit burn. But even after this partial success, Young was conscious that they'd been just a decision away – had he chosen to activate the back-up flight software – from losing flight control of *Columbia*. *And that*, he thought with characteristic understatement, *would have been very bad for us*.

Nor was the drama of STS-9 quite over yet. Four and a half minutes from touchdown on the dry lake, while flying a revised approach into Edwards from the north, the temperature in one of the auxiliary power units rose sharply. Unknown to the crew, a hydrazine leak, exposed to higher levels of oxygen as they descended through 40,000 feet, had ignited. As Young touched down on Runway 17, two of the three APUs that powered the Orbiter's flight controls were on fire. Then, as the nosewheel came down heavily as a result of a late change to the flight control software, one of the main computers failed again. Within twenty-five minutes of landing, the two burning APUs had been destroyed by fire.

The headline news around STS-9, Young's sixth and final spaceflight, was Spacelab's successful debut. The failure of the computers and APU fires – both invisible to television viewers – went largely unnoticed by the public. They had been told the Shuttle was now operational and, without conspicuous drama or catastrophe, they had no reason to believe otherwise. Where was the interest in job done?

The Shuttle's success in doing what she was designed to do was the very thing that, in the end, inured the public to her extraordinary and unique capabilities. It simply wasn't possible for each flight to be special. And so the Space Shuttle, the high-water mark of the US aerospace industry – the last American flying machine built to fly higher and faster than everything that had come before – began to attract less attention. And she became familiar quickly. As early as the fifth mission – the first 'operational' mission – Pilot Bob Overmyer, the fifth MOL veteran to fly, was asked how he felt about being a space trucker. He could live with that, he told the interviewer. So too, it turned out, could

Chuck Yeager. Having spent twenty years making plain his view that astronauts were riding rather than flying their spacecraft, he'd changed his tune now graduates of his own Edwards AFB Aerospace Research Pilots School had a winged spaceplane to fly. When asked what he made of Overmyer's reply, the legendary test pilot said, with a touch of desire in his voice, 'to be called a Space Shuttle Pilot, man – to be called a trucker – I'd be called *anything* to fly it . . .'

But the question he'd been asked got to the heart of it. If successful, the Shuttle programme became one of utility rather than adventure. Instead of exploring and striking out in new directions and discovering new worlds, the Shuttle – the most remarkable flying machine ever built – was no more than a tool; a means to an end. And, from inside NASA's Washington HQ, Hans Mark campaigned for three years to make sure that that end was the construction of a space station.

In January 1985, his effort appeared to have succeeded. Sitting in the gallery of the Capitol Building, a short walk from the NASA office block, the agency's Deputy Administrator listened as the President delivered his State of the Union address, anticipating the moment when Reagan would announce news of his decision.

After opening the speech with news of the country's economic growth, Reagan moved on to what he described as 'America's new frontier'. In words designed to echo the resolve and optimism of Jack Kennedy's decision to shoot for the moon, Reagan announced: 'We can follow our dreams to distant stars, living and working in space for peaceful, economic and scientific gain. Tonight, I am directing NASA to develop a permanently manned space station and to do it within a decade.'

The space station was to be named *Freedom*, and the Shuttle, as first laid out by Nixon's 1969 Space Task Group, would be both the means of its construction and its connection with Earth on completion.

Ultimately, plans to build *Freedom* were to falter before eventually evolving into the International Space Station following the fall of the Berlin Wall. Until then, the Orbiter itself, the last great achievement of NASA's Apollo generation, would serve as a means through which we learned to live and work in space. With each new mission, the growing Shuttle fleet – three-strong with the addition of *Discovery* in the summer of 1984, and joined by *Atlantis* the following year – grew into the role, expanding the range of jobs they could perform, from, with Spacelab on board, mini-space station through satellite launch, repair and retrieval to early on-orbit construction. The promise of launching probes to distant planets and the space station lay ahead.

It would be too late for Max Faget. He could see trouble ahead for the *Freedom* programme in any case. At the age of sixty, soon after Engle and Truly brought the second Shuttle mission to a successful conclusion, Houston's little Cajun engineering dynamo decided to call it a day. The truth was that Faget, who'd first resisted then embraced the Shuttle zealously, felt burned out. He'd lived through it all. Part of the original NACA Space Task Group that had left Langley and set up home in Houston, his story *was* that of the American manned space programme, from Mercury through to the Shuttle. But with *Columbia*'s successful first flight he saw his world was changing. The adventure was over.

You could see, he thought, *Camelot fading into the fog.*

EPILOGUE

Brought Down

'A lot of folks said the Shuttle was a lemon. But my mom said, "It's not a lemon, it's a peach."'

John Young

On 16 MAY 1985, Bob Crippen wrote to aspiring artist Tim Gagnon with his thoughts on the design of a patch for his next mission in command of the Shuttle. Since the early seventies, Crip had been encouraging the young artist in his ambition to see his work reach space. For his upcoming flight, Crip provided Gagnon with a few guidelines: 'I'm partial to round patches, but other shapes are okay if there are a minimum of protuberances. Also, simple is good. Too much detail in a patch doesn't work. As for colors, I'm partial to red, white and blue, but that is not a constraint. However, the maximum of colors should be about ten. It would be desirable, if the first launch from Vandenberg was symbolized somehow. One idea is we commonly use the term V1 when talking about that flight. In addition, a polar orbit indicator would be appropriate. Crew names are Crippen, Gardner, Mullane, Gardner, and Ross. We will add any Payload Specialist(s) names at the bottom.'

Five months later, Crippen, who throughout his time as an astronaut had remained a serving naval officer, was introduced to the press, alongside the rest of his crew, in the brand-new Orbiter Processing Facility (OPF) at Vandenberg Air Force Base. Only one of them was not wearing military uniform, and he was hardly less military than the rest of them. Pete Aldridge, flying with the crew as a Department of Defense Payload Specialist, was Hans Mark's successor as Under-Secretary for the Air Force and Director of the

National Reconnaissance Office. The last member of the crew was Payload Specialist Brett Waterson, the first of Mark's cadre of Air Force manned spaceflight engineers to be assigned a spaceflight. The mission aboard the third Orbiter in the fleet, *Discovery*, was designated STS-62A under a new numbering system in which '6' denoted the year of launch, '2' a West Coast launch, and 'A' that it was the first mission, from that site, of the year. And it was to be the most conspicuously military manned spaceflight yet launched by the United States – a tentative embrace of the Space Shuttle by the Air Force.

In early 1985, in preparation for the mission, the Orbiter prototype *Enterprise* had been flown to Vandenberg on the back of the Shuttle Carrier Aircraft for a series of tests to prove the facilities at the USAF's new spaceport were ready.

For Crippen, it marked a return to Slick Six, the launch complex first developed in support of the Manned Orbiting Laboratory. As well as the new OPF, the old MOL Launch Control Center was renovated and fitted with the same state-of-the-art computer system used at Kennedy. The MOL launch mount and flame trench were adapted to accommodate the different configuration of the Shuttle, while the railway line that ran through the base now, rather than being an unwelcome complication, came into its own as a means of delivering the solid rocket booster segments from Thiokol in Utah. Building 8505, the old MOL crew quarters, was refurbished, ready once again to become home to Bob Crippen and his fellow military spacemen.

Crip felt his career as an astronaut had come full circle. At first the Air Force had resisted NASA's selection of a naval aviator as Commander on what they regarded as a bluesuit mission, but Crip himself lobbied hard for it. And, as the experienced Commander of three Shuttle missions since STS-1 and a veteran of the Air Force's own cancelled manned space programme, the case for assigning him the mission was too strong. Having won command of the flight, though, Crip never then got the chance to wear the STS-62A mission patch he'd briefed so carefully.

In January 1986, Crip led a flight of three T-38s down to Albuquerque, New Mexico. From there, the STS-62A crew were flown in a light aircraft to the Los Alamos Laboratories, where they were being given instruction on one of their two mission payloads, an experiment codenamed CIRRIS, designed to collect data on levels of background radiation in space which might then be used to inform the development of future military hardware.

On the morning of the 28th, the crew took time out from the classroom to watch the launch of *Challenger*, the second Orbiter, from Kennedy Space Center. For less than thirty seconds they watched *Challenger*'s roaring ascent before, with Shuttle launches beginning to seem routine, coverage broke off in favour of more standard morning programming.

'Let's see if they're covering it on one of the other channels,' someone said, and Crip spun the dial in search of an alternative station. No joy. As the crew prepared to leave for work, Dale Gardner suggested 'Well, let's try it one more time' and turned on the TV set again. As the screen blinked on, dreadful images took shape in front of them. From the back of the room, Pete Aldridge saw violent tendrils of white smoke corkscrewing away from a massive, boiling explosion. *Jesus*, he thought. Ahead of him, sitting closer to the screen, he saw Bob Crippen's head drop. Any uncertainty about what they'd just witnessed was swept aside by his Commander's obvious and instant devastation. *Challenger* was lost, and with her the lives of her seven-strong crew.

While each member of the STS-62A crew tried to process what had happened, Aldridge called the Air Force. 'I need a flight,' he told them. 'We need seats for seven people.' He knew the astronauts had to return to Houston immediately. There were the T-38s waiting at Albuquerque, but he didn't think Crippen was in the right frame of mind to be piloting an aeroplane.

'No,' Crip told him, 'I'm going to fly. I need to think a bit.'

He'd lost a friend – Dick Scobee, *Challenger*'s Commander, had flown as Pilot on Crip's third Shuttle mission. But as a senior member of the Astronaut Office, Crip was close to all seven members of the crew. Alongside them, he also mourned an eighth loss: the Orbiter herself. For over a decade he'd invested everything in bringing the Space Shuttle to life. Following his historic first flight with John Young, Crip had been given command of *Challenger*'s second flight. He was as familiar with her inside and out, her qualities and foibles, as anyone alive, and now she had been smashed to smithereens in an instant, the cause of the destruction unknown. *That beautiful machine*, he thought as he replayed the indelible image of the explosion. *What the heck happened to her?*

There was no radio chatter between the three T-38s as Crippen led the formation home to Ellington, nor between Crip and Mike Mullane in the back seat, but with each change of frequency as they flew cross-country back to Houston, air traffic controllers offered 'NASA Flight' their condolences.

By nightfall, George Abbey had already asked Crip to act as deputy head of the Review Board investigating the loss of *Challenger* and her crew of seven. The next day, he once again strapped himself into the cockpit of a T-38 to fly to Kennedy where he got to work on the cause of the accident, the recovery of the vehicle, and of the crew's remains, a task led by another MOL veteran, Bob Overmyer.

Had Pete Aldridge flown aboard STS-62A, Crippen thought, he'd have seen what the Shuttle was capable of. She'd have convinced him. He'd have become, like his predecessor Hans Mark, a supporter. But with the loss of *Challenger*, the arranged marriage between NASA and the Air Force was on the rocks before it was properly consummated. DoD had always had reservations about putting all their eggs in the Shuttle basket. From within NASA, as Deputy Administrator, Mark had continued driving towards that goal but, left with a fleet reduced to just three vehicles, *Columbia*, *Discovery* and *Atlantis*, it became impossible to sustain the flight rates and availability required by the Air Force launch schedule. And so, after DoD had spent at least $2.8 billion on redeveloping Slick Six – including $79.5 million on a windscreen to shield the Shuttle from the strong winter gusts that whistled through the surrounding hills – Pete Aldridge took the decision to pull the plug on Vandenberg. While maintaining that a West Coast spaceport remained, he said, 'essential', he reduced the site to operational caretaker status. In theory, the facility could be reactivated, but it never happened. No Shuttle would ever launch from Vandenberg. And it remained the biggest regret of Bob Crippen's career that he never had the opportunity to do so.

Within weeks of the tragedy, Dick Truly was brought back to NASA as Associate Administrator for Manned Spaceflight, chairing an agency task force set up to work alongside the Rogers Commission investigating the disaster. While they looked into the cause of the failure of the solid rocket boosters that had led to the loss of *Challenger*, Truly assigned Bob Crippen the job of leading an internal review of the Shuttle programme's management structure. When Crip submitted his report two months later, in August 1986, recommending that there be a NASA Shuttle Programme Deputy Director, based at Kennedy, with responsibility for Shuttle operations, Truly responded with a challenge.

'Crip,' he told his friend, 'if you really believe that, you'll hang up your flying boots and come take that position.'

He did. And, after assuming his new role at the Cape, Bob Crippen never flew the Orbiter again. Instead, he and Truly led the effort to get the vehicle itself airborne once more.

Alongside modifications to the solid rocket boosters, thousands of other changes, deemed critical, were made post-*Challenger* to try to ensure the Shuttle's safety. After STS-26, dubbed the 'Return to Flight', launched in 1988 with Rick Hauck in command, the Shuttle fleet flew for another fifteen years without major incident. Then, in 2003, the appearance that NASA had succeeded in making manned spaceflight routine was shattered. And this time it was *Columbia*, the ship in which Young and Crippen had so spectacularly ushered in the Shuttle era, that was lost.

Eighty-two seconds after launching from Pad 39A, *Columbia* was travelling at nearly two and a half times the speed of sound. As she climbed through 66,000 feet, a chunk of insulating foam tore away from the external tank and punched a hole in the reinforced carbon-carbon leading edge of her port wing. It was a fatal blow, but one which would wait to reveal itself. The only conceivable hope for the astronauts was that it might be discovered. The tragedy was that it so nearly was.

As in 1981, there had been concern, following the launch, about the integrity of the heatshield. But twenty-two years after the Shuttle's maiden flight, the Kranz doctrine appeared to be in danger of being forgotten. While those technicians responsible for examining footage of the Shuttle launch were concerned about possibly serious damage as a result of a foam strike to the wing, the prevailing view among the programme's management was that this was a familiar problem rather than an event that demanded an exceptional response. A request made to the Air Force for high-resolution photographs of *Columbia* on orbit was rescinded and further efforts to enlist DoD help were dead-ended through failed bureaucracy. With that failure, *Columbia* and her crew were condemned.

On re-entry, hot plasma gas roared in through the hole in the leading edge and whipped around the aluminium compartments within the wing, burning and melting whatever was in its path. Inside Mission Control, telemetry downlinked from *Columbia* began to record a series of anomalous readings that couldn't be attributed to any single failure: first a higher than usual strain on the wing spar, then 'off-scale low' readings from hydraulic sensors. At 13.59:15 GMT, fifteen minutes after the Orbiter's entry interface at 400,000 feet over the Pacific, tyre pressure in the port-side main landing gear was lost.

As the air thickened through 200,000 feet, the integrity of the port wing began to fail, its structure devoured from inside as if by an aggressive cancer. With greater aerodynamic pressure pressing against the unsupported skin of the lower wing, its carefully calculated curve buckled and collapsed, creating asymmetric levels of lift between port and starboard that tried to flip her on to her back. At the same time the increased aerodynamic drag from the ruined wing pulled the nose hard to the left. *Columbia*'s computers commanded the RCS thrusters in the tail to fire in a last-ditch effort to try to hold her stable, but she was overwhelmed, first rolling on to her back, then tumbling uncontrollably through the high Texas skies.

Columbia had struggled mightily to save herself and her crew. Labelled as fragile, NASA's fleet of Orbiters turned out to be anything but. In thirty years of operations, not one of them failed her astronauts. They proved themselves capable of absorbing, containing and enduring whatever faults developed in their own systems. Instead, on the two occasions when an Orbiter was lost, she fell victim to an assault: *Challenger* to a blowtorch of burning rocket fuel that destroyed the stack to which she was attached; *Columbia* to the percussive attack of a sharp-edged lump of moulded foam that might as well have been a sledgehammer. Subjected to such precisely applied violence, neither machine stood a chance.

Six days after the loss of *Columbia*, Bob Crippen, now retired, returned to the Kennedy Space Center where, standing in front of the workforce, gathered together on the facility's Shuttle runway, he delivered a eulogy to the lost spacecraft and her crew. Clearly emotional, occasionally pausing to control his voice, Crip spoke slowly, in a rich, cracked baritone:

> *Columbia* was a fine ship. She was named after Robert Gray's exploration ship that sailed out of the Boston harbour in the eighteenth century. *Columbia* and the other orbiters were all named after great explorer ships, for that is their mission: to explore the unknown. *Columbia* was hardly a thing of beauty, except for those of us who loved and cared for her. She was often bad-mouthed for being a little heavy round the rear end. But many of us can relate to that. Many said she was old and past her prime. Still she had only lived barely a quarter of her design life. In years she was only twenty-two. *Columbia* had a great many missions ahead of her. She, along with the crew, had her life snuffed out in her prime.
>
> I was here at the Shuttle runway in March of 1978 [*sic*] when *Columbia* first arrived at Kennedy Space Center. She came in on the back of a seven-forty-seven escorted by Deke Slayton in a T-38. Readied for launch by the loving care of the Kennedy team – the same care they've given to all twenty-eight of her flights. She was ready to fly in April of 1981. John Young and I were privileged to take her on that maiden flight. She performed magnificently. 'The world's greatest electric flying machine' was what John described her as. Because she was a little heavy, she didn't get some of the more glamorous missions, but she was our leader at doing science on orbit, just as she was doing with this crew in Spacelab on STS-107; microgravity scientific

exploration was her bag. She carried Spacelab numerous times, studying material processing and life sciences, all of which were focused at giving us a better life here on Earth. *Columbia* also helped us better understand the heavens and learn about the origins of the universe with several missions including Astro, also by deploying the most advanced X-ray observatory ever built, the Chandra Space Telescope, and by her very recent Hubble Space Telescope servicing mission.

Just as her crew has, *Columbia* has left us quite a legacy. There's heavy grief in our hearts, which will diminish with time, but it will never go away. And we'll never forget.

Hail Rick, Willy, KC, Mike, Laurel, Dave and Ilan.

Hail *Columbia*.

GLOSSARY

AFB	Air Force Base
AFS	Air Force Station
ALT	Approach and Landing tests
Angle of Attack	angle between the oncoming air and a reference line on the aircraft, often a line connecting the leading edge with the trailing edge of the wing
AOA	abort once around
APU	auxiliary power unit
ARPS	Aerospace Research Pilots School
ATO	abort to orbit
BFS	back-up flight system
Blue Cube	Air Force Satellite Control Facility
CAP	crew activity plan
CapCom	Capsule Communicator
CCD	charge-coupled device
CIA	Central Intelligence Agency
CMP	Command Module Pilot
CORONA	codename for the KH-1, KH-2, KH-3 and KH-4 reconnaissance satellites
CRT	cathode ray tube
CSM	Command and Service Module

DFI	development flight instrumentation
DMI	deployment mapping instrument
DoD	Department of Defense
DORIAN	codename for the KH-10 camera system designed for the Manned Orbiting Laboratory
EST	Eastern Standard Time (US)
ET	external tank
EVA	extra-vehicular activity
FBI	Federal Bureau of Investigation
FIDO	Flight Dynamics Officer
FRSI	felt reusable surface insulation
FSL	Flight Simulation Laboratory
g	unit of acceleration
GAMBIT	codename for the KH-7 and KH-8 reconnaissance satellites
GAO	General Accounting Office
GMT	Greenwich Mean Time
GPC	general purpose computer
GVA	grizzled veteran astronaut
HAC	heading alignment circle
HEXAGON	codename for the KH-9 reconnaissance satellite
hydraulic oleo	a shock absorber used on landing gear
hydrazine	rocket fuel
hypergolic rocket	one where the propellants spontaneously ignite upon coming into contact
ICBM	intercontinental ballistic missile
IP	initial point
IR	infra-red
IRIS	infra-red imagery of Shuttle

JSC	Johnson Space Center
KAO	Kuiper Airborne Observatory
KENNEN	codename for the KH-11 reconnaissance satellite
KEYHOLE	codename for US spy satellites
KSC	Kennedy Space Center
LACROSSE	codename for a family of NRO radar reconnaissance satellites. Properly known, according to some sources, as LACROS
LLTV	lunar landing training vehicle
LM	Lunar Module
LMP	Lunar Module Pilot
LOR	lunar orbit rendezvous
LOS	loss of signal
LST	large space telescope
Mach	measurement of speed in relation to the speed of sound
Max Q	maximum dynamic pressure
ME	main engine(s)
MECO	main engine cut-off
MET	mission elapsed time
MIT	Massachusetts Institute of Technology
MMU	manned manoeuvring unit
MOL	Manned Orbiting Laboratory
MPAD	Mission Planning and Analysis Division
MSC	Manned Spacecraft Center
MSE	manned spaceflight engineer
MSFC	Marshall Space Flight Center
NACA	National Advisory Committee for Aeronautics
NAS	Naval Air Station

NASA	National Aeronautics and Space Administration
NORAD	North American Aerospace Defense Command
NRO	National Reconnaissance Office
OMB	Office of Management and Budget
OMS	orbital manoeuvring system
OPF	Orbiter Processing Facility
OPS	operational sequences
OTF	orbital test flight
PAD	pre-advisory data
PASS	primary avionics software system
PIO	pilot-induced oscillation
PJ	pararescue jumper
PRO	proceed
RAE	Royal Aircraft Establishment
RAF	Royal Air Force
RAT	ram air turbine
RCS	reaction control system
RFP	request for proposal
RID	review item discrepancy
ROTC	Reserve Officers Training Corps
RTCC	Real Time Computer Complex
RTLS	return to launch site
RTV	room temperature vulcanization
SAIL	Shuttle Avionics Integrated Laboratory
SALT	Strategic Arms Limitation Talks
SCA	Shuttle Carrier Aircraft
SCAPE	self-contained atmospheric protection ensemble
SDS	Satellite Data System. A network of communication satellites used to relay

	information from KH-11 and LACROSSE back to US ground stations
sep	short for separation
SimSup	simulator supervisor
SIP	strain isolation pad
SMS	Shuttle Main Simulator
SRB	solid rocket booster
SSME	Space Shuttle main engines
STA	Shuttle Training Aircraft
STS	space transportation system
TACAN	tactical air navigation
TAL	transatlantic abort landing
TEOS	Tetraethyl Orthosilicate
TFNG	The F***ing New Guys
TIG	time of ignition
TPS	thermal protection system
USAF	United States Air Force
VAB	Vehicle Assembly Building
VOS	USAF Satellite Control Facility Shuttle operations office

BIBLIOGRAPHY

It would have been impossible even to contemplate writing *Into the Black* without drawing on the work of others who preceded me. While a complete bibliography follows, I am particularly indebted to a handful of fine writers who require a special mention. Dennis Jenkins' book *Space Shuttle: The History of the National Space Transportation System* and T. A. Heppenheimer's two-volume *History of the Space Shuttle* were both vital and comprehensive references with respect to the story of the Shuttle itself. For providing some insight into the black world of the NRO and KEYHOLE satellite programme, William E. Burrows' book *Deep Black*, Jeffrey T. Richelson's *America's Secret Eyes in Space* and numerous outstanding articles by Dwayne A. Day from the website thespacereview.com were indispensable. I recommend the work of all five gentlemen without reservation.

Books

Arnold, David Christopher, *Spying from Space* (Texas A & M University Press, 2005)

Baker, David, *The History of Manned Spaceflight* (Outlet, 1986)

—— *International Space Station* (Haynes Publishing, 2012)

—— *NASA Space Shuttle 1981 Onwards (All Models)* (Haynes Publishing, 2011)

Bugos, Glenn, *Atmosphere of Freedom* (NASA, 2000)

Burrows, William E., *Deep Black* (Bantam Press, 1988)

—— *Exploring Space* (Random House, 1990)

—— *This New Ocean* (Random House, 1998)

Butrica, Andrew J., *To See the Unseen* (NASA, 1996)

Cernan, Eugene, and Davis, Don, *The Last Man on the Moon* (St Martin's Press, 1999)

Chaikin, Andrew, *A Man on the Moon* (Michael Joseph, 1994)

Clarkson, Jeremy, *I Know You Got Soul* (Michael Joseph, 2004)

Collins, Michael, *Carrying the Fire* (Farrar, Straus and Giroux, 1974)

Compton, W. David, and Benson, Charles D., *Living and Working in Space: A History of Skylab* (NASA, 1983)

Cooper Jr, Henry S. F., *Before Lift-Off* (The Johns Hopkins University Press, 1987)

Crickmore, Paul F., *Lockheed SR-71* (Osprey, 1993)

Crossfield, A. Scott, and Blair Jr, Clay, *Always Another Dawn* (Hodder & Stoughton, 1961)

Dethloff, Henry C., *Suddenly, Tomorrow Came . . .* (NASA, 1993)

Doyle, Steven E. (ed.), *History of Liquid Rocket Engine Development in the United States 1955–1980* (American Astronautical Society, 1992)

Dunar, Andrew J., and Waring, Stephen P., *Power to Explore* (NASA, 1999)

Dyson, George, *Project Orion* (Allen Lane, 2002)

Ellis, Warren, and Doran, Colleen, with Stewart, Dave, *Orbiter* (DC Comics, 2003)

Evans, Ben, *Space Shuttle Columbia* (Springer-Praxis, 2005)

Gainor, Chris, *Arrows to the Moon* (Apogee Books, 2001)

Gatland, Kenneth, *The Illustrated Book of Space Technology* (Salamander, 1981)

Gertner, Jon, *The Idea Factory* (Penguin Books, 2012)

Gleick, James, *Isaac Newton* (Fourth Estate, 2003)

Godwin, Robert (ed.), *Space Shuttle STS Flights 1–5: The Nasa Mission Reports* (Apogee Books, 2001)

Gordon, Yefim, and Gunston, Bill, *Soviet X-Planes* (Midland, 2000)

Graham, Richard H., *SR-71 Revealed* (MBI, 1996)

Gray, Mike, *Angle of Attack* (Norton, 1992)

Grey, Jerry, *Enterprise* (William Morrow, 1979)

Gunston, Bill, *F-4 Phantom* (Ian Allan, 1977)

—— *Faster than Sound* (Patrick Stephens Limited, 1992)

Hallion, Richard P., and Gorn, Michael H., *On the Frontier* (Smithsonian Books, 2003)

Hansen, James R., *First Man* (Simon & Schuster, 2005)

Hendrickx, Bart and Viz, Bert, *Energiya-Buran: The Soviet Space Shuttle* (Springer-Praxis, 2007)

Heppenheimer, T. A., *Development of the Space Shuttle 1972–1981* (Volume Two) (Smithsonian Institution Press, 2002)

—— *The Space Shuttle Decision 1965–1972* (Smithsonian Institution Press, 2002)

Higham, Robert (ed.), *Flying American Combat Aircraft: The Cold War* (Stackpole Books, 2005)

Hill, C. N., *A Vertical Empire* (Imperial College Press, 2012)

Illiff, Kenneth W., and Peebles, Curtis L., *From Runway to Orbit* (NASA, 2004)

Jenkins, Dennis R., *Space Shuttle: The History of the National Space Transportation System* (Dennis R. Jenkins, 2010)

—— and Landis, Tony R., *Hypersonic* (Speciality Press, 2003)

Joels, Kerry Mark, and Kennedy, Gregory P., *The Space Shuttle Operator's Manual* (revised edition) (Ballantine, 1988)

Jones, Tom, *Skywalking* (Smithsonian Books, 2007)

Kaplan, Marshall H., *Space Shuttle* (Aero Publishers, 1983)

Kraft, Chris, *Flight* (Dutton, 2001)

Langewiesche, William, *Aloft* (Penguin, 2010)

Launius, Roger D., and Jenkins, Dennis R., *Coming Home* (NASA, 2012)

Lewis, Richard S., *The Voyages of Columbia* (Columbia University Press, 1984)

Lipartito, Kenneth, and Butler, Orville R., *A History of Kennedy Space Center* (University Press of Florida, 2007)

Logsdon, John M. (ed.), *Exploring the Unknown* (NASA, 1999)

Lovell, Jim, and Kluger, Jeffrey, *Apollo 13* (30th Anniversary Edition) (Houghton Mifflin, 2000)

Macknight, Nigel, *Shuttle* (Macknight International, 1984)

Mark, Hans, *The Space Station: A Personal Journey* (Duke University Press, 1987)

Marrett, George J., *Contrails Over the Mojave* (Naval Institute Press, 2008)

Miller, Jay, *The X-Planes* (Midland, 2001)

Muenger, Elizabeth A., *Searching the Horizon* (NASA, 1985)

Mullane, Mike, *Riding Rockets* (Scribner, 2006)

Nelson, Craig, *Rocket Men* (John Murray, 2009)

Neufeld, Jacob, et al. (eds), *Technology and the Air Force* (USAF, 1997)

Peebles, Curtis, *Dark Eagles* (Presidio, 1995)

—— *High Frontier* (Air Force History and Museums Program, 1997)

Perry, Robert L., *A History of Satellite Reconnaissance* (NRO, 1973)

Pressel, Phil, *Meeting the Challenge* (American Institute of Aeronautics and Astronautics, 2013)

Reichardt, Tony, *Space Shuttle: The First 20 Years* (Smithsonian/Dorling Kindersley, 2002)

Rich, Ben R., and Janos, Leo, *Skunk Works* (Little, Brown, 1994)

Richelson, Jeffrey T., *America's Secret Eyes in Space* (HarperCollins, 1990)

Rose, Bill, *Secret Projects Military Space Technology* (Midland, 2008)

Rumerman, Judy A. (compiler), *NASA Historical Data Book Volume VI* (NASA, 1999)

Sheehan, Neil, *A Fiery Piece in a Cold War* (Vintage, 2009)

Sivolella, Davide, *To Orbit and Back Again* (Springer-Praxis, 2014)

Slayton, Donald K., with Cassutt, Michael, *Deke!* (Forge, 1994)

Stafford, Thomas P., with Cassutt, Michael, *We Have Capture* (Smithsonian Books, 2002)

Stock, William, and Noble, John, *Space Liner* (Times Books, 1981)

Taylor, John W. R., and Morley, David, *Spies in the Sky* (Ian Allan, 1972)

Thompson, Milton O., *At the Edge of Space* (Smithsonian Institution Press, 1992)

Tomayko, James E., *Computers in Spaceflight: The NASA Experience* (NASA, 1988)

—— *Computers Take Flight: A History of NASA's Pioneering Digital Fly-by-wire Project* (NASA, 2000)

Tregaskis, Richard, *X-15 Diary: The Story of America's First Space Ship* (Dutton, 1961)

Turnill, Reginald, and Reed, Arthur, *Farnborough: The Story of RAE* (Robert Hale, 1980)

van Pelt, Michel, *Rocketing into the Future* (Springer-Praxis, 2012)

Weiner, Tim, *Legacy of Ashes* (Allen Lane, 2007)

Westman, Paul, *John Young* (Dillon Press, 1981)

Whittle, Richard, *The Dream Machine* (Simon & Schuster, 2010)

Wolfe, Tom, *The Right Stuff* (Jonathan Cape, 1980)

Wright, David, et al., *The Physics of Space Security: A Reference Manual* (American Academy of Arts and Sciences, 2005)

Yeager, General Chuck, and Janos, Leo, *Yeager* (Bantam, 1985)

Young, John, *Forever Young* (University Press of Florida, 2012)

Magazines, journals and newspapers

Achenbach, Joel, 'NASA Gets Two Military Spy Telescopes for Astronomy', *Washington Post*, 2012

Air and Space, 'Astronaut Stories: The World's First Spaceplane', 2011

Aviation Week and Space Technology

Bamford, James, 'America's Supersecret Eyes in Space', *New York Times*, 1985

Bartholomew, Dana, 'Space Shuttle Breathed New Life Into Rocketdyne', *LA Daily News*, 2011

Brown, David, 'Home of the Right Stuff', *Air International*, 1985

Cassutt, Michael, 'The Manned Space Flight Engineer Programme', *Spaceflight*, 1989

—— 'Mr Inside', *Air and Space*, 2011

—— 'Secret Space Shuttles', *Air and Space*, 2009

Cooper Jr, Henry S. F., 'Shuttle – I', *New Yorker*, 1981

—— 'Shuttle – II', *New Yorker*, 1981

—— 'We Don't Have to Prove Ourselves', *New Yorker*, 1991

Cooper, Paul A., and Holloway, Paul F., 'The Shuttle Tile Story', *Astronautics and Aeronautics*, 1981

Easterbrook, Gregg, 'Beam Us Out of This Deathtrap, Scotty', *Washington Monthly*, 1980

Flight International

Glanz, James, and Schwartz, John, 'Dogged Engineer's Effort to Assess Shuttle Damage', *New York Times*, 2003

Kenden, Anthony, 'Was "Columbia" Photographed by KH-11?', *BIS Space Chronicle*, 1983

Launius, Roger D., 'Astronaut Envy', *Space and Defense*, 2010

Lewis, Richard S., 'Skylab Brings NASA Down to Earth', *New Scientist*, 1978

Little, Geoffrey, 'Spaceman', *Air and Space*, 2005

McCrum, Kirstie, 'Flying High', *Wales Online*, 2013

Middlecamp, David, 'Air Force Space Shuttle Astronauts Introduced at Vandenberg', *Tribune*, 2013

National Geographic, October 1981

Oberg, James, 'Max Faget: Master Builder', *Omni*, 1995

—— 'Skylab's Untimely Fate', *Air and Space*, 1992

Overmyer, Bob, 'The Gulfstream II in the Shadow of the Shuttle', *AOPA Pilot*, 1997

Posey, Carl, 'Meet Fitz Fulton', *Air and Space*, 2014

—— 'A Sudden Loss of Altitude', *Air and Space*, 1998

Recer, Paul, 'Astronomer is Flying High Over Peek at Mercury', *Los Angeles Times*, 1995

Salisbury, David, 'An Airborne Eye on the Cosmos', *Christian*

Science Monitor, 1983

Weiss, Rick, 'Satellite View of Shuttle Unsought; Columbia Orbited in Camera Range', *Washington Post*, 2003

Wilford, John Noble, 'New Test, Open Hatch Suspected in Space Shuttle Fatality', *New York Times*, 1981

—— 'Shuttle Rockets Into Orbit on First Flight', *New York Times*, 1981

—— 'Spy Satellite Reportedly Aided in Shuttle Flight', *New York Times*, 1981

Archive documents

The Air Force in Space Fiscal Year 1962 (USAF)

Air Force Satellite Control Facility Historical Brief and Chronology 1954–Present (USAF, 1984)

Apollo XVI Lunar Surface Journal (NASA)

Ben Franklin Gulf Stream Drift Mission Report, 1969

Berger, Carl, *A History of the Manned Orbiting Laboratory (MOL) Program* (USAF, 1970)

Contingency Aborts 21007/31007 (United Space Alliance, 2007)

DARPA Technical Accomplishments: An Historical Review of Selected DARPA Projects Vol. 1 (Institute for Defense Analyses, 1990)

Dr Hans Mark interviewed by Gerald Haines (NRO, 1997)

Eilertson, W. H., *Two-Stage, Fully Reusable Space Shuttle Lunar Mission Capability* (Bellcomm. Inc., 1969)

Eppley, Charles V., *History of the USAF Experimental Flight Test Pilot School* (USAF, 1963)

Ferguson, Matthew J., and May, Chester B., *Use of Ben Franklin Submersible as a Space Station Simulator* (NASA, 1970)

The Gambit Story (NRO, 1991)

Geiger, Clarence J., *Strangled Infant: The Boeing X-20A Dyna-Soar* (USAF)

Haynes, Davy A., *Feasibility of Cislunar Flight Using the Shuttle Orbiter* (NASA, 1991)

The Hexagon Story (NRO, 1992)

Houston, Robert S., Hallion, Richard P., and Boston, Ronald G., *Transiting from Air to Space: The North American X-15* (USAF, 1987)

Infrared Imagery of Shuttle (IRIS) (NASA/Martin-Marietta, 1977)

Infrared Imagery of Shuttle (IRIS) Experiment IRIS/STS-3 Engineering Report (NASA, 1982)

Interview with Dr William O. Baker (Center for History of Physics of the American Institute of Physics, 1996)

Interview with Vice Admiral Richard H. Truly (US Navy, 2005)

Johnson Space Center News Release Logs, 1973–81

Jones, J. J., *Use of the Shuttle Orbiter as a Research Vehicle* (AIAA, 1981)

Lambert, John V., and Kissell, Kenneth E., *The Early Development of Satellite Characterization Capabilities at the Air Force Laboratories, Landbased Instrumentation Handbook* (Headquarters Western Space and Missile Center, Air Force Systems Command, USAF, 1981)

Lunar Expedition Plan (USAF, 1961)

Man in MOL (USAF, 1970)

Milestones in Airborne Astronomy: From the 1920s to the Present (American Institute of Aeronautics and Astronautics)

Moser, T. L., and Schneider, W. C., *Strength and Integrity of the Space Shuttle Orbiter Tiles* (AIAA, 1981)

NASA Johnson Space Center Oral History Project:
    Aldridge Jr, Edward C. 'Pete'
    Algranti, Joseph S.
    Allen, Joseph P. 'Joe'
    Blackburn, Gerald
    Bobko, Karol J.
    Brand, Vance
    Brandenstein, Daniel C.
    Coats, Michael L.
    Cohen, Aaron

Covey, Richard O.
Creighton, John O.
Crews Jr, Albert H.
Crippen, Robert L.
Cuzzupoli, Joseph W.
Fabian, John M.
Faget, Maxime A.
Fullerton, C. Gordon
Greene, Jay H.
Gregory, Frederick D.
Haines, Charles R.
Haise Jr, Fred W.
Hart, Terry J.
Hartsfield Jr, Henry W. 'Hank'
Hauck, Frederick H.
Hawley, Steven A.
Hodge, John D.
Hoffman, Jeffrey A.
Hooks, Ivy
Hutchinson, Neil B.
Kiker, John W.
Kraft Jr, Christopher C.
Lee, Dorothy B.
Lenoir, William B.
Lousma, Jack R.
Lunney, Glynn S.
McBride, Jon A.
Mattingly II, Thomas K.
Moser, Thomas L.
Mullane, Richard M.
Myers, Dale D.
Nagel, Steven R.
Nelson, George D.
Odom, James B.

Peterson, Donald H.

Ried Jr, Robert C.

Ross, Jerry L.

Schaffer, Phillip C.

Seddon, Margaret Rhea

Sheehan, Gerald D.

Shriver, Loren J.

Silvera, Milton A.

Stafford, Thomas P.

Sullivan, Kathryn D.

Thompson, Robert F. 'Bob'

Van Hoften, James D. A.

Williams, Donald E.

Yardley, John F.

Young, Kenneth A.

NASA's Kuiper Airborne Observatory, 1971–1995 (NASA, 2013)

National Reconnaissance Office Mission Ground Station Declassification 'Questions and Answers'

OL-AG Phillips Laboratory Malabar Test Facility User Manual, c/o fas.org

Project Hexagon Overview (NRO)

Report of the Space Task Group, 1969 (NASA)

Mr Robert Crippen Oral History (Kennedy Space Center, 2002)

Space Shuttle Orbiter Approach and Landing Test Final Evaluation Report (NASA, 1978)

Space Shuttle Transoceanic Abort Landing (TAL) Sites (NASA, 2006)

STS-1 Operational Flight Profile (NASA, 1980)

STS-1 First Space Shuttle Mission Press Kit (NASA, 1981)

STS-1 In-Flight Anomaly Report (NASA, 1981)

STS-1 Transcript (NASA, 1981)

System Failure Case Studies: Tough Transitions (NASA, 2011)

Weingarten, Norman C., *History of In-Flight Simulation and Flying Qualities Research at Calspan* (AIAA, 2005)

Websites

aiaahouston.org – Jackson, Dr A. A., *The Ugly Spaceship*, 2012

americaspace.com – Covault, Craig, *Top Secret KH-11 Spysat Design Revealed by NRO's Twin Telescope Gift to NASA*

amyshirateitel.com – *A History of the Dyna-Soar*

astronautix.com

astronomycafe.net – *Observing with the Kuiper Airborne Observatory*

c141heaven.info – NASA 714

collectspace.com

fas.org – Lacrosse/Onyx; *The SLC-6 Saga*

flightjournal.com – *747 Pilot Comments About Flying the Shuttle*

globalsecurity.org – Lacrosse/Onyx

hankbrandli.com – *The Rescue of Apollo 11*

jamesmskipper.us – SMEAT

johnwyoung.com

launiusr.wordpress.com – *NASA's Space Shuttle and the Department of Defense*

mach25.nl – STS-62A

nbcnews.com – Windrem, Robert, and O'Dea, Tierney, *Peeking at the Shuttle from Space*

nf104.com russianspaceweb.com

www.satobs.org – Molczan, Ted, *Could Columbia Have Been Imaged by a Keyhole? / STS-1 and KH-11-2 Encounters / STS-2 and KH-11-3 Encounters*

sony.net – *Determination Drove the Development of the CCD 'Electronic Eye'*

spacefacts.de

spaceflightnow.com

  – *In His Own Words: Pentagon Space Official Pete Aldridge*

  – Ray, Justin, *From Shuttles to Rockets / Space Launch Complex 6*

thespacereview.com

  – Day, Dwayne A., *Astronauts and Area 51: the Skylab Incident* (2007) / *A Bat Outta Hell* (2010) / *The Big Bird and the Turkey*

(2011) / *Big Black and the New Bird* (2010) / *Blue Skies on the West Coast* (2007) / *General Power vs Chicken Little* (2005) / *The Hexagon and the Space Shuttle* (2011) / *The Hour of the Wolf* (2012) / *Ike's Gambit* (2009) / *Those Magnificent Spooks and Their Spying Machine* (2013) / *The Sounds of Distant Thunder* (2012) / *The Spooks and the Turkey* (2006) / *Theft, The Sincerest Form of Flattery* (2012)

– Oberg, James, *Academic Honors for a Spaceflight Prophet* (2005)

thuleab.dk

utexas.edu – Mark, Hans, *The Education of the Guardians* (2011)

Audio

https://archive.org – NASA Audio Collection STS-1 mission transcript

www.youtube.com – Hans Mark: Origin Story of Carl Sagan's plaque on *Pioneer 10*

Video

www.youtube.com
– Bob Crippen STS-107 eulogy, 2003
– Hail Columbia!, IMAX Corporation, 1982
– STS-1 Post-mission Press Conference, NASA, 1981
– Astrospies, NOVA, 2008
– MIT Aircraft Systems Engineering Course, 2005
– The Space Shuttle's Last Flight, Darlow Smithson Productions, 2011

APPENDIX

Space Shuttle Cutaway

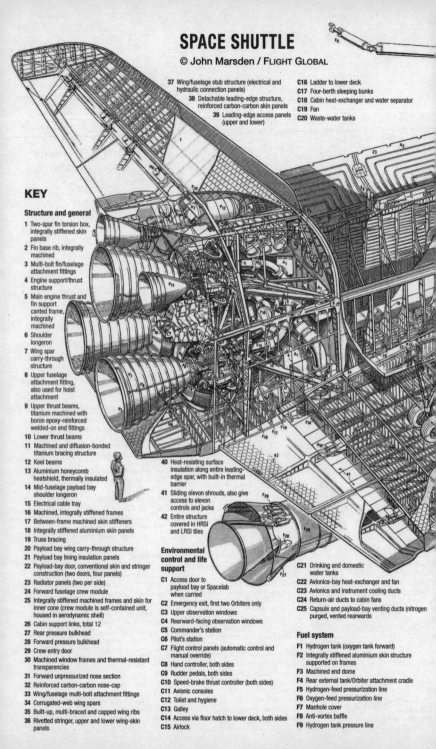

SPACE SHUTTLE

© John Marsden / FLIGHT GLOBAL

37 Wing/fuselage stub structure (electrical and hydraulic connection panels)
38 Detachable leading-edge structure, reinforced carbon-carbon skin panels
39 Leading-edge access panels (upper and lower)

C16 Ladder to lower deck
C17 Four-berth sleeping bunks
C18 Cabin heat-exchanger and water separator
C19 Fan
C20 Waste-water tanks

KEY

Structure and general

1 Two-spar fin torsion box, integrally stiffened skin panels
2 Fin base rib, integrally machined
3 Multi-bolt fin/fuselage attachment fittings
4 Engine support/thrust structure
5 Main engine thrust and fin support canted frame, integrally machined
6 Shoulder longeron
7 Wing spar carry-through structure
8 Upper fuselage attachment fitting, also used for hoist attachment
9 Upper thrust beams, titanium machined with boron epoxy-reinforced welded-on end fittings
10 Lower thrust beams
11 Machined and diffusion-bonded titanium bracing structure
12 Keel beams
13 Aluminium honeycomb heatshield, thermally insulated
14 Mid-fuselage payload bay shoulder longeron
15 Electrical cable tray
16 Machined, integrally stiffened frames
17 Between-frame skin stiffeners
18 Integrally stiffened aluminium skin panels
19 Truss bracing
20 Payload bay wing carry-through structure
21 Payload bay lining insulation panels
22 Payload-bay door, conventional skin and stringer construction (two doors, four panels)
23 Radiator panels (two per side)
24 Forward fuselage crew module
25 Integrally stiffened machined frames and skin for inner cone (crew module is self-contained unit, housed in aerodynamic shell)
26 Cabin support links, total 12
27 Rear pressure bulkhead
28 Forward pressure bulkhead
29 Crew entry door
30 Machined window frames and thermal-resistant transparencies
31 Forward unpressurized nose section
32 Reinforced carbon-carbon nose-cap
33 Wing/fuselage multi-bolt attachment fittings
34 Corrugated-web wing spars
35 Built-up, multi-braced and capped wing ribs
36 Rivetted stringer, upper and lower wing-skin panels

40 Heat-resisting surface insulation along entire leading-edge spar, with built-in thermal barrier
41 Sliding elevon shrouds, also give access to elevon controls and jacks
42 Entire structure covered in HRSI and LRSI tiles

Environmental control and life support

C1 Access door to payload bay or Spacelab when carried
C2 Emergency exit, first two Orbiters only
C3 Upper observation windows
C4 Rearward-facing observation windows
C5 Commander's station
C6 Pilot's station
C7 Flight control panels (automatic control and manual override)
C8 Hand controller, both sides
C9 Rudder pedals, both sides
C10 Speed-brake thrust controller (both sides)
C11 Avionic consoles
C12 Toilet and hygiene
C13 Galley
C14 Access via floor hatch to lower deck, both sides
C15 Airlock

C21 Drinking and domestic water tanks
C22 Avionics-bay heat-exchanger and fan
C23 Avionics and instrument cooling ducts
C24 Return-air ducts to cabin fans
C25 Capsule and payload-bay venting ducts (nitrogen purged, vented rearwards)

Fuel system

F1 Hydrogen tank (oxygen tank forward)
F2 Integrally stiffened aluminium skin structure supported on frames
F3 Machined end dome
F4 Rear external tank/Orbiter attachment cradle
F5 Hydrogen-feed pressurization line
F6 Oxygen-feed pressurization line
F7 Manhole cover
F8 Anti-vortex baffle
F9 Hydrogen tank pressure line

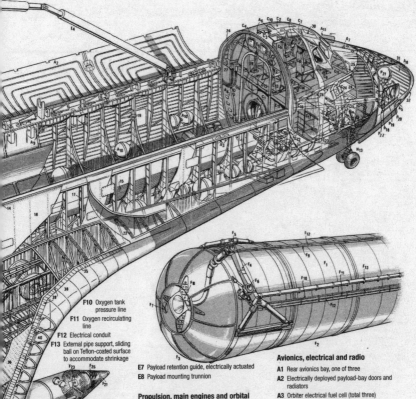

F10 Oxygen tank pressure line
F11 Oxygen recirculating line
F12 Electrical conduit
F13 External pipe support, sliding ball on Teflon-coated surface to accommodate shrinkage

F14 Orbiter oxygen pull-off umbilical connection (hydrogen on other side)
F15 Hydrogen main fuel-feed pipe
F16 Oxygen umbilical panel (hydrogen on other side)
17 Oxygen tank fill and drain pipe (hydrogen on other side)
18 APU fuel tanks (three)
19 Helium tanks (four) used for pneumatic systems, valve-actuation and tank-purging
20 Solid-fuel rocket booster, one each side
21 Ribbon-type drogue chute (four)
22 Recovery parachute (three)
23 Navigation stowage and thrust-vector system access
24 Cable tunnel
25 Forward separation motors, 15,000lb thrust each
26 Rear separation motors, 15,000lb thrust each
27 Rear skirt and launch support cone
28 Contoured expansion nozzle, 8° omni-axial gimbal capability

Payload-handling equipment

E1 Payload-handling station
E2 Rendezvous and docking station
E3 Payload specialist's station
E4 Payload manipulating arm
E5 Mating adaptor
E6 Manipulator arm rest

E7 Payload retention guide, electrically actuated
E8 Payload mounting trunnion

Propulsion, main engines and orbital manoeuvre system

P1 Rocketdyne liquid-propellant, throttleable main engine, one of three
P2 Gimbal actuator (two per engine) ± 10.5° pitch, ± 8.5° yaw
P3 Low-pressure turbopump (liquid hydrogen) mounting
P4 High-pressure hydrogen turbopump
P5 High-pressure oxygen turbopump
P6 Engine controller
P7 Thrust nozzle
P8 Nozzle cooling pipes
P9 Heatshield
P10 Orbital manoeuvring system engine packs (both sides inter-connected)
P11 OMS main engine (6,000lb thrust in vacuum) one of two, electro-mechanical gimbal-actuated
P12 OMS propellant tanks containing nitrogen tetroxide and mono-methyl hydrazine
P13 Helium bottle to pressurize OMS propellants
P14 Orbiter reaction-control system (RCS)
P15 Primary bipropellant thrusters (870lb in vacuum), 12 per side
P16 Vernier thrusters (25lb in vacuum), one each side
P17 RCS propellant tanks containing nitrogen tetroxide and mono-methyl hydrazine
P18 Forward RCS propulsion units (three)
P19 Fixed thrusters (total 14)
P20 Vernier thrusters (total2)
P21 RCS forward propellant tanks
P22 Ground servicing receptacle

Avionics, electrical and radio

A1 Rear avionics bay, one of three
A2 Electrically deployed payload-bay doors and radiators
A3 Orbiter electrical fuel cell (total three)
A4 Crew module electrical connecting panels to payload bay
A5 Static and temperature probes (normally retracted except for approach and landing)
A6 S-band antenna
A7 L-band antenna
A8 Ku-band antenna
A9 Floodlight (three per side)
A10 Hydrogen/oxygen fuel cells
A11 Transparency cover for star-tracker

Hydraulics

H1 One of three hydraulic reservoirs serving three independent systems
H2 Pump (three) 3,000lb/sq in variable delivery
H3 Sundstrand APU (one of three)
H4 APU exhaust ducts
H5 Water boilers and coolers (one of three)
H6 Boiler blow-off outlets
H7 Rudder/speed-brake actuator unit
H8 Rotary actuator (one of four)
H9 Body-flap actuators (total three) −11.7° +22.5°
H10 Eleven actuators (two per side, total four) −40° +15°
H11 Main engine gimbal actuating jacks (two per engine) and control valves (three)
H12 Hydraulically actuated anti-skid device, main undercarriage, and up-lock (all three legs of free-fall design)
H13 Nose undercarriage, hydraulically steered

PICTURE ACKNOWLEDGEMENTS

Although every effort has been made to trace copyright holders and clear permission for the photographs in this book, the provenance of a number of them is uncertain. The author and publisher would welcome the opportunity to correct any mistakes.

First section

A young Bob Crippen: © Bob Crippen; Dick Truly and Vought F-8 Crusader: © Richard Truly; Bob Crippen and A-4 Skyhawk: © Bob Crippen.

Bob Crippen at the Aerospace Research Pilots School: © Bob Crippen; John Young with F-4H-1 Phantom II; Project HIGH JUMP, Phantom: *both* © US Navy; John Young and Gus Grissom in *Gemini III*: © NASA; Joe Engle in North American X-15: © USAF; X-15: © NASA; X-20 Dyna-Soar; Project LUNEX spacecraft and lander: *both* © USAF.

MOL diagram: © NRO; MOL mission patch; Dick Truly with scale model of space station; MOL pilots in 1966; Dick Truly, Mike Adams and Jack Finley at Wright-Patterson Air Force Base: *all* © USAF; USAF Titan IIIC rocket: © NASA; KH-9 HEXAGON reconnaissance satellite: © NRO; re-entry bucket and C-130: © USAF.

The MOL guys; John Young on the moon; Max Faget; DC-3 in wind tunnel; models for a delta-winged Orbiter; Richard Nixon and James Fletcher; Dr Hans Mark; ceramic tiles being tested; *all* © NASA.

Shuttle solid rocket booster; RS-25 Space Shuttle main engine: *both* © NASA; Bob Crippen after Skylab Medical Experiment Altitude Test: © Bob Crippen; John Young at Johnson Space Center: © NASA.

Second section

Spock, Sulu, Bones and Scotty at *Enterprise* roll-out ceremony; VIRTUS carrier aircraft: *both* © NASA; Short-Mayo Composite: © Crown Copyright; Lockheed C-5 Galaxy: © NASA.

NASA Boeing 747; scale model of 747/Shuttle: *both* © NASA; modified Gulfstream business jet: © David W. Whittle courtesy of NASA JSC Oral History Project; *Enterprise* astronauts 1977; *Enterprise* flies free; Approach and Landing test mission patch; *Enterprise* banks over the Mojave Desert; *Enterprise* lands on Rogers Dry Lake Bed: *all* © NASA.

Haise, Fullerton and *Enterprise*; airborne salute to *Enterprise*; Joe Engle and Dick Truly on Shuttle's fourth flight; fifth ALT test flight; NASA introduces crews for Shuttle test flights; Fred Haise in simulator; 'Buck Rogers' mission; Orbiter in wind tunnel: *all* © NASA.

Shuttle on back of SCA; F-104 and F-15 jet fighters flying at high speed; tiles being re-attached to *Columbia*; Dick Truly using on-orbit tile repair kit; *Columbia* in the Vehicle Assembly Building; *Columbia* on journey to Pad-39A; STS-1 Mission patch; *Columbia* on Pad-39A: *all* © NASA.

John Young in T-38; Bob Crippen and John Young study STS-1 flight plan; Bob Crippen descends from Shuttle Training Aircrafts; Bob Crippen and John Young in test lab: *all* © NASA.

Third section

Casselli's watercolour of John Young; STS-1 crew; Young and Crippen in simulation flight; T-38s of the Chase team: *all* © NASA.

Columbia takes off; *Columbia* in flight; *Columbia* ascends past 150,000 feet; *Columbia*'s discarded external tank; Crippen and Young in aft crew station; tiles missing from heatshield; Gene Kranz, Chris Kraft and Max Faget; Dr Hans Mark: *all* © NASA.

William Twinting: © NASA; Blue Cube: © USAF; Hubble Telescope; Cape Cod pictured from *Columbia*; Joe Engle and Dick Truly at Capcom console; painting showing temperature and pressure wave created by *Columbia* on re-entry; *Atlantis* re-entering; Kuiper Airborne Observatory: *all* © NASA.

Columbia and Chase team glide towards Edwards; *Columbia* over Rogers Dry Lake; John Young brings *Columbia* in to land; *Columbia* on Runway 23; view from inside *Columbia* on landing; Young and Crippen leave *Columbia*; *Columbia* post-flight: *all* © NASA.

Columbia returns to Edwards; Vandenberg AFB: *both* © USAF.

INDEX

Nomex felt 178, 295
NORAD air defence radar 246
North American Aviation 141, 311
  Apollo programme 141–2
  B-25 Mitchell 141
  B-45 Tornado 141
  F-86 Sabre 141
  P-51 Mustang 141
  RA-5C Vigilante 259
  T-6 Texan 141
  X-15 179, 180, 230
    air-launching 168
    engines 113
    history 141
    influence on the Shuttle design 69, 93
    Joe Engle and 31, 33–5, 204, 208, 506
    maximum altitude 367
    pilot ejection 311
    X-15 programme 32–5, 40, 59, 96, 103
  XB-70 Valkyrie 141, 189, 230
North American Rockwell, see Rockwell International
Northrop
  B-2 Spirit 284
  T-38 Talon 180, 205, 214, 256, 269, 282, 338, 405
    at Aerospace Research Pilots School 28
    Challenger disaster 523–4
    Chase Air Force 288, 313–14, 329, 461, 468, 470, 473, 489–91, 492, 494–5
    Columbia leaves Rockwell 281
    Shuttle Approach and Landing tests 221, 224, 225, 234–5
    STS-1 launch 350
NPO Energiya 189
Nygren, Rick 153, 156, 157

Oceana Naval Air Station 21
Odom, James 192
Odyssey (Command Module) 103
O'Neill, Gerald 217
Operation EAGLE CLAW 301–2
orbital manoeuvring system (OMS) 369, 371, 372, 375–8, 393, 424
orbital velocity 130–2
orbiter manoeuvring systems (OMS) pods 398, 404, 405, 440, 482

Orbiter Processing Facility (OPF) 296, 297, 319, 521–2
Orbiters, *see Orbiters by name, and components*
Order of the Elephant membership 26
Orion, see Lunar Modules
Orroral Valley ground station 409, 417, 422, 423, 466
Overmyer, Bob: Apollo-Soyuz Test Project 183, 184, 185, 189
  Challenger disaster 524
  MOL programme 40
  NASA selection 90
  Shuttle Approach and Landing tests 204, 221, 222, 224, 233
  STS-5 397

Packard, David 78
Page, George 356
Pan-Am International 73
Panama Canal Zone 57
pararescue jumpers (PJs) 484, 499
PASS, *see* computer systems
Patrick Air Force Base 356, 357, 367, 414, 415
Patuxent River Naval Air Station 14, 21, 83, 259
Payload Specialists 218, 286, 521
Peenemunde rocket research facility 186
Perkin-Elmer (manufacturers) 44, 81, 390
Peterson, Bruce 167
Peterson, Don: MOL programme 40, 56, 74, 82
  NASA selection 90
  Space Shuttle programme 161, 303
Phillips, Hewitt 170
Phillips, Major General Sam 89
pilot-induced oscillation (PIO) 211, 212, 235, 237
Pioneer 343, 458
Pioneer 10 216–17
Pioneer 11 216–17
Pluto 217, 458
Pompidou, George 123
Pontiac Catalina V8 102
Pratt and Whitney: J52 turbojets 260
  JT8D-15 turbofans 180
  Space Shuttle engines 113, 114, 116
  TF-33-P-7 turbofans 464
  XLR-129 113